Engineering Response to Global Climate Change

Planning a Research
and Development Agenda

Edited by
Robert G. Watts

National Institute for Global Environmental Change
Tulane University
New Orleans, Louisiana

LEWIS PUBLISHERS

Boca Raton   New York

Acquiring Editor:	Joel Stein
Project Editor:	Joan Moscrop
Marketing Manager:	Greg Daurelle
Direct Marketing Manager:	Arline Massey
Cover design:	Dawn Boyd
Manufacturing:	Sheri Schwartz

Library of Congress Cataloging-in-Publication Data

Engineering response to global climate change : planning a research and development agenda /
 edited by Robert G. Watts
  p.   cm.
 Includes bibliographical references and index.
 ISBN 0-56670-234-8
  1. Climate changes--environmental aspects. 2. Environmental engineering--
 Environmental aspects.   3. Ecological engineering.   4. Global environmental
 change.   5. Greenhouse gases--Environmental aspects.   I. Watts, Robert G.
 QC981.8.C5E56 1997
 363.738'745--dc20   96-47401
                                                                                        CIP

ABOUT THE AUTHOR/EDITOR

Robert G. Watts is the Cornelia and Arthur L. Jung Professor of Mechanical Engineering at Tulane University. After receiving an undergraduate degree in Mechanical Engineering from Tulane University, he received a M.S. in Nuclear Engineering from the Massachusetts Institute of Technology and a Ph.D. in Mechanical Engineering from Purdue University in 1965. He spent a post-doctoral year at Harvard University. During the academic year 1977, he was a senior scientist at the International Institute for Applied Systems Analysis in Austria, where he studied the effects of climatic change on agriculture. His current research is related to paleoclimatology, recent climatic change and its effects on humans, and energy systems. He is the regional director of the South Central Region of the National Institute for Global Environmental Change, which is located at Tulane University. He is the author with Terry Bahill of *Keep Your Eye on the Ball: The Science and Folklore of Baseball* (1990).

ACKNOWLEDGMENTS

This book grew out of a workshop that took place in Palm Coast, Florida from June 1–6, 1991. I would first like to acknowledge the Engineering Foundation for first suggesting that I organize the workshop. Financial support for a planning workshop was provided by the National Science Foundation and the U.S. Environmental Protection Agency. Major funding for the workshop itself was provided by the U.S. DoE through the National Institute for Global Environmental Change (NIGEC) Southern Regional Office located at Tulane University.

Through much of the planning as well as post-workshop organizing I benefited greatly from conversations with Drs. Gregg Marland, Brian Flannery and Martin Hoffert, as well as my colleague at NIGEC, Mr. Juan Parke. Ms. Susan Williams, my very efficient secretary, along with Juan Parke, were so well organized that even my limited organizational skills were not too detrimental.

The workshop was ably facilitated by a team from the Institute for Cultural Affairs: Ms. Jean Watts, Mr. Jim Wiegel and Mr. Larry Henschen. I owe them gratitude for keeping the workshop on track.

Leslie Morantine, Susan Williams, Jean Watts and Juan Parke kept the paper flowing and the PCs operating well into the night, preparing for the next day's session, and taking care of those little details that are so important to providing a good and efficient working environment. Dr. Michael Morantine helped in this task in addition to participating in the Geoengineering workshop. Ms. Pat Smith was very helpful in putting the final touches on the book. This book is gratefully dedicated to them.

Also, for his computer and layout expertise, special thanks to Gary Reggio, who produced the entire book on Macintosh PowerPCs™, using QuarkXPress™ for the layout, with PhotoShop™ and FreeHand™ for various illustrations.

Finally, I thank all of those who participated in the workshop for their hard work both in Florida and afterward.

Robert G. Watts
Tulane University
New Orleans, 1996

TABLE OF CONTENTS

Chapter 1
THE FIFTH REVOLUTION
Robert Watts

Chapter 2
EMISSIONS AND BUDGETS OF RADIATIVELY IMPORTANT ATMOSPHERIC CONSTITUENTS
Author: Don Wuebbles
Co-Authors: Jae Edmonds, Jane Dignon, William Emanuel,
Donald Fisher, Richard Gammon, Robert Hangebrauck,
Robert Harris, M.A.K. Khalil, John Spence, Thayne M. Thompson

Chapter 3
ENERGY DEMAND REDUCTION
Authors: Arthur Rosenfeld, Barbara Atkinson,
Lynn Price, Bob Ciliano, J.I. Mills, Kenneth Friedman
Co-Authors: Ed Flynn, Mary Hopkins,
Henry Shaw, John Wilson, Francis Wood
Contributors: Ruth Reck, Eric Larson

Chapter 4
ENERGY SUPPLY
Authors: Martin Hoffert, Seth D. Potter
Contributors: Jerry Delene, Peter E. Glaser, Michael Golay,
Harold M. Hubbard, Murali Kadiramangalam, Alfred Perry,
Myer Steinberg, Carl-Jochen Winter

Chapter 5
WATER RESOURCES
Authors: William H. McAnally
Co-Authors: Phillip H. Burgi, Darryl Calkins, Richard H. French,
Jeffery P. Holland, Bernard Hsieh, Barbara Miller, Jim Thomas
Contributors: William D. Martin, James R. Tuttle

Chapter 6
SEA LEVEL RISE AND COASTAL HAZARDS: AN ASSESSMENT OF IMPACTS AND COASTAL ENGINEERING RESEARCH NEEDS

Authors: Ashish Mehta
Contributors: Robert Dean, Hans Kunz, Victor Law,
Say-Chong Lee, Zal Tarapore

Chapter 7
AGRICULTURE AND BIOLOGICAL SYSTEMS
Authors: Norman R. Scott
Contributors: John N. Walker, Gerald F. Arkin,
James A. DeShazer, Gary R. Evans, Glenn J. Hoffman,
James W. Jones

Chapter 8
GEOENGINEERING CLIMATE
Authors: Brian P. Flannery, Haroon Kheshgi,
Gregg Marland, Michael C. MacCracken
Contributors: Hioshi Komiyama, Wallace Broecker,
Hisashi Ishatani, Norman Rosenberg, Meyer Steinberg,
Tom Wigley, Michael Morantine

Chapter 9
**ENGINEERING RESPONSE TO
GLOBAL ENVIRONMENTAL CHANGE**
Robert Watts

• • •

Chapter 1

THE FIFTH REVOLUTION

Author: Robert G. Watts

1.1 INTRODUCTION

Certainly one of the most important events in the long history of mankind's development was his mastery of fire. In his book, "The Next Million Years," Charles Galton Darwin (1953) suggests that the history of mankind has seen four occasions when man made a step forward that was essentially irreversible in the sense that the progress afforded was never lost. He refers to these as four revolutions. The first of these was the discovery of fire. It was, of course, not the discovery of fire, but its mastery, that was so important. The use of fire for cooking and warmth, and eventually for the creation of tools out of bronze and iron changed the nature of life on Earth irreversibly. The second revolution was the invention of agriculture. The people who participated in organized agricultural practices were able to provide themselves with food, both animal and vegetable, more readily than hunter/gatherers. The third of Darwin's revolutions was the urban revolution. The invention of cities brought the advantages of the division of labor and the establishment and regular practice of trade, through which people in different locations could obtain goods not available locally. According to Darwin, the fourth revolution was the scientific revolution, which he describes as "the discovery that it is possible consciously to make discoveries about the fundamental nature of the world, so that by their means man can intentionally and deliberately alter his way of life."

It is useful to contemplate what Darwin termed the fifth revolution. This will occur when mankind exhausts the store of the source of the first revolution: fire.

The sources of fire (heat) that powered the industrial revolution have been mainly coal, oil, and natural gas: the fossil fuels. Both

1

Darwin and Haldane (1923) before him point out that resources of these fossil fuels are but centuries from exhaustion. In the long run we shall have to find alternatives to the use of fossil fuels to produce energy, and both Darwin and Haldane went on to suggest the usual possibilities of harnessing energy from the sun directly, or indirectly through the wind, the tides, or rivers. The possibility of atomic power was discussed by both Darwin and Haldane, and Darwin also discussed geothermal heat and the use of the vertical temperature gradient of the ocean as well as the use of plants to produce alcohol as a convenient fuel.

Neither Darwin nor Haldane appears to have understood (although it had been previously suggested by other scientists) that another potential problem concerning the use of fossil fuels would soon arise. The burning of fossil fuels releases carbon dioxide into the atmosphere, and carbon dioxide strongly absorbs infrared radiation. The result is that solar radiation is allowed to penetrate the atmosphere, while infrared radiation is selectively impeded from leaving the Earth–atmosphere system. If the atmospheric loading of carbon dioxide becomes large enough, the climate of the Earth will change; the Earth will become warmer. The consequences of a warmer Earth are largely unknown, but they will surely include shifting of climatic zones, with resulting changes in the regions where agriculture can be successfully practiced; changes in the flows of rivers and streams; and, quite possibly, changes in sea level. This means that we may be faced with Darwin's fifth revolution sooner than either Darwin or Haldane suspected.

Darwin clearly believed that the fifth revolution. brought about by the exhaustion of fossil fuels, would differ fundamentally from the first four. While the first four revolutions led to the ability to sustain large increases in population, or to improve the comfort of those populations, the fifth would have the opposite effect. Technological optimists, however, believe (e.g., Weinberg, 1966) that we are far from finished with Darwin's fourth revolution. The essence of the scientific revolution lies in the discovery that nature can be understood, and that mankind can, by intelligent manipulation, use this understanding to create better conditions for human life. It is in this spirit that this document has been prepared.

1.2 THE NATURE OF THE PROBLEM

The atmospheric concentrations of a number of gases that are chemically or radiatively active are increasing. At least part of the increase is the result of human activities. Many of these gases absorb more strongly in the infrared part of the radiation spectrum than at the shorter wavelengths characteristic of solar radiation. The sun emits radiation principally in the wavelength band between 0.2 and 4.0 microns, while the Earth and its atmosphere emit in the range of about 4 to 100 microns. The atmosphere absorbs about 23% of the incoming solar radiation, principally by ozone and water vapor. On the other hand, water vapor, carbon dioxide (CO_2), methane (CH_4), nitrous oxide (N_2O), chlorofluorocarbons (CFCs), and a number of other atmospheric gases exhibit very strong absorption of radiation in the infrared range associated with terrestrial radiation. The trapping of outgoing terrestrial radiation by the atmosphere is referred to as the greenhouse effect. Without the presence of these greenhouse gases, the Earth would be much colder than it is (see Section 1.3.1). If the concentrations of greenhouse gases increase substantially as a result of human activities, most scientists now believe that the Earth will become warmer. In other words, a manmade greenhouse effect will be created that enhances the natural greenhouse effect.

A number of the activities that support modern societies result in the production of greenhouse gases. For example, the best known greenhouse gas, carbon dioxide, is produced by the burning of fossil fuel, by deforestation, and to a lesser extent by industrial processes such as cement manufacture. Methane is produced by activities associated with energy production (e.g., coal mining and natural gas production and distribution) and with food production (ruminant animals and rice farming). These and other sources of greenhouse gases are discussed in Chapter 2.

In light of the increasing atmospheric concentrations of greenhouse gases we need to speculate about what the future holds. How serious is the concern about greenhouse warming?

In order to estimate future greenhouse gas-induced climatic change it is necessary to know future atmospheric concentrations of greenhouse gases as well as the effect of these gases on the climate.

A group of scenarios for estimating future greenhouse gas exchange has been developed by the Intergovernmental Panel on

Climate Change (IPCC, 1990). The scenarios are based on specific assumptions regarding economic and population growth during the next century. It was assumed that the population will approach 10.5 billion in the second half of the next century. Economic growth was assumed to be 2–3% annually in OECD countries and 3–5% in Eastern European countries and developing nations during the coming decade, decreasing thereafter. Four growth scenarios, covering projected emissions of carbon dioxide, methane, nitrous and other nitrogen oxides, chlorofluorocarbons, and carbon monoxide, have been presented. We quote directly from the IPCC report:

- "In the *Business-as-Usual Scenario (Scenario A)* the energy supply is coal intensive and on the demand side only modest efficiency increases are achieved. Carbon monoxide controls are modest, deforestation continues until the tropical forests are depleted and agricultural emissions of methane and nitrous oxide are uncontrolled. For CFCs the Montreal Protocol is implemented albeit with only partial participation."
- "In *Scenario B* the energy supply mix shifts towards lower carbon fuels, notably natural gas. Large efficiency increases are achieved. Carbon monoxide controls are stringent, deforestation is reversed and the Montreal Protocol implemented with full participation."
- "In *Scenario C* a shift towards renewables and nuclear energy takes place in the second half of next century. CFCs are now phased out and agricultural emissions limited."
- "For *Scenario D* a shift to renewables and nuclear in the first half of the next century reduces the emissions of carbon dioxide, initially more or less stabilizing emissions in the industrial countries. The scenario shows that stringent controls in industrial countries combined with moderated growth of emission in developing countries could stabilize atmospheric concentrations. Carbon dioxide emissions are reduced to 50% of 1985 levels by the middle of the next century."

The results of these scenarios are shown in Figure 1.1. It is important also to note that some potentially important feedback processes have been omitted in these scenarios. The climatic change that will accompany an increase in the greenhouse effect will very likely cause

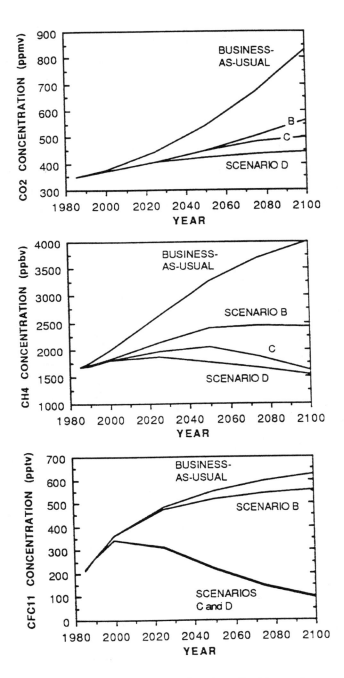

Figure 1.1   Atmospheric concentrations of carbon dioxide, methane and CFC-11 resulting from the four IPCC emissions scenarios. (From Houghton, J.T., G.J. Jenkins, and J.J. Ephraums, *Climate Change: The IPCC Scientific Assessment*, 1990.)

a change in the net flow of carbon dioxide, methane, and perhaps other gases between the atmosphere and terrestrial ecosystems. These feedback processes are currently poorly understood, but it seems likely that they will act to further increase the greenhouse gas content of the atmosphere as the Earth becomes warmer. The effective greenhouse forcing is different for different greenhouse gases because some absorb infrared radiation more strongly than others and because they have different lifetimes in the atmosphere. For example, although the concentrations of both methane and CFC-11 are far smaller than that of carbon dioxide, an incremental molecule of methane in the atmosphere is about 21 times and of CFC-11 12,400 times as effective in absorbing infrared radiation as is a molecule of carbon dioxide.

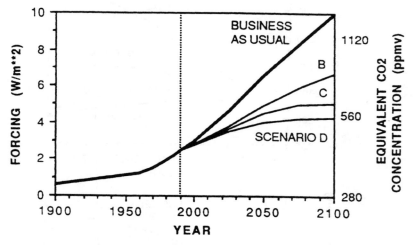

Figure 1.2   Increase in radiative forcing since the mid-18th century, and predicted to result from the four IPCC emissions scenarios, also expressed as equivalent carbon dioxide concentrations. (From Houghton, J.T., G.J. Jenkins, and J.J. Ephraums, *Climate Change: The IPCC Scientific Assessment,* 1990.)

By combining the radiative effects of all the greenhouse gases, the four scenarios can be combined as shown in Figure 1.2. The right-hand coordinate shows the atmospheric concentration of CO_2 that would produce the same net radiative forcing do all of the greenhouse gases combined. The value of that forcing is shown on the left-hand coordinate. Note that even in Scenario D the effective CO_2 concentration nearly doubles its pre-industrial value by the end of the next century.

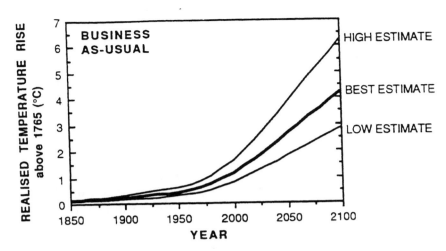

Figure 1.3   Simulation of the increase in global mean temperature from 1850-1990 due to observed increases in greenhouse gases, and predictions of the rise between 1990 and 2100 resulting from the Business-as-Usual emissions. (From Houghton, J.T., G.J. Jenkins, and J.J. Ephraums, *Climate Change: The IPCC Scientific Assessment,* 1990.)

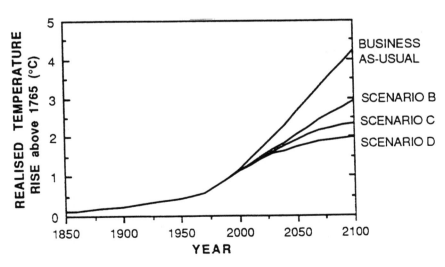

Figure 1.4   Simulations of the increase in global mean temperature from 1850-1990 due to observed increases in greenhouses gases, predictions of the rise between 1990 and 2100 resulting from the IPCC Scenario B, C and D emissions, with the Business-as-Usual case for comparison. (From Houghton, J.T., G.J. Jenkins, and J.J. Ephraums, *Climate Change: The IPCC Scientific Assessment,* 1990.)

These scenarios do not include consideration of the indirect effects that result from chemical reactions in the atmosphere, i.e., the CFCs are greenhouse gases, but through chemical reactions result in the destruction of stratospheric ozone, another greenhouse gas. These indirect effects are just beginning to be understood. For a given greenhouse gas scenario, the direct radiative forcing can be calculated with fair confidence. As we discuss more fully in Section 1.3.3, far less confidence can be placed in the resulting climatic change estimates. For this reason the IPCC working group has presented three scenarios for globally averaged temperature resulting from the Business-as-Usual emissions scenario. These are shown in Figure 1.3. They reflect different estimates of the "climate sensitivity," the extent to which climate will change in response to changes in the radiative forcing. (For details of the assumptions involved in these scenarios the interested reader is referred to the IPCC Report.) The "high estimate" results in a global temperature increase of more than 6°C by the end of the next century. Even for the "low estimate" the temperature increase is nearly 3°C.

The IPCC report also includes an estimate of the global temperature increase for each of the four emissions scenarios with climatic sensitivity set at the "best estimate" level. These are shown in Figure 1.4. Even if the rather strong prevention strategies of Scenario D are implemented a substantial change in the climate of the Earth by the end of the 21st century is implied.

1.3 THE GREENHOUSE EFFECT AND CLIMATE CHANGE

1.3.1 The Natural Greenhouse

A primary reason for concern about the increasing concentrations of these gases is that they are strong absorbers of thermal radiation in the infrared wavelength range while absorbing very weakly in the short wavelength range of solar radiation (see Section 1.2). They therefore allow solar radiation to pass through the atmosphere, but selectively absorb terrestrial radiation emitted by the Earth and its atmosphere. The resulting trapping of heat in the Earth atmosphere system is generally referred to as the greenhouse effect, and the gases as greenhouse gases (GHGs). If no such gases were present in the Earth's atmosphere the Earth would be considerably colder than it is today. It would, in fact, be too cold to sustain life. Conversely, if the concentrations of GHGs were larger, the Earth would be a warmer planet. Suppose, for example, that the atmosphere (including clouds) did not participate at all in the radiation balance of the Earth. A global radiation balance wherein the solar radiation absorbed by the Earth, αSA_c, is set equal to the infrared radiation emitted by the Earth, $\sigma \varepsilon T_{av}^4 A$, gives:

$$\alpha SA_c = \sigma \varepsilon T_{av}^4 A$$

(Eq 1.1)

where S is the solar constant (the radiant heat flux from the sun at the average distance of the Earth from the sun, taken here as 1360 w/m^2), α is the absorptivity (co-albedo) of the Earth, σ is the Stefan Boltzman constant, ε is the emissivity of the Earth, A_c is the cross section area of the Earth, A is the total surface area of the Earth, and T_{av} is the average surface temperature. Satellite measurements indicate that the average value of α is about 0.7, while most of the surface materials of the Earth have emissivities greater than 0.95. The ratio A/A_c is 4. Using these values in the above equation gives $T_{av} = 258K$ ($-18°C$). The actual spatially and annually averaged surface temperature is approximately 33 degrees higher than this. This difference is a manifestation of the greenhouse effect of the present atmosphere.

If one performs the same type of calculation for Mars and Venus one finds that nature has provided us with an empirical proof of the existence of the greenhouse effect. The temperature of the surface of Mars in the absence of an atmosphere would be approximately $-57°C$. Its atmosphere contains CO_2, but it is so thin that there is essentially

Table 1.1   Nature's proof of the greenhouse effect.

	Surface Pressure (Relative to Earth)	Main Greenhouse Gases	Surface Temperature In Absence of Greenhouse	Observed Surface Temperature	Warming Due to Greenhouse Effect
Venus	90	>90% CO_2	$-46°C$	477°C	523°C
Earth	1	~0.04% CO_2 ~1% H_2O	$-18°C$	15°C	33°C
Mars	0.007	>80% CO_2	$-57°C$	$-47°C$	10°C

no greenhouse effect at all. On the other extreme is Venus, whose atmosphere is dense and contains large amounts of CO_2. The surface temperature of Venus without an atmosphere would be approximately -46°C. The presence of greenhouse gases results in an actual surface temperature of 477°C. These results are summarized in Table 1.1.

1.3.2 The Man-Made Greenhouse

There is no question that the greenhouse effect, warming due to the presence of GHGs in the atmosphere, is real. Therefore, there is little doubt that increasing the concentrations of GHGs in the atmosphere will lead to more warming. There is a good deal of uncertainty, however, about the magnitude, distribution, and timing of the resulting climate change with gradually increasing GHG concentration as well as the impacts of this change on the Earth and its animal and vegetable inhabitants. Nevertheless, a consensus has developed among climatologists that the globally and annually averaged equilibrium temperature of an Earth with a radiative forcing equivalent to doubling of the CO_2 concentration would be between 1.5°C and 4.5°C warmer than the present (IPCC, 1990), warmer than at any time during which mankind has been its inhabitant. Atmospheric concentrations of GHGs equivalent to a doubling of carbon dioxide are currently predicted to occur at least by the second half of the next century (IPCC, 1990).

If currently prevailing human activities do not change in the next few decades we shall likely be faced with a climate substantially different from that of today. In addition, the *rate* of change of climate will in all likelihood be unprecedented in recent history. The rate of change may well prove to be large enough to make it very difficult for many species of plant and animal life (including humans) to adapt to the new climate as they sometimes have in the past (see Section 1.5.2). It would, therefore, seem prudent to consider our alternatives. There are two general ways to prevent GHG-induced climate change. The first is to somehow cause the concentrations of atmospheric GHGs to reach steady state values not too different from today's. This can be accomplished either by decreasing the sources of these gases or by increasing the sinks. The second method is by controlling the climate itself, for example, by using huge orbiting solar reflectors to reflect solar radiation away before it reaches the Earth's atmosphere. This may sound like science fiction, but such ideas should not be rejected out of hand. Landing a human on the moon seemed like science fiction only a few decades ago.

If climate change cannot be prevented, and it probably cannot be entirely prevented given our current heavy commitment to burning fossil fuel and given the present uncertainty as to the sources of many of the other greenhouse gases (e.g. methane), we shall have to adapt to the changing climate. Since most adaptive methods will be more or less regionally specific, it would be most useful to know regional details of climate change and the resulting impacts. Unfortunately, current climate models have not proved successful at predicting even the present climate on regional scales (Grotch and MacCracken, 1991; Willmott and Legates, 1993). Hence, in planning adaptive strategies we shall be faced with the age old engineering problem of designing in the face of uncertain constraints.

The bulk of this book is not about climate change. It is about the engineering response to climate change. Nevertheless, any response to climate change must be undertaken with an understanding of the situation we seek to avert along with its seriousness and its certainty. We will need to decide between adaptation and mitigation and will have to judge responses in terms of their cost, effectiveness, potential risk, etc. Engineering responses must be envisioned within the context of the problem. We provide a brief summary of the problem as currently conceived; more detailed discussion is available in monographs such as IPCC (1990 and 1992). We begin with a discussion of climate sensitivity and climatic feedbacks and use this as a background for describing the results of climate change scenarios predicted by general circulation models of the Earth's atmosphere and oceans.

1.3.3 Climate Sensitivity and Climatic Feedbacks: Qualitative Results from Simple Models

The simplest of climate models are the energy balance models. Energy balance models are conceptually useful because they can be used to illustrate and at least qualitatively evaluate the roles of some of the various internal feedbacks on the response of the Earth's temperature to external changes, such as changes in the GHG concentration of the atmosphere. Illustrating with a couple of basic equations can help us appreciate the relationships among atmospheric chemistry, clouds, ice cover, water vapor, etc., and the Earth's surface temperature.

1.3.3.1 Climatic Feedbacks

In a state of thermodynamic equilibrium, the solar radiation absorbed by the Earth and its atmosphere, cryosphere, and hydrosphere are balanced by the infrared radiation emitted to space. This is expressed in the equation:

$$\alpha_{av}Q - R_{av} = 0$$

(Eq. 1.2)

where $Q = S/4$ is the average solar radiation flux incident on the outer edge of the atmosphere, α_{av} (the co-albedo) is the fraction of this radiation that is absorbed, and R_{av} is the average infrared radiant flux leaving the atmosphere. All the quantities in this equation are, of course, globally and annually averaged quantities. Equation 1.2 leaves unanswered questions concerning the spatial and temporal distribution of climatic variables. Nevertheless, one can draw some inferences and even make some quantitative estimates about climatic change using global energy balance models.

If the GHG content of the atmosphere were to suddenly change to a new value, the co-albedo and the infrared flux term R_{av} in Equation 1.2 would, after some transients, change to new values such that Equation 1.2, a steady state equation, would again be true. If we denote the GHG content of the atmosphere by y_c then following an incremental change in y_c to $y_c + dy_c$, the co-albedo changes to $\alpha_{av} + d\alpha_{av}$ and the infrared flux to $R_{av} + dR_{av}$. Once steady state is achieved

$$Qd\alpha_{av} - dR_{av} = 0$$

(Eq. 1.3)

The co-albedo and the infrared flux are both functions of many variables, including y_c. For example, the infrared radiation is a function of the average effective temperature of the Earth's surface and atmosphere, T_{av}, and its chemical and physical properties. Among these are the concentration of GHGs, y_c, the water vapor concentration, y_w, the fraction of the surface covered by clouds, $\bar{N}$, the average height of the clouds, $\bar{h}$, and perhaps other quantities, y_j. The co-albedo is a function of the reflective properties of the atmosphere (together with clouds) and the Earth. The presence of GHGs increases the absorption of infrared radiation by the atmosphere. The radiative balance is also affected by the presence of clouds and changes in the surface properties of the Earth such as, for example, changes in the extent of ice and snow cover and changes in the distribution of vegetation. Mathematically, the changes in R_{av} and α_{av} are written as follows:

$$dR_{av} = \frac{\partial R_{av}}{\partial y_c}dy_c + \frac{\partial R_{av}}{\partial y_w}\frac{dy_w}{dT_{av}}dT_{av} + \frac{\partial R_{av}}{\partial \overline{N}}\frac{d\overline{N}}{\partial T_{av}}dT_{av} + \sum_i \frac{\partial R_{av}}{\partial y_i}\frac{dy_i}{dT_{av}}dT_{av}$$

(Eq. 1.4)

$$d\alpha_{av} = \frac{\partial \alpha_{av}}{d\overline{N}}\frac{d\overline{N}}{dT_{av}}dT_{av} + \frac{\partial \alpha_{av}}{\partial \overline{h}}\frac{d\overline{h}}{dT_{av}}dT_{av} + \frac{\partial \alpha_{av}}{\partial x_s}\frac{dx_s}{dT_{av}}dT_{av} + \sum_j \frac{\partial \alpha}{dy_j}\frac{dy_j}{dT_{av}}dT_{av}$$

(Eq. 1.5)

The variable x_s stands for the average latitudinal extent of snow and ice. The variables y_i and y_j here stand for any extra variables, perhaps yet unknown, including, for example, vegetation distribution. Using Equations 1.3, 1.4, and 1.5 and solving for dT_{av}, we find that

$$dT_{av} = \frac{-\left(\frac{\partial R_{av}}{\partial y_c}\right)\partial y_c}{\left\{\left[\frac{\partial R_{av}}{\partial T_{av}} + \frac{\partial R_{av}}{\partial y_w}\frac{dy_w}{dT_{av}} + \frac{\partial R_{av}}{\partial \overline{N}}\frac{d\overline{N}}{dT_{av}} + \frac{\partial R_{av}}{\partial \overline{h}}\frac{d\overline{h}}{dT_{av}} + \sum_i \frac{\partial R_{av}}{\partial y_i}\frac{dy_i}{dT_{av}}\right] - Q\left[\frac{\partial \alpha_{av}}{\partial \overline{N}}\frac{d\overline{N}}{dT_{av}} + \frac{\partial \alpha_{av}}{\partial \overline{h}}\frac{d\overline{h}}{dT_{av}} + \frac{\partial \alpha_{av}}{\partial x_s}\frac{dx_s}{dT_{av}} + \sum_j \frac{\partial \alpha_{av}}{\partial y_j}\frac{dy_j}{dT_{av}}\right]\right\}}$$

(Eq. 1.6)

This is a very instructive equation, and is well worth dwelling upon. The quantity λ

$$\lambda = \frac{1}{\left\{\left[\frac{\partial R_{av}}{\partial T_{av}} + \frac{\partial R_{av}}{\partial y_w}\frac{dy_w}{dT_{av}} + \frac{\partial R_{av}}{\partial \overline{N}}\frac{d\overline{N}}{dT_{av}} + \frac{\partial R_{av}}{\partial \overline{h}}\frac{d\overline{h}}{dT_{av}} + \sum_i \frac{\partial R_{av}}{\partial y_i}\frac{dy_i}{dT_{av}}\right] - Q\left[\frac{\partial \alpha_{av}}{\partial \overline{N}}\frac{d\overline{N}}{dT_{av}} + \frac{\partial \alpha_{av}}{\partial \overline{h}}\frac{d\overline{h}}{dT_{av}} + \frac{\partial \alpha_{av}}{\partial x_s}\frac{dx_s}{dT_{av}} + \sum_j \frac{\partial \alpha_{av}}{\partial y_j}\frac{dy_j}{dT_{av}}\right]\right\}}$$

(Eq. 1.7)

is usually called the sensitivity parameter in the climate modeling literature. It describes the relationship between the change in the concentration of greenhouse gas and the change in temperature. The denominator of the sensitivity parameter contains the feedback terms, which we now briefly discuss.

A change in the GHG concentration affects the infrared radiation, changing the average temperature T_{av} of the system directly. This temperature change affects climatic variables such as atmospheric water vapor, cloud cover, ocean temperature, ice and snow cover, and perhaps other variables y_i and y_j. Changes in these variables cause a further change in R_{av} and α_{av}, and thus T_{av}, through feedback processes. If the result is an enhancement in the original change in T_{av}, the feedback is referred to as a positive feedback. If it acts to reduce T_{av}, it is called a negative feedback. At least some of the feedbacks are positive. For example, an increase in T_{av} causes an increase in the water vapor content of the atmosphere, increasing the infrared

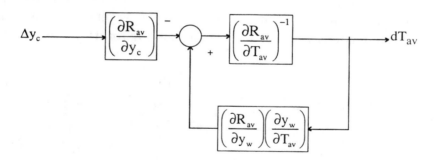

Figure 1.5   An illustration of climatic feedback: the water vapor feedback.

absorption, trapping more of the infrared radiation that would otherwise leave the system, and further increasing T_{av}. This is illustrated in Figure 1.5 for the case of water vapor feedback.

The situation with clouds is less certain. Most clouds are reflectors of solar radiation and absorbers of infrared radiation. If cloudiness increases with increased T_{av} (which we note may or may not be the case), the first of these is a negative feedback, since it tends to reflect energy away from the system when T_{av} increases. The second, absorption of infrared radiation, is a positive feedback. Mitchell et al. (1989) have shown that changes in the state of cloud water (ice crystal vs. water droplets) could provide a substantial negative feedback.

Surface snow and ice cover change provides a positive feedback. An increase in T_{av} leads to a decrease in snow and ice cover. Since snow and ice are efficient reflectors of solar radiation their disappearance leads to increased absorption of solar energy and a further increase in T_{av}. Little is known about vegetation changes that might result from a warming or cooling of the Earth, but Cess (1978) has indicated that this might be a positive feedback. In the three-dimensional world the effect of clouds will also depend on the nature, the height, and the geographic location of the clouds.

Let us now examine the results of the feedbacks in a more quantitative way. The terms in the numerator of Equation 1.6 represent the rate at which the net radiative flux at the top of the atmosphere changes with GHG content, holding all other variables constant, multiplied by the change of GHG content, dy_c. The value of this term has been determined by a number of investigators (IPCC, 1990) to be approximately -4.0 w/m^2 for a doubling of CO_2. It is generally felt that this result is accurate to within at least $\pm 20\%$.

The first term in the denominator is the rate at which the infrared flux changes with temperature, all other variables remaining constant. Its value may be estimated by supposing that the Earth–atmosphere system radiates as a black body at some equivalent temperature $T_{av,e}$. Thus, the appropriate temperature to be used is the temperature "seen" by a viewer in space, and is approximately 258K. This gives $\partial R_{av}/\partial T_{av}$ as about 3.5 w/m^2°C. With no feedbacks at all, a doubling of the atmospheric CO_2 might therefore be expected to increase the average temperature of the atmosphere by about $4.0/3.5 = 1.1$°C.

It is likely that the largest of the feedback terms is the water vapor feedback. An increase in the temperature of the atmosphere produces an increase in the equilibrium water vapor pressure in the atmosphere. It seems plausible, and is borne out at least qualitatively by model studies, and recently by satellite observations (Ravel and Ramanathan, 1989), that the relative humidity of the atmosphere remains approximately constant during a climate change. The total water vapor content of the atmosphere, the absolute humidity, therefore increases as the temperature increases. This increases the infrared absorptivity of the atmosphere, decreasing the infrared flux to space, and further increasing the atmospheric temperature. According to the IPCC study, water vapor and lapse rate feedbacks reduce the denominator of the sensitivity parameter to 2.4 ± 0.1, implying that

$(\partial R_{av}/\partial y_w)/(\partial y_w/\partial T_{av})$ is approximately -1.1 ± 0.1 w/m^2 C. The sensitivity as given by Equation 1.6 is thus substantially increased. The predicted temperature change resulting from a doubling of CO_2 is now $4.0/(3.5 - 1.1) = 1.67°C$.

The cloud amount and cloud height feedback terms are probably the least well known of all the feedbacks. Ramanathan (1977) has shown that if the cloud-top and surface temperatures increase or decrease together, the net feedback is zero. Cess (1978) also suggested that cloud amount feedback may be small. It is possible to find both positive and negative values of cloud amount feedback in the literature, however, and this feedback effect remains the most questionable of those discussed here. In Section 1.3.3.1 we will draw attention to the great range of cloud effects in general circulation climate models. For the present, we place its value at -0.3 ± 0.7w/m^2°C, as indicated in the IPCC (p. 139).

Some of the early energy balance climate models, e.g., Budyko (1969) and Sellers (1969) indicated that the ice-albedo feedback is a rather large positive feedback, perhaps as large as -0.7w/m^{2}C. It is now believed that the albedo feedback is a moderate positive feedback, perhaps -0.3±0.2 w/m^{2}C. (IPCC, page 139).

We are now in a position to fill in some of the most important terms in Equation 1.6. Dropping for the present the "unknown" feedbacks involving y_i and y_j and filling in the others as estimated above, we find that

$$\Delta T_{av} = \frac{4.0}{3.5 - (1.1 \pm 0.1 + 0.3 \pm 0.7 + 0.3 \pm 0.2)}$$

The upper and lower limits on our estimate of the atmospheric temperature change due to a CO_2 doubling become 5°C and 1.4°C. The uncertainty is seen to be fairly large. It is important to note that in making the calculation above we assumed that the errors in the estimates of the different terms are independent of each other and that the feedbacks themselves are not interactive. We also note that "unknown" feedbacks may make the sensitivity either larger or smaller, a point often not sufficiently emphasized by the climate modeling community. Detractors claim that unknown feedbacks may reduce climate sensitivity even further, but it is important to note that it is equally likely that the climate may be even more sensitive than predicted. The answer lies in the collection of terms represented in the denominator of Equation 1.6.

1.3.3.2 The Delaying Effect of the Oceans

All of the equations discussed so far have been steady state equations. The GHG content of the atmosphere will not change instantaneously of course, nor will the atmospheric temperature respond instantaneously. The strong interaction between the atmosphere and the ocean surface keeps the two more or less in equilibrium at some time scale, but the response of the ocean will lag behind any input forcing its change because of its large heat capacity. This thermal lag has been studied by several investigators.

Hunt and Wells (1979) used a radiative-convective atmospheric model coupled with a simple ocean consisting of a mixed layer of uniform temperature and fixed depth to estimate the response of the ocean-atmosphere system. Using a mixed layer 300 m deep, they determined that the response would be delayed by 8 to 10 years. Cess and Goldenberg (1981) coupled the mixed layer to the deep ocean through a diffusive mechanism, while Hoffert et al. (1980) used diffusion plus an imposed overturning to couple the two ocean layers. These studies indicate that time delays of 10 to 20 years might be expected due to ocean heat capacity. Each of these studies was concerned with the transient response of the surface temperature to an instantaneous doubling of the greenhouse forcing. In the real world the greenhouse effect is in the process of increasing gradually, and Thompson and Schneider (1982) pointed out that the transient response in this case might be quite different from the response to an abrupt change in the greenhouse forcing. Morantine and Watts (1990, 1993), Kim et al. (1992), and Watts and Morantine (1994) have studied the response to linear radiative forcing using energy balance climate models. Coupled atmosphere/ocean general circulation models are in the initial stages of studying the response to gradual increases in greenhouse forcing (see Section 1.3.4.2). The results indicate that time delays can be very large, with the transient lagging behind the predicted steady state response (that is, the response if it were immediate) by 50 to 100 years.

Schneider and Thompson (1981) examined the effect of the nonuniform latitudinal distribution of the mixed layer depth and its interaction with the deep ocean and found that the transient response of climate may vary substantially with latitude. This is borne out both by the energy balance model results of Kim et al. (1992) and by recent GCM experiments (see Section 1.3.4.2).

All these studies indicate that the transient response is a major uncertainty in the greenhouse gas-climate problem and that any observed change in climate will lag considerably behind observed changes in atmospheric chemistry.

1.3.4 General Circulation Models and Their Sensitivity

While energy balance climate models have proved to be very useful in understanding the transient response and feedback processes, they cannot be used to study the regional distribution of many important climatic variables such as precipitation and soil moisture, for example. For this purpose three dimensional models similar to weather prediction models are used. These models, which are called general circulation models, or GCMs, incorporate the momentum, energy, and continuity equations together with complex codes for computing radiative transfer in numerical models. They use a numerical grid that is typically about 400 to 500 km in the horizontal with 2 to more than 10 levels in the vertical direction. Obviously, many phenomena take place on scales considerably smaller than the numerical grid scale (clouds, precipitation, hydrologic phenomena, biota). These are usually parameterized in rather simplistic ways. Although numerical weather predictions diverge from observed conditions within several days, climate models, in which the simulations are extended over several years, appear to broadly resemble the actual climate when averaged, for example, over a month or a season.

We may speak of model sensitivity from any of several points of view: sensitivity of model prediction to various parameterization schemes, to changing boundary conditions, or to external forcing. In the first case, for example, we might be interested in whether "improved" parameterizations lead to substantially improved prediction. In the second case we might be interested in how deforestation will affect the local or global climate. The sensitivity of the model to external forcing is of interest for studying the response of the climate to greenhouse forcing. In this section we first compare predictions of the current climate by four widely cited GCMs that use a variety of resolutions and have been used to simulate climatic change due to increased atmospheric CO_2 concentrations. Next, we discuss several attempts to determine the sensitivity of individual models to the inclusion of more sophisticated parameterizations. We then examine the effects of changing environmental conditions (such as deforestation) on climate. Finally, we look briefly at CO_2-induced climate change as predicted by several models.

1.3.4.1 Modeling the Current Climate

If we wish to compare the climate simulated by a model to the actual climate, we must first have accurate data on the current climate. We will use the global climatology of mean monthly surface temperatures developed by Legates (1987) and Legates and Willmott (1990) from at least 20 years of data, as reported by Willmott and Legates (1993). Clearly, surface air temperature is only one measure of climatology, but it is the most easily accessible climate measure and probably the most reliable. We compare only the zonally averaged temperatures and refer the interested reader to the paper by Willmott and Legates (1993) for details of the fields. Figures 1.6 and 1.7 have been prepared from the data reported in that paper. Each of the models uses a slab ocean.

The Geophysical Fluid Dynamics Laboratory (GFDL) model predicts globally averaged temperatures that are far too low both in January (3.4°C) and in July (2.2°C) even though the solar constant used is far too high (1443.7w/m² compared to an actual value of about 1360 w/m²). Temperatures are far too low almost everywhere except at high latitudes in the southern hemisphere. This is especially true of mid-latitude continents in the northern hemisphere, where

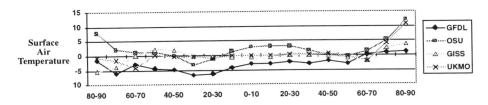

Figure 1.6   Difference between observed zonally average surface air temperature for Jan. and four GCM simulations. (Adapted from Willmott et al., 1993.)

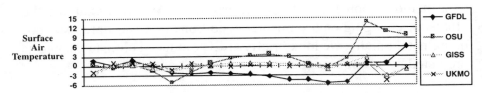

Degrees Longitude with North Pole on far Left and South Pole on far Right

Figure 1.7   Difference between observed zonally average surface air temperature for July and four GCM simulations. (Adapted from Willmott et
al., 1993.)

temperatures in Asia are more than 10°C too low. The model performs
somewhat better in July, but tropical temperatures are still far too low.

In the Oregon State University (OSU) GCM the globally averaged
temperature is overestimated by 1.4°C in January and 1.5°C in July.
Tropical and high latitude temperatures are generally too high and
mid-latitude temperatures too low in both hemispheres and both seasons. Temperatures over most continents are far too high, while the
model underpredicts ocean temperatures in the vicinity of western
boundary currents.

Globally averaged temperatures are well simulated by the Goddard
Institute of Space Studies (GISS) GCM. Mid-latitude January temperatures are higher than observed in the northern hemisphere and
lower than observed in the southern hemisphere. Southern high latitude temperatures are too high and northern high latitude temperatures are too low. In July, temperatures in the low latitudes are slightly overestimated, while in middle and high northern hemisphere latitudes they are generally underestimated. High latitude continental
regions, especially northern Canada, northeast Asia and Antarctica are
much better simulated by the GISS model than by either the GFDL or
the OSU model, and this probably results from the fact that the GISS
model incorporates a prescribed poleward ocean heat transport, while
the other models do not.

The United Kingdom Meteorological Office (UKMO) GCM simulation is also close to the observed temperatures. High southern hemisphere latitudes are too warm in January and high northern latitudes

are too cold in July. On a regional scale, the Tibetan Plateau and northern Europe are too cold in January. The predicted ocean temperatures agree almost as well with actual climatology as the GISS model, probably because the ocean component also included prescribed ocean poleward heat transport.

The comparisons presented here are based on model results that were readily available. Most of them have probably been updated with more sophisticated treatments of atmospheric boundary layer and land surface processes, and perhaps clouds also. The GFDL and OSU models used in these simulations are somewhat more crude than the GISS and UKMO models. For example, both the GISS and the UKMO models incorporate diurnal cycles and both include prescribed ocean poleward heat transport. The limited horizontal resolution of all the models results in inadequate representation of surface elevations. This may be why all of the models predict lower than observed temperatures in the Tibetan Plateau region and higher than observed values over Antarctica.

1.3.4.2 Sensitivity to Land Surface Parameterization

Sato et al. (1989) studied the effects of replacing a conventional land surface hydrological model with a simple biophysically based model of terrestrial vegetation in a GCM. The control model had three thermally active soil layers and the hydrology used a simple "bucket" 15 cm deep as in Manabe (1969). The "improved" model uses the Simple Biosphere Model (SiB) as described by Sellers et al. (1986). It addresses the effect of vegetation on land surface–atmosphere interactions by modeling biophysical processes that affect heat, momentum and mass transfer.

The control model is a rather sophisticated version of the Numerical Weather Center (NMC) GMC, which is an 18 layer model (5 or 6 in the planetary boundary layer) with spectral wave number 40. It is described in more detail in Sato et al. (1989). Surface roughness in the control model is prescribed and varies with location but is time invariant. In the SiB-GCM the surface roughness length, albedo, and heat and moisture exchange depend upon vegetation type and state, and thus vary with both space and time.

The SiB-GMC produced a much more realistic partitioning of energy flux on land surfaces, both globally and regionally, than did the control GCM. Generally, the SiB-GCM produced more sensible heat

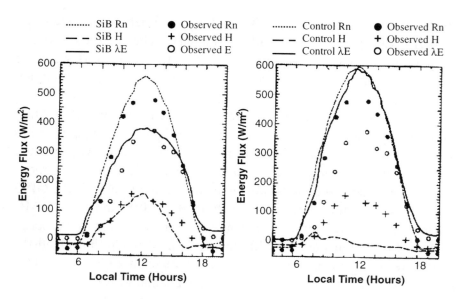

Figure 1.8   Simulated (SiB on the left, Ctl-GCM on the right) and observed
mean surface energy balance for June 15 – July 14 at 3°S and 59°W.
(From Sato et al., *J. Atmos. Sci.*, 46, pp. 4058–4097, 1989. With permission.)

flux, less latent heat flux and reduced precipitation rates over
continents. As one example, we reproduce in Figure 1.8 the simulated
heat flux partitioning in both models in a grid square near Manaus in
the Amazon basin (59 W, 3 S) compared to observations made over
the Amazon forest by Shuttleworth et al. (1984) during the period
June 15 through July 14. In the figure, R_n is net radiation, H is sensible heat, and λE is latent heat.

A somewhat simpler model (the SECHIBA model) has been
employed by Docoudre et al. (1993) within a GCM and compared to
results from the same model without the vegetation parameterization.
The results are quite comparable to those reported by Sato et al. It is
certainly not clear how much complexity the vegetation model must
contain in order to faithfully simulate regional climatology.

Marengo et al. (1994) examined the effect of an even simpler land
surface parameterization (Abramopoulos et al. 1988) and a river routing scheme on the hydrology of the Amazon basin. The general circulation model used was a version of the GISS GCM. The control
experiment incorporated a two layer soil storage model similar to the
bucket model. The "improved" version included vegetative resistance,
evaporation from precipitation intercepted by a canopy, and the

partition of soil water due to transpiration according to root density. Runoff from a box enters a river channel and moves downstream according to a river direction file developed by Miller et al. (1995). The control experiment predicted evaporation rates for the Amazon Basin that were far higher than the measured rates and this resulted in river runoffs that were very low. The "improved" parameterization of the hydrology resulted in river runoff rates and evaporation rates that were close to the observed values, but also predicted surface air temperatures that were too warm. The authors suggest that improved parameterization of spatially heterogeneous rainfall in each grid box should improve simulations of spatial and temporal variations of evaporation and runoff. Wood and Lakshmi (1993) investigated this problem and suggested that surface fluxes can be scaled, and that macroscale models based on effective parameters can be used to account for at least some small scale herogeities.

The sensitivity of climate to local albedo changes was probably first explored systematically by Charney (1975) and Charney et al. (1977). In semi-arid regions, an increase in albedo results in a decrease in the temperature of the atmospheric column and therefore a decrease in ascent and decreased precipitation. This is, of course, a positive feedback, tending to maintain desert areas. Garratt (1993) reviewed many sensitivity experiments and concluded that whether or not models include interactive soil moisture, they consistently predict that an increase in albedo always results in decreased evaporation over land, precipitation over land and increased precipitation over the sea.

An increase in surface roughness (for example, as modeled by increased surface roughness length) increases the drag coefficient and tends to increase potential evaporation. We note here that surfaces with large roughness lengths tend also to have relatively low albedo. The effects of including a canopy model in a GCM (as compared to bare soil) include a decrease in albedo, an increase in roughness length, the interception in reevaporation of water from vegetation, and the physiological control of evaporation through stomatal resistance (Garratt, 1993). Several studies that focus on the impact of tropical deforestation on climate have been published (Dickinson and Henderson-Sellers, 1988; Lean and Warrilow, 1989; Shulka et al., 1990; and Nobre et al., 1991). In all of these studies temperatures in Amazonia increase by 2–3°C, with decreases in annual precipitation of 500–800 mm and decreases in evaportranspiration of 200–400 mm.

These results imply that it might be difficult to reestablish tropical forests once deforestation occurs.

1.3.4.3 Sensitivity to Model Grid Resolution

The sensitivity of atmospheric GCM (AGCM) simulations to horizontal resolution is of considerable interest because many atmospheric processes have characteristic length scales that are much smaller than the grid spacing of current models (or of any future models within a reasonable time). For example, convective clouds have characteristic horizontal lengths on the order of a few hundred meters, and ground surface and vegetation of perhaps one kilometer. On the other hand, the kinetic energy spectrum of the atmosphere peaks at several thousand kilometers, the length scale of baroclinic waves. Therefore, there is some question as to how much can be gained through higher resolution as opposed to using the computational power for more refined modeled physics. In fact, Held and Phillips (1993), using a simple zonally symmetric GCM, found that the increase in eddy momentum fluxes and zonal winds that occurs when resolution is increased is principally due to meridional rather than zonal resolution.

Chen and Tribbia (1993) found that increasing the resolution of the NCAR CCM 1 produced stationary long waves that were *increasingly unrealistic*. Phillips et al. (1995) simulated summer and winter climates with a GCM at a variety of horizontal resolutions to examine the effects of horizontal resolution on simulation of moist processes. Progressing from low to moderate resolutions produced qualitative changes, while increasing the resolution further yielded smaller changes. Large scale moist processes, especially in the tropics, were simulated with progressively *less* realism as the resolution increased.

A similar study was performed with an ocean GCM (OGCM) described by Semtner and Chervin (1988) by Covey (1992). The model was run with horizontal resolutions of $4° \times 4°$, $2° \times 2°$, $1° \times 1°$, and $1/2° \times 1/2°$. The effective model viscosity was increased as the resolution increased in order to keep the computation numerically stable. Lower resolution cannot adequately capture western boundary currents. For resolutions $2° \times 2°$ or finer, however, the barotropic stream function changes little, leading the author to conclude that "that there is much to be gained by making the horizontal resolution of ocean GCMs finer than $4°–5°$ grid spacings used in traditional climate models, but diminishing returns set in as grid spacing is

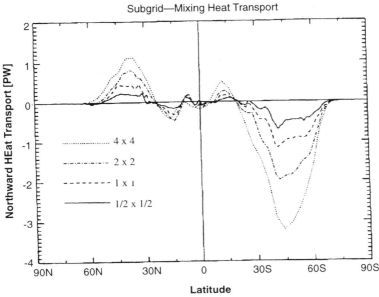

Figure 1.9   Sub-gridscale diffusion component of ocean heat transport, integrated over longitude and depth, for four resolution cases. (From Covey, C., *PCMDI Report No. 4, 1992.*)

decreased below about 1°." We note that the poleward transport of heat *decreases monotonically* with higher resolution in runs reported, and continues to change dramatically from the 1° x 1° run to the 1/2° x 1/2° run, indicating that this important variable is far from converged (Figure 1.9).

1.3.4.4 Carbon Dioxide Doubling Experiments

Until very recently many, but not all climate models used in transient experiments represented the ocean as a single well-mixed layer, usually with fixed depth. Some models allow for (fixed) horizontal heat transport by the deep ocean, but some neglect this heat transport entirely. Some have fully interactive deep oceans. In determining surface moisture, precipitation and snowmelt are accumulated at each grid point in a "bucket" until overflow occurs. Overflow is diagnosed as runoff and is transported to nearby grid points. The GISS model uses a somewhat more sophisticated model with two levels of soil moisture. Clouds are usually parameterized simplistically also. When a certain region attains a certain relative humidity, clouds form at that level. In early models cloudiness was fixed at current values. Most models now use model generated cloudiness.

Table 1.2   Globally and annually averaged changes in three CO_2-doubling studies.

Model	Surface Temperature Change (°C)	Precipitation Change (%)
CCC	3.5	4
GFHI	4.0	8
UKHI	3.5	9

A number of review articles in the literature compare the responses of various GCMs to a CO_2 doubling. We limit our discussion to a brief description of the comparative results. In CO_2 doubling model experiments, the models are first run until they reach steady seasonal predictions of the present climate. The CO_2 content of the atmosphere is then suddenly doubled and the model run to a new steady state. Typical globally averaged surface temperature and precipitation changes are shown in Table 1.2.

CCC refers to the Canadian Climate Centre Model, GFHI to the high resolution geophysical fluid dynamics model, and UKHI to the United Kingdom Meteorological Office high resolution model. The models agree on some general global features of a 2 x CO_2 world. Neither temperature nor precipitation changes are uniform spatially, however. The spatial distributions of two important parameters, surface temperature and soil moisture changes, for the summer season are shown in Figures 1.10 and 1.11.

There is considerable variation among models, especially for soil moisture changes, within individual continents. For example, the CCC model predicts dry conditions in the central U.S. for both summer and winter, while the Geophysical Fluid Dynamics Laboratory model predicts wetter winters and dryer summers. The United Kingdom Meteorological Office model predicts wetter winters in the western U.S., dryer winters in the eastern U.S., and dryer summers throughout the country. Clearly, the models are not yet very helpful in determining regional scenarios. Rind (1988) performed a series of numerical experiments with the GISS model in which the grid size and the ocean boundary conditions were varied in order to assess the sensitivity of the hydrologic cycle. He found that under doubled CO_2 mid-latitude drying over land is very sensitive to high latitude temperature amplification, sea surface temperature gradient, and model

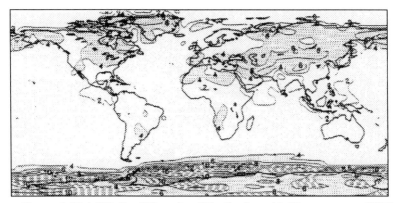

Figure 1.10a) JJA 2 x CO_2 - 1 x CO_2 Surface Air Temperature: CCC

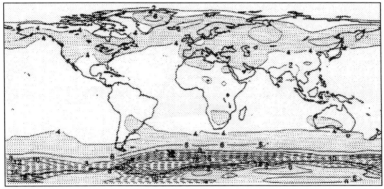

Figure 1.10b) JJA 2 x CO_2 - 1 x CO_2 Surface Air Temperature: GFHI

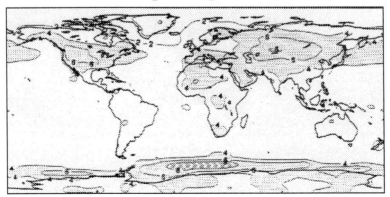

Figure 1.10c) JJA 2 x CO_2 - 1 x CO_2 Surface Air Temperature: UKHI

Figure 1.10   Change in surface air temperature (10–year means) due to doubling CO_2, for months July–August, as simulated by three high resolution models: a) CCC, b) GFIH, and c) UKHI. (From Houghton, J.T., G.J. Jenkins, and J.J. Ephraums, *Climate Change: The IPCC Scientific Assessment*, 1990.)

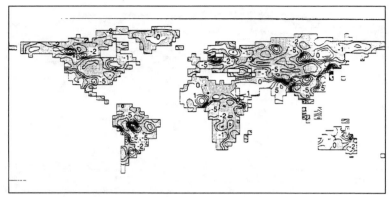

Figure 1.11a) JJA 2 x CO_2 - 1 x CO_2 Soil Moisture: CCC

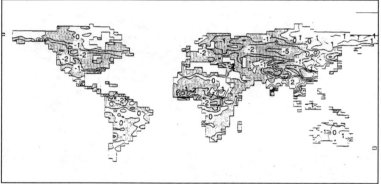

Figure 1.11b) JJA 2 x CO_2 - 1 x CO_2 Soil Mixture: GFHI

Figure 1.11c) JJA 2 x CO_2 - 1 x CO_2 Soil Mixture: UKHI

Figure 1.11   Change in soil moisture (smoothed 10-year means) due to doubling CO_2, for months July–August, as simulated by three high resolution models: a) CCC, b) GFHI, and c) UKHI. Note that a) has a geographically variable soil capacity whereas the other two models have the same capacity elsewhere. Contours at +0, 1, 2, 5 cm, areas of decrease stippled. (From Houghton, J.T., G.J. Jenkins, and J.J. Ephraums, *Climate Change: The IPCC Scientific Assessment*, 1990.)

resolution, and he concluded that precipitation and soil moisture are probably very sensitive to the way the hydrologic cycle is parameterized.

Lest the reader get the wrong impression about the substantial agreement between the globally averaged model results in Table 1.2, we should point out that these results, and not those from other model calculations, are listed for convenience because they correspond to the results shown in Figures 1.10 and 1.11, which are used to demonstrate the regional differences predicted by different models.

The results of many other models are given in the IPCC report. Some details about the differences among the various models are given in the IPCC (1990). The predicted globally averaged temperature increase for a CO_2 doubling ranges from 1.9°C (Mitchell et al., 1989) to 5.2°C (Mitchell et al., 1987). In the first of these the precipitation change is 3%, while in the second it is 15%. The only major difference in the two analyses by Mitchell et al. was in the way clouds and their radiative properties were modeled.

To illustrate the sensitivity of the various models to the treatment of clouds, Cess et al. (1990) performed a set of numerical experiments with 19 atmospheric general circulation models in which the global sea surface temperature was raised by 2°C and then lowered by 2°C while holding sea ice extent fixed. A perpetual July simulation was

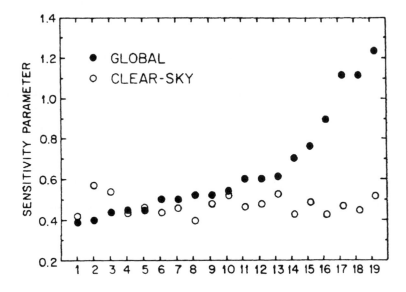

Figure 1.12   The clear sky and global sensitivity parameters for 19 general circulation models. (From Cess, R.D. et al., *J. Geoph. Res.*, 95, 1990.)

used in order to keep albedo feedback at a minimum. Cloud effects were isolated by averaging the top-of-the-atmosphere radiative fluxes over clear sky regions and thereby evaluating the sensitivity parameter for an equivalent "clear sky" Earth for comparison with the Earth as a whole. The results are shown in Figure 1.12. The "clear sky" sensitivities are remarkably consistent among the various models, while the global sensitivities vary by a factor of nearly three, suggesting that disagreements among model results can be traced largely to differences in cloud feedback.

The parameterization of clouds in GCMs is closely tied to relative humidity. The magnitude and distribution of relative humidity in climate models changes much less than the absolute humidity as the climate warms or cools. Seasonal data suggest that the real climate also behaves in this manner (Raval and Ramanathan, 1989; Rind et al., 1991). Most climate models with interactive cloud predictions show reductions in total cloudiness as global temperature rises, but an increase in clouds near the tropopause (Wetherald and Manabe, 1986; Mitchell and Ingram, 1992). Both lead to positive feedbacks. High clouds exhibit a lower solar reflectance than lower clouds. They are also cooler. Therefore, an increase in high clouds leads to a loss of reflected solar radiation to space and also a reduced infrared loss to space. A reduction in low and middle level clouds warms the troposhere and the surface because absorption of solar radiation is larger (relative to the surface) and the cooler clouds emit less infrared radiation.

It is important to realize that, while these results seem physically reasonable, they may or may not be accurate. We also point out that other feedbacks related to clouds might also be of considerable importance. Cess et al. (1989) have suggested that an increase in cloud water content that might accompany a tropospheric temperature increase could induce either a positive or a negative feedback. Mitchell et al. (1989) have shown that a change in state of the water in high clouds might produce a large negative feedback.

Much remains to be learned about the nature and magnitude (and even the sign) of the cloud feedback.

1.3.4.5 The Transient Response of GCMs

Studies of the response of the climate to a gradual increase in GHG concentration using GCMs have only recently begun. The first transient response models were step response studies. That is, the CO_2

concentration was instantaneously doubled and the subsequent approach to equilibrium studied. Such studies were performed, for example, by Schlesinger et al. (1985), Washington et al. (1989), and by Manabe et al. (1990). Model responses to gradually increasing CO_2 concentrations have been studied by Hansen et al. (1988), Washington et al. (1989), Stouffer et al. (1989), and more recently by Manabe et al., (1991), Meehl et al. (1993), Cubasch et al. (1992), and Manabe and Stouffer (1993). The ocean model used by Hansen et al. consisted of a mixed layer with diffusive heat flux at the bottom and a horizontal flux convergence within the mixed layer, all fixed at present conditions. The others used fully coupled atmosphere-ocean models.

Washington and Meehl (1989) performed a GCM model experiment in which the concentration of CO_2 was increased linearly at a rate of 1% per year. The 30% increase in CO_2 achieved after 30 years resulted in a 0.7°C increase in the globally averaged temperature. The warming was most evident over Northern Hemisphere continents and the Southern Hemisphere oceans between 30 and 60 S latitude, especially in winter. There were large areas of cooling over the high latitude Northern Hemisphere oceans, possibly reflecting a weakening of the thermohaline circulation. Geographic patterns of surface air temperature are in general quite different from those predicted by the same model with instantaneous CO_2 doubling.

The results of the numerical experiment by Stouffer et al. are described in considerable detail in the IPCC report. The more recent papers by Manabe et al. (1991, 1993) represent extensions of that work. The models were subjected to an exponentially increasing CO_2 content at a rate of 1% per year (compounded), leading to a doubling of the CO_2 content after 70 years. This rate was chosen because it approximates the current rate of increase of greenhouse gas concentration as CO_2 equivalent. The period between 60 and 80 years after the radiative forcing began was chosen for detailed analysis. In the steady state CO_2 doubling experiments the model surface temperature increase was very large during the winter in the Arctic and surrounding regions and in the coastal Antarctic. During the summer season it was somewhat smaller. In the transient models, vertical mixing of heat between the upper layer of the ocean and deeper layers in high southern latitudes and in the high latitude North Atlantic led to smaller temperature increase (i.e., a slower response) in these regions.

Nevertheless, the distribution of temperature increase in the transient experiment resembles that of the steady state CO_2 doubling experiment in most of the Northern Hemisphere and also in the Southern Hemisphere at low latitudes. Summer soil moisture is reduced over most continental regions except for the Indian sub-continent. Winter soil moisture is increased in large regions of the mid- and high latitude Northern Hemisphere and decreased in the sub-tropics.

Cubasch et al. (1992) and Meehl et al. (1993) show very similar climate change features during transients. Meehl et al. used the Community Climate Model (CCM) of the National Center for Atmospheric Research, subjected to an increase of greenhouse gas concentration of 1% per year (linear), while Cubasch et al. subjected a modified version of the European Center for Medium Range Weather Forecasts model to a greenhouse concentration forcing of Scenario A, as described in the IPCC report. This is forcing close to that used by Manabe et al. (1991) and increased the concentration by a factor of two after about 60 years. In each experiment the globally averaged temperature increased slowly at first, reaching a rate of about 0.03 to 0.036°C per year after several decades, and reaching about 2.5°C after 100 years. Generally speaking, the models showed that (1) the Northern Hemisphere warms faster than the Southern Hemisphere; (2) land masses warm faster than oceans; (3) high northern latitudes outside the North Atlantic contribute most to the warming; and (4) high southern latitudes contribute least. Cubasch et al. predict a rise in sea level of about 15 cm after 100 years while Manabe and Stouffer show a rise of about 20 cm, increasing to nearly 2 m after 500 years if the forcing is turned off after quadrupling. In both cases the rise is due only to the thermal expansion of the ocean; possible melting of land ice is not included in either model.

It appears that the slow response of the high latitude oceans to greenhouse forcing is the result of changes in the thermohaline circulation of the ocean (Manabe et al., 1993). The fact that relatively small variations in the thermohaline circulation can cause measurable variations in the ocean surface temperature was pointed out earlier by Watts (1985). Decadal variability of the GFDL GCM as reported by Lau (1992) supports this. Another interesting and important possibility is suggested by Manabe and Stouffer. When their model was run until the CO_2 concentration was doubled and then held stable, the thermohaline circulation in the North Atlantic decreased for several decades after the CO_2 stopped increasing, but eventually returned to its previous level. However, when the CO_2 concentration increased to four times the current level and then leveled off, the thermohaline circulation essentially disappeared after about 200 years and had not returned when the model run was completed after 500 years. The possibility of a profound impact on the biogeochemistry of the ocean cannot be rejected.

All globally averaged surface temperature predictions in the transient models show a steady rise after a brief initial phase. The IPCC report states that the response is linear but with a smaller slope than the equilibrium response would indicate, in agreement (they state) with results from upwelling-diffusion models. However, analytical solutions of upwelling-diffusion models (Morantine and Watts, 1990) and the GCM results of Cubasch et al. (1992) show that the lag time approaches a constant after 100 years or so, with the temperature response slope equal to that of the equilibrium models.

A close examination of the Stouffer et al. results (IPCC, 1990) shows the same upward curvature and the approach to the asymptotic response obtained by Cubasch et al. This is an important distinction, because it means that the rate of temperature increase is not constant at about 68% of the equilibrium rate as estimated in the IPCC report, but gradually increases with time until it approaches the equilibrium rate. According to the IPCC report the global warming that we should have experienced since pre-industrial times (defined there as 1765) falls between 0.6°C and 1.3°C, with a best estimate of 0.9°C. As we will see in the next section, global surface temperature data sets show that a global temperature increase of approximately 0.5°C has taken place during the last century. This falls just below the low end of the IPCC estimate. Does this mean that even the least sensitive of the models is too sensitive? Could it imply that the models are missing important negative feedbacks? Perhaps the time delay due to the ocean is larger than anticipated. In the next section we briefly describe the efforts to detect recent climate change.

1.4 DETECTING CLIMATE CHANGE

There is some controversy over whether a climate signal resulting from GHG increases has already occurred. If a temperature change is not yet scientifically demonstrable, what will it require to make such a demonstration? In the transient model of Hansen et al. (1988), GHG loading was increased according to three scenarios, one for which the growth of GHGs continues at typical rates of the 1980s, another with a moderate decrease in the rate of increase, and a third with drastic decreases such that the GHG forcing ceases to increase after 2000. In all cases the globally averaged temperature had risen above the noisy signal of the control run by the year 2000. Hansen concluded that if the real climate behaves in approximately this way it will constitute a detection of greenhouse warming with very high confidence.

In the meantime it is instructive to examine the results of Hansen's control run. Prior to the transient experiment, a 100-year run was performed with atmospheric composition fixed at 1958 levels. The globally averaged temperature departed from the long-term average by as much as 0.3 to 0.4°C during the run. Similar results are shown in Figure 1.13 from a 100-year unforced run with a coupled ocean-atmosphere GCM reported by the IPCC report and in Manabe et al. (1991). The result was not unexpected. It has been known for a long time (Lorenz, 1969) that the climate signal can display large internal natural variability. This variability is clearly displayed in actual temperature records. Figure 1.14 shows hemispherically and globally averaged temperature departure from the 1951–1980 mean values for

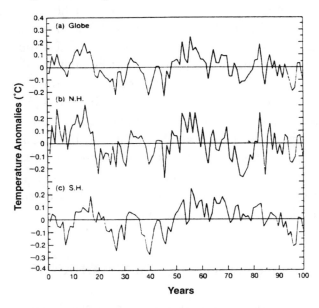

Figure 1.13   Global temperature variability in an unforced GCM run. (From Manabe, S., et al., *J. Climate*, 4, 1991.)

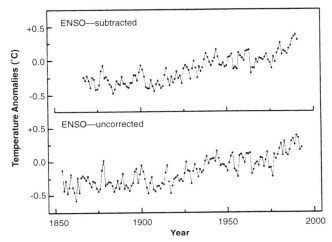

Figure 1.14a Global annual temperature anomalies, 1854–1993. (From Jones, P.D. et al., Global and hemispheric temperature anomalies—land and marine instrumental records. In T.A. Boden et al., eds, *Trends '93,* 1994.)

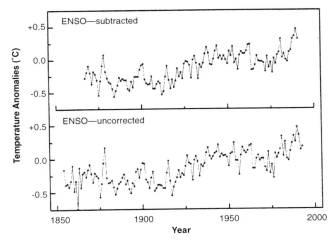

Figure 1.14b Northern Hemisphere annual temperature anomalies, 1854–1993. (From Jones, P.D. et al., Global and hemispheric temperature anomalies—land and marine instrumental records. In T.A. Boden et al., eds, *Trends '93,* 1994.)

years 1854–1993 *(Trends '93,* 1994). These global and hemispheric temperature anomaly estimates were compiled by Jones et. al. (1994) based on corrected land and marine data. Two series of global and hemispheric temperature anomaly estimates are shown, one adjusted for the influence of the El Niño Southern Oscillation (ENSO) events and the other uncorrected for ENSO events. The ENSO influence can be extracted by using regression techniques (Jones, 1988; Angell, 1990). According to Jones (1988), the Southern Oscillation Index (SOI) explains 20–30% of the high-frequency temperature variance. Large annual and decadal variability is evident. A linear trend fitted between 1890 and 1989 shows an increase of 0.47°C per 100 years for the Northern Hemisphere, 0.53°C for the globally averaged value.

Even the filtered data is clearly not increasing monotonically, however. Over the 30-year period between about 1940 and 1970 the global and Southern Hemisphere data show a leveling off, and the Northern Hemisphere data show a cooling. If the global temperature rise of roughly half a degree Centigrade is the result of GHG warming, how does one account for the large variability?

There have been a few attempts to account for the variability by supposing the external forcing by the Sun and by atmospheric particulates has changed (Hansen et al., 1981; Gilliland, 1982). The recent Marshall Institute Report (Seitz et al., 1989, 1992) appears to hold the position that the entire recent temperature data set might be explained by changes in solar activity. However, a critical analysis of satellite data between 1980 and 1989 by Foukal and Lean (1990) implies that the variations in solar activity are far too small to account for the observed temperature variations. Kelly and Wigley (1990) and Wigley and Raper (1990) also estimate that the effect of solar

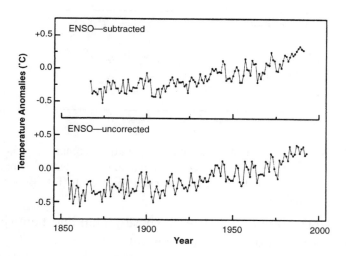

Figure 1.14c   Southern Hemisphere annual temperature anomalies 1854–1993. (Jones, P.D. et al., Global and hemispheric temperature anomalies— land and marine instrumental records. In T.A. Boden et al., eds, *Trends '93*, 1994.)

variability can account for only a very small fraction of the observed warming.

If we write $\Delta T_{OBS} = \Delta T_{GHG} + \Delta T_{VAR}$ in which ΔT_{OBS} is the observed global temperature anomaly, ΔT_{GHG} is the change due to greenhouse forcing, and ΔT_{VAR} is the part of the anomaly due to all other causes, we observe that the greenhouse warming signal can be obtained by subtracting ΔT_{VAR} from ΔT_{OBS}, or $\Delta T_{GHG} = \Delta T_{OBS} - \Delta T_{VAR}$. The problem is then reduced to finding the magnitude of the climate variability, here defined as all global temperature variation except for the greenhouse signal. (This is a bit of a misnomer since other anthropogenic inputs, such as sulfur from fossil fuel burning might also be involved.) Observe that ΔT_{VAR} may come from either internal or external variability. External variability is regarded here as variability that is not an integral part of the climate system itself. Solar constant and orbital parameter variations and volcanic or anthropologically produced aerosols are examples (e.g., Hansen et al., 1981; Charlson et al., 1990). Internal variability is here regarded as that which is somehow internal to the climate system. Variations in cloudiness on interannual or decadal time scales is one clear example of internal variability. (We note that variations in cloudiness might be entirely uncorrelated with surface temperature changes, and may not be associated with atmospheric feedback. We nevertheless consider it as internal variability.) This has recently received considerations by, among others, Wigley et al. (1990), who forced an energy balance (upwelling-diffusion) climate model with white noise radiation with an interannual standard deviation of approximately 1 w/m^2 (adjusted to match the high frequency variations in the global temperature). The low frequency characteristics of the output of the climate model were then examined. The fluctuations with timescales of 10–100 years were substantial, but not large enough to account for the recent global warming.

Charlson et al. (1987) have pointed out that sulfate aerosols originating from dimethylsulfide produced by planktonic algae may be a major source of cloud condensation nuclei over the oceans. The presence of sulfate aerosols is an important determinant of cloud albedo and thus may have an effect on climate. If the rate of dimethylsulfide emissions by the ocean is also affected by the climate, this may be an important feedback that can produce, amplify, or damp variability.

A third example of internal variability is the transfer of heat between reservoirs such as the atmosphere and the mixed layer and deep oceans. It is clear that the ocean can absorb very large amounts of heat without showing a substantial temperature increase because of its large heat capacity. It has long been recognized that the greenhouse gas-climate signal will be delayed by several decades due to heat leaking into the deep ocean from the mixed layer, which itself serves as a somewhat smaller capacitance. On the other hand, it has only recently been pointed out that variations of the dynamical state of the ocean can cause rather large variations in the ocean surface temperature (Roemmich and Wunsch, 1984; Watts, 1985; Levitus, 1989; Antonov, 1990; Watts and Morantine, 1991; Rahmstorf, 1995; Parrilla et al., 1994). Both Levitus and Antonov have shown that during at least parts of the period (about 1940–1979) during which the globally averaged temperature was decreasing, the temperatures at mid-depth in the North Atlantic ocean were warming. Antonov has since shown that temperatures in the South Atlantic have also increased at mid-depths (private communication, P. Groisman, 1991). Watts and Morantine estimated from the data of Levitus that the rate of heat transfer to the mid-depth ocean during the periods between 1955–59 and 1970–74 was approximately 36 million Mw. Averaged over the surface of the Earth this amounts to 0.072 w/m^2. If a similar amount of heat was transferred to the intermediate water in the South Atlantic, as implied by Antonov, the resulting 0.144 w/m^2 would be more than enough to completely offset the greenhouse forcing during this time period.

It was suggested by Rooth (1982), and rendered at least plausible by a number of experiments with numerical ocean models (e.g., Weaver et al., 1991 and the references therein), that the ocean can show considerable variability on decadal and longer time scales. Covey (1991) further suggests that this implies that much of the variability of the climatic "noise" on time scales of one or more decades might come from the variability of oceanic flows (see also, Lau, 1982). Delworth et al. (1993) found 40–60 year cycles in an unforced 600-year integration of a general circulation model that are localized in the northwest Atlantic and are associated with variations of the ocean thermohaline circulation. The deep ocean is a very large thermal reservoir whose temperature variations have not been carefully monitored.

There are, of course, other types of data that can be used as bellwethers for the detection of GHG-induced global warming. For example, all CO_2 doubling experiments show a cooling of stratospheric temperatures above the 50 mbar pressure level, suggesting that stratospheric temperature might be an important variable for detection of GHG-induced climatic warming. However, natural variability, the poor quality and short duration of the measurements, and possible effects of other signals such as ozone depletion and the presence of volcanic aerosols make it difficult to interpret the data.

Tropospheric water vapor should increase as the climate warms, and it seems to have done so (Raval and Ramanathan, 1989). The short duration and inhomogeneity of the available data, however, again makes the record difficult to interpret in terms of long-term trends.

Increasing global temperature might be expected to cause sea level rise due both to thermal expansion of the ocean and to melting of land ice. There has apparently been a long-term retreat of small glaciers (Wood, 1988) but the record is noisy. Both Barnett (1988) and Gornitz and Lebedeff (1987) report a 10–12 cm rise in sea level over the last 100 years. It has also been recently reported (Gloersen and Campbell, 1991) that the sea ice extent in the Arctic has decreased by about 2% over the years 1978 to 1987. Once again the duration is short and the record noisy. Both Northern Hemisphere snow extent and the sizes of many glaciers appear to be decreasing (IPCC, 1992).

It is also worth noting that a recent analytical model predicts that a moderate warming will produce a net accumulation of ice in Antarctica and therefore a decrease in sea level (Huybrechts and Oerlemans, 1990; Huybrechts et al., 1991).

It has been proposed that detecting a greenhouse-induced climate signal (that is, detecting a climate signal and attributing it to the greenhouse effect) might be possible with higher confidence through observation of a multivariate signal. This simply means that if several predicted variable changes are borne out by observational data, confidence in a cause and effect relationship increases. This presupposes, of course, that at least some signals predicted by models are reasonably accurate. This approach is generally referred to as the multivariate or fingerprint method (see MacCracken and Moses, 1982; IPCC, p. 252).

An example of the use of the fingerprint method has recently been presented by Karoly et al. (1992). The recent variation of the zonal mean atmospheric temperature as a function of height and latitude was compared with the difference between control and doubled CO_2 simulations in three climate models. Although there was a significant increase in the greenhouse signal in the observational data during the period examined (1963–1988), the authors point out that the results must be treated with caution both because the period of observation of the upper air data is short and because the fingerprint used may not be unique to an enhanced greenhouse effect. The method should prove to be more useful in the future as climate models improve, as a longer time series of upper air data becomes available, and as more variables are identified as important components of the fingerprint. Karl (1994) points out the importance of improving the monitoring of a number of important environmental parameters.

Singular spectrum analysis (SSA) is a data analysis method that can be used to describe the main physical phenomena reflected in a time series. It was first used in the analysis of paleoclimatic records by Vautard and Ghil (1989). The ability of SSA to identify the principal climatic oscillations and the regime changes in the amplitude of the records was demonstrated.

Ghil and Vautard (1991) used SSA to analyze a 135-year global surface air temperature record (Jones, Wigley and Wright, 1986) and showed the existence of interannual and interdecadal oscillations and a secular warming trend. The interannual oscillations were of periods 6 and 5 years, and were attributed to global aspects of the El Niño Southern Oscillation (ENSO) phenomenon. The interdecadal oscillations periods were 21 and 16 years, and were ascribed to changes in the extra-tropical ocean circulation.

Elsner and Tsonis (1991) reanalyzed different lengths of the temperature record and 5 other datasets and concluded that there was no support for the presence of the bidecadal oscillations reported by Ghil and Vautard (1991). They attributed the presence of possibly faulty data before 1880 as the cause of the oscillations, as the oscillations were nonexistent in analyses of subsets of the record that excluded the period prior to 1880.

Allen et al. (1992), in another examination of the results from Ghil and Vautard (1991), argued that if the power of the pair of empirical orthogonal functions corresponding to the bidecadal oscillation were confined to the early part of the time series, then the amplitude of the 20-year oscillation would have diminished over time, and not maintained its strength as reported.

Schlesinger and Ramankutty (1994) used a simple hemispherically resolved climate/ocean model (Schlesinger et al., 1992; Schlesinger and Ramankutty, 1992) to determine global mean temperature changes for prescribed GHG and atmospheric sulfate aerosol radiative forcing. This was subtracted from the IPCC-observed global mean surface temperature changes over 1858–1992 to construct a residual temperature change series that ostensibly reflects what is not caused by atmospheric sulfate aerosol or GHG radiative forcing. They analyzed this residual series and showed the existence of a 65–70 year oscillation in the global temperature record. This oscillation was found to be the statistical consequence of 50- to 88-year oscillations for the North Atlantic Ocean and its bounding Northern Hemisphere continents.

Mahasenan, Watts and Dowlatabadi (1996) performed SSA on 27 proxy and historical temperature records that range in length from 173 to 1481 years and found significant variability on the century time scale with the most predominant modes being oscillations with periods 150 to 170 and approximately 80 years, similar to those reported by Broecker (1975). The oscillations in the Northern Hemisphere peak around 1950, suggesting that the natural variability now shows a tendency toward decreasing temperature. The warming over the past few decades might therefore suggest that the climate sensitivity to greenhouse gas forcing is greater than generally acknowledged, since the recent warming is partially offset by the cooling of the long-term cycle.

Important new evidence concerning climatic variability has recently been presented by project members of the Greenland Ice Core Project (GRIP). The ice core from Summit, Greenland, provides evidence that during the last interglacial period, the Eemian, approximately 120,000 years ago, the variability was considerably larger than it has been during the present interglacial. The average temperature during the Eemian is currently thought to have been somewhat warmer than the present interglacial. Another ice core near the GRIP core does not show high variability during the Eemian (Groots et al., 1993). However, evidence from rock magnetism (Thouveny et al., 1994) and a European pollen record (Field et al., 1994) do show high variability. It is well documented that the climate during the glacial period was also highly variable. This raises sobering questions about changes in the variability of the climate that might result from future warming (Dansgaard et al., 1993).

1.5 HAS THE TIME COME TO RESPOND?

It can safely be said that the climate of the Earth has warmed by approximately 0.5°C during the last 100 years. It is also clear that the atmospheric loading of a number of GHGs has increased and continues to increase. It is arguable, however, that this does not automatically establish cause and effect. What with the uncertainties in both modeling and detection of greenhouse-induced climatic change, one must take seriously the question of whether it is time to respond. In this section we briefly outline and respond to some of the arguments in favor of waiting for much greater certainty before we respond.

1.5.1 Dissenting Opinions

Ellsaesser et al. (1986) point out that the Earth has been warmer than the present during periods such as the altithermal and the medieval optimum, presumably without additional atmospheric CO_2. There are, in fact, those who feel strongly that climate change due to the greenhouse effect will not likely be large (e.g., Lindzen, 1991).

Budyko (1991) has argued that after a certain point a general global warming will benefit humans. Relying for the most part on the study of past historical warm periods, Budyko contends that, while a slight global warming will produce dryer summers in both the U.S. and the Soviet Union, once the warming reaches 3 to 4°C precipitation will be enhanced everywhere. He and his colleagues in the State Hydrological Institute in Leningrad estimate that the increased precipitation and elevated CO_2 levels will increase world agricultural yield by some 40%, saving perhaps 2 billion people from starvation. However, as the catastrophic flooding in the Southeastern U.S. and in China can attest, more precipitation is not good by definition. It must occur at the right time and in the right place.

Ausubel (1991), on the other hand, argues that humans have adapted to climate change or to unfavorable local climatic regions successfully in the past, and that technological advances of the recent past have rendered human society even less vulnerable to climate variation. Why, then, should we begin to adopt possibly expensive and disruptive policies to limit GHG emissions?

1.5.2 Responses to the Dissenters

First of all, the evidence supporting the idea that increased GHGs will lead to global warming is very strong indeed. There is widespread, practically unanimous agreement among climatologists that a doubling of the atmospheric concentration of CO_2 would lead to an increase of the surface air temperature of the Earth of between 1.5 and 4.5°C under equilibrium conditions. The range of these numbers reflects the admitted uncertainties of the model predictions. The value is as likely to be outside this range on the high side as it is on the low side. If we choose to allow the buildup of the GHG concentration to continue unabated, then we are presumably counting on climatic sensitivity being much smaller than that implied by most climate models (as suggested by Lindzen), or that any change will be either benign or an improvement over the present climate (as suggested by Budyko), or that we will be able to easily adapt to the changes (as suggested by Ausubel). Losing such a bet could prove to be very expensive, both in terms of environmental impact and human suffering, and in terms of real money. It cannot be overemphasized that the cost of adaptation once the impact of climatic change has been realized may be substantially larger than the cost of avoiding climatic change.

The suggestion by Budyko that a warmer climate will be better deserves careful consideration. During the Little Ice Age a few hundred years ago the globally averaged surface air temperature of the Earth was perhaps a degree or so colder than the present. If colder temperatures are worse, then why are warmer temperatures not better? The answer lies in the fact that we have adapted our infrastructure, our population distribution, the regions where agriculture is successfully practiced, etc., to the current climate. In a very real sense, therefore, *any* substantial change from the current climate will very likely stress the system in a negative way. Moreover, the increasing world population will make matters significantly more difficult. In many areas of the world water resources are scarce. If climatic change makes this situation worse, as is likely in the U.S. midwest, for example, it will be very difficult to adapt to this change. The increasing population makes adaptation increasingly difficult because resources, such as water, are finite.

Although, as Ausubel states, humans have in the past been able to adapt successfully to various climates, the rate of climate change caused by the present rate of GHG increase is likely to be unprecedented in human history, and might therefore severely tax our ability to adapt successfully. This is especially true of unmanaged ecosystems. An example might help to emphasize this point. Trees migrate by spreading their seeds. The maximum distance that almost any tree can spread its seeds during a season is perhaps 2 km. A rough estimate of the minimum rate of change of temperature that would allow a stand of trees to migrate can be obtained from the approximate differential relationship $dT/dt = (dy/dt)(dT/dy)$ where dy/dt is the maximum rate of migration, dT/dy is the poleward gradient of temperature and dT/dt is the rate of change of temperature. Taking dy/dt as 2 km per year and dT/dy as about 0.005°C per km, approximately the present value, gives a maximum rate of temperature change of 0.01°C per year, or 0.1°C per decade. This estimate is in agreement with a number recently given by Vellinga and Swart (1991). A more rapid temperature change will very likely prove very damaging to natural ecosystems. Pielou (1991) points out that rapid climatic change might also have an adverse effect on mammals. At the end of the last glaciation, between about 12,000 and 9,000 years ago, between 35 and 40 species of large mammals became extinct. Although the reasons for these extinctions are far from clear, the possibility that rapid climatic change might be very destructive to mammals as well as plant species looms too large to be ignored. Kessler (1991) has recently pointed out that even if we can adapt, and even if we are likely to do so, this does not mean that the best policy is to wait for even an uncertain global warming to occur and adapt to it. Waiting and adapting may sacrifice overall economic improvement in the long run.

In any case, if current predictions are nearly correct, some adaptation will be necessary. Even if the greenhouse gas forcing were abruptly stabilized, an unlikely event, at least for the near future, the global temperature would continue to rise for many years (IPCC, 1990). As Ausubel (1991) points out, humans have been adapting to climate change and regional variability for thousands of years. New technologies may make adaptation easier, but the stress of rising populations in the face of already scarce regional resources (such as water) may, in fact, make it more difficult. Furthermore, adaptation works best when it follows from careful planning. In developing countries, where the infrastructure for delivering energy is just beginning to be put into place, planning for the future by incorporating more energy efficient and environmentally friendly technologies could prevent the mistakes of the past and avoid the necessity for expensive changes on the future. Many adaptive strategies are also robust, in the sense that they will be worthwhile even in the absence of large scale climatic change. For example, the development of drought-resistant crops would benefit regions where severe droughts now occur frequently.

If we discover at some future time that climatic change and its impacts are more severe than we have anticipated, or that they will occur more rapidly than expected, we shall have to consider whether to attempt large scale human control of either biogeochemical cycles or the climate itself. This is referred to as geoengineering. Many schemes have been proposed for study. No one proposes that we actually do these things anytime in the near future, of course. We discuss a number of possibilities in Chapter 8. In virtually every case there are possibilities of unforeseen side effects which could, of course, make the cure worse than the disease. Until we understand the climate system very thoroughly, one should be very wary of disturbing it in any way that might lead irreversibly to undesirable side effects. Nevertheless, research along this line should be pursued vigorously.

There are many demand-side responses and some supply-side responses that are cost effective even in the short term and even if climatic change is not going to be severe. Many of these involve increased efficiency on both the demand-side and the supply-side. Chapter 3 presents an analysis of a number of options. At some time in the not too distant future there will be a shortage of fossil fuel, because the amount of fossil fuel is not infinite. Presently identified reserves of coal, oil, and natural gas stand at about 1000 gigatons, with unproven reserves estimated optimistically at perhaps 5000 gigatons (Rotty and Masters, 1985; Hammond, 1990). It is therefore axiomatic that at some time in the not too distant future we will need to have available some alternative energy source. Given that current global energy use is about 346×10^{15} Btu/yr and that it is increasing at 1.5% per year (averaged over the past 10 years), we shall run out of fossil fuel within the span of several centuries. It is quite likely that it will become a good deal more expensive long before that time. It would seem sensible that we consider our options and make them ready.

We believe that it is sound policy to act now. Many of the actions that we recommend take the form of intensified research, but some are more immediate, involving changes in the way we produce and consume energy, and in the way we design and build structures, for example.

1.6 ABOUT THIS BOOK

In the remainder of this book we set the stage by discussing the separate issues of emissions of radiatively important atmospheric constituents, energy demand, energy supply, agriculture, water resources, coastal hazards, and geoengineering.

In June 1991, we convened the Workshop on the Engineering Response to Global Climate Change: Planning a Research and Development Agenda. Approximately 70 scientists and engineers spent 4 days discussing what research was necessary in order to identify the causes and the extent of climatic change, to assess its consequences, and to prepare for mitigative and adaptive measures. Seven working groups were established: Sources and Sinks of Greenhouse Gases; Energy Demand, Energy Supply; Agricultural and Biological Systems; Water Resources; Coastal Hazards; and Geoengineering. Each working group was asked to identify a set of goals and a set of approaches to accomplish these goals. Common goals and approaches were then identified, and small, interdisciplinary groups were convened to expand upon the various approaches. Chapters 2 through 8 were written by the seven working groups largely after the workshop, so that their views were enriched by discussions with the other working groups.

This book is divided into nine chapters. In the first chapter we have discussed the background and nature of the problem. We have included as an appendix to Chapter 1 a brief history of important developments in the greenhouse gas problem, mainly for the purpose of elucidating the difficulties of the problem, pointing to the development of ideas, and showing how scientists have come to erroneous conclusions in the past. It follows, of course, that we may be missing important aspects of the problem even now. In Section 1.3 we outlined the problem of climatic change and the concept of feedbacks. We defined the climate sensitivity and reviewed the results of climate models of various complexity. In Section 1.4 we discussed the important problem of detection. In Section 1.5 we asked the important question "has the time come to respond" and reviewed briefly and responded to the various dissenting opinions.

Chapter 2 consists of a detailed review of what is known about the sources of greenhouse gases along with suggested areas for further research.

A review of various demand-side strategies for reducing the rate of anthropogenic production of atmospheric carbon dioxide through more efficient energy use is presented in Chapter 3.

Although the discussion in Chapter 3 makes it clear that current industrialized societies are wasteful of energy and that a comfortable and vibrant society can be maintained with much less energy use in these societies, Chapter 4 makes it equally clear that lesser developed countries can only prosper if their energy use is greatly increased.

Chapter 5 presents an overview of the problems and issues likely to result from changes in the magnitude and distribution of water as a result of global climate change. Potential response strategies required to cope with current hydrologic variability, as well as with potential changes, are examined. Also outlined are the research and development needed to allow the flexibility, resiliency, and robustness of water resource systems necessary to deal with projected, but uncertain climatic changes and the present variability that stresses existing systems.

If global warming results in a substantial sea level rise, the effects on many low-lying communities could be devastating. Chapter 6 emphasizes the current uncertainties in the magnitudes and timing of projected coastal hazards, including flooding of lowlands, beach erosion, storm hazards and ecological change. Research is suggested that will provide policy makers with rational assessments of likely future hazards, their impacts on the coastal zone, and the engineering responses required to meet future targets fixed by society.

Chapter 7 is concerned with agricultural and biological systems. An attempt is made in this chapter to identify the effects of climatic change on the agricultural sector as well as the role of agricultural activity on greenhouse gas emissions. Engineering issues and approaches that might be pursued to reduce emissions or to adapt to anticipated changes in climate are discussed. The major focus of this chapter is on intensively managed systems for food, fiber, and forest products. Agricultural practices that might lead to a better managed ecosystem are discussed. Where the knowledge base as it relates to managed systems is inadequate to define corrective strategies, specific research goals are suggested.

The idea of controling the climate through purposeful intervention in planetary processes, either by controling the sources and sinks of greenhouse gases or by regulating the climate system itself, is not new. In Chapter 8 we emphasize that although technically and even economically feasible options may soon become available, it would be quite foolish to attempt to implement them with the current lack of understanding of the climate system itself. Intentional actions of this sort might well have unintended and unforeseen consequences, which might even be disastrous and irreversible. One might do well to think of the Sorcerer's apprentice in the Disney film *Fantasia* being over-whelmed by his own creation. The purpose of this chapter is emphatically not to suggest that the methods examined actually be employed, but only that we begin to think about and evaluate the possibilities. The chapter should, therefore, be considered as exploratory.

Chapters 2 through 8 mainly address technological issues. In Chapter 9 we discuss some of the practical consequences of the fact that engineering must be practiced within the context of the real world in which technology inevitably interacts with people and their values, that is, within a social context. Lasting change, at least in a democratic world, can occur only with the permission and cooperation of society. In the case of the global problems addressed in this book, cooperation between nations, both democratic nations and those with centrally planned economies, is a problem that must be dealt with in the social, as well as the political arena. Practical solutions that address the problems of every nation must be considered. In the long term, nations must form a partnership in dealing with global concerns. Opportunities exist for global cooperation. Our hope is that we have presented a construct within which global cooperation can begin.

APPENDIX:

A BRIEF HISTORY OF THE GREENHOUSE PROBLEM

A look at the history of ideas about greenhouse gases and their effect on the climate of the Earth provides some very interesting insights. For example, although the potential effects of increased carbon dioxide on the climate of the Earth and the possibility that carbon dioxide from fossil fuel burning could affect the climate was at least suspected as early as 1861, it was not until quite recently that this was seen as an unfavorable occurrence. Early scientists appear to have felt, as do a few scientists today, that a warmer world would benefit humans. There have also been a few early clear misconceptions. For example, some papers reported that the lifetime of carbon dioxide in the atmosphere was as short as 10 years, and that any excess carbon dioxide produced by fossil fuel burning would be quickly absorbed into the ocean or the biosphere. One fairly recent report (SMIC, 1971, p. 242) stated that "because CH_4 has no direct effect on the climate or the biosphere it is considered to be of no importance for this report." Four years later the World Meteorological Organization's report on climate modeling came to the same conclusion. Methane has since been discovered to be a very important greenhouse gas (Wang et al., 1976; Ramanathan, 1985). The point here is that it is by no means impossible that we are still failing to recognize important aspects of the greenhouse problem. A few scientists have pointed out that we may be overestimating the extent and impact of greenhouse warming. Fewer have emphasized the opposite extreme that we may also be underestimating it.

The greenhouse effect emerged as a potential policy issue in 1963 when the Conservation Foundation expressed concern about the implications of increasing atmospheric CO_2. They reported that the estimated global temperature rise due to a doubling of CO_2 was 3.9°C and that his might have serious biological, geographical and economic consequences, including possible flooding of coastal areas due to the melting of glaciers.

The issue was again raised two years later in the Report of the Environmental Pollution Panel of the President's Scientific Advisory Committee. Since that time concern about the greenhouse effect has gained the attention of scientists and, to some extent, policy makers through the 1971 SMIC Report, the 1977 National Research Council Report on "Energy and Climate," and many similar reports that have followed.

A.1 CARBON DIOXIDE

The first recognition that slight variations in the Earth's atmospheric composition could lead to climatic change may have come from Tyndale (1861). Fourier (1824) had previously described the greenhouse effect of the Earth's atmosphere, comparing it to a glass covering a container. Some 35 years after Tyndale's papers the Swedish scientist Arrhenius (1896) stated that "the selective absorption of the atmosphere is, according to the researches of Tyndale, Lecher and Pernter, Rontgen, Heine, Langley, Angstrom, Paschen, and others, of a wholly different kind. It is not exerted by the chief mass of air, but in a high degree by aqueous vapor and carbonic acid, which are present in the air in small quantities." Arrhenius was seeking a possible explanation of the occurrence of ice ages, being convinced that the theory then advanced by Croll (now referred to as the Milankovitch theory) was not correct. Both water vapor and ice albedo feedbacks were discussed in this remarkable paper, as was the role of clouds. Each was dismissed as small, however. Arrhenius predicted that the temperature increase due to doubled atmospheric carbon dioxide would be 5–6°C.

Chamberlin (1899), an American geologist, again presenting an alternative to the Croll theory of ice ages, referred to carbon dioxide as a potential cause. Chamberlin also referred in passing to the possible effect on climate of the continued use of fossil fuel in cities, with the attendant production of carbon dioxide. In that same year, Tolman (1899) demonstrated the importance of the ocean on the global carbon cycle.

In a series of papers Callendar (1938, 1940, 1949, 1958, 1961) suggested that fossil fuel burning could increase the atmospheric loading of CO_2 enough to change the climate and reported many previous measurements of the carbon dioxide content of the atmosphere. By estimating the amount of carbon dioxide produced, Callendar noted that the increase in the atmosphere compared well with the total carbon dioxide produced by fossil fuel burning. There is, however, no indication of how he estimated the total fossil fuel burned. He appears to have assumed that all of the fossil fuel carbon dioxide stays in the atmosphere. Callendar used data from the World Weather Records of the Smithsonian Institution to estimate a recent increase in the global temperature, which he attributed to increasing carbon dioxide. He regarded the increase in global temperature as beneficial to mankind and stated that there "is no danger that the amount of CO_2 in the air will become uncomfortably large because as soon as the excess pressure in the air becomes appreciable, say about 0.0003 atmospheres, the sea will be able to absorb this gas as fast as it is likely to be produced." Eriksson and Welander (1956), on the other hand, implied that injection of fossil carbon dioxide would produce changes in the biosphere, but practically no increase in the atmosphere. Revelle and Seuss (1957) examined the ratios of carbon 14 to carbon 12 and carbon 13 to carbon 12 in wood and marine material and concluded that most of the carbon dioxide produced by fossil fuel burning since the beginning of the industrial revolution must have been absorbed by the oceans. Both Eriksson and Welander, and Revelle and Seuss concluded that the increase in atmospheric carbon dioxide was likely due to natural variability of the Earth system. Arnold and Anderson (1957) came to a similar conclusion.

The data reported by Callendar appear to have convinced the scientific community of the importance of obtaining more accurate measurements. During a conference on atmospheric chemistry at the University of Stockholm in May 1954, K. Buch proposed establishing a network of sampling stations in Scandinavia. The station network began operation in November 1954, reporting data regularly to the journal *Tellus*. In reporting the first Scandinavian data, Fonselius, Koroleff and Warme (1956) suggested that it would be highly desirable if similar measurements were made in other countries.

The geophysical observatory at Mauna Loa in the Hawaiian Islands, was established as part of the U.S. contribution to the International Geophysical year in 1957 (Machta, 1972). A similar program had been initiated at the South Pole a few months previously. Early results from Mauna Loa were published by Pales and Keeling (1965) and results from the South Pole were published by Brown and Keeling (1965). Since that time many observing stations have been added. The most recent results show that the atmospheric loading of carbon dioxide is at 353 ppmv (parts per million by volume) and is increasing at approximately 0.5% per year (Keeling et al., 1989).

A.2 OTHER TRACE GASES

The atmosphere contains a large number of trace gases that are radiatively and chemically active. The history of our understanding of the contribution of anthropologically caused increases in these gases is relatively short. Ramanathan (1975) first pointed out that chlorofluorocarbons in very small concentrations can contribute significantly to the greenhouse effect. CFCs are both strongly absorbing in the infrared and are long-lived. Wang et al. (1976) used a radiative-convective atmospheric model to estimate that doubling the N_2O, CH_4, and NH_3 concentrations of the atmosphere would cause increases in the surface temperature of 0.7, 0.3, and 0.1°C, respectively. Ramanathan et al. (1985) later studied the potential warming effects of about 30 trace gases that are known to be radiatively active and whose atmospheric concentrations are known to be increasing. Based upon projections of observed trends, they concluded that other trace gases will more than double the CO_2 warming by the year 2030. The major contributors in their study were methane, chlorofluorocarbons, tropospheric ozone, nitrous oxide, and carbon tetrachloride.

A.3 MODELING CLIMATIC CHANGE

Over the past several decades a hierarchy of mathematical climate models has been developed to study climate change. They range from simple energy balance climate models (EBMs), which use only the energy balance of the Earth, to complex general circulation models of the atmosphere and the ocean (GCMs), which incorporate the energy and momentum equations with the conservation of species such as water vapor and trace gases in the atmosphere and salt in the oceans. Of intermediate complexity are radiative-convective models (RCMs) which treat the vertical structure of the atmosphere by assuming stable stratification, dividing the atmosphere into layers vertically, but ignoring horizontal motion.

The use of modern radiative-convective models to study climatic change associated with a doubling of atmospheric carbon dioxide began in the 1960s. (The importance of the convective adjustment had been understood for some time. In a discussion of Callendar's 1938 paper, Sir George Simpson and Prof. D. Brunt pointed out to Callendar that the atmosphere is not in a state of radiative equilibrium, but that the temperature distribution is governed almost entirely by convection.) Manabe and Moller (1961) developed a radiative model that included the effects of absorption of both solar and long-wave radiation by carbon dioxide, water vapor and ozone. The predicted temperature of the tropopause was much too low because of the neglect of moist convection and the vertical transport of heat by large scale eddies. Moller (1963) subsequently used this model to obtain an estimate of the surface temperature rise due to a doubling of carbon dioxide. Assuming the absolute humidity of the atmosphere was independent of temperature, Moller obtained a very large CO_2 doubling temperature. Manabe and Wetherald (1967) added a convective adjustment, i.e., used a radiative-convective model with fixed relative humidity, a much more realistic condition, to obtain a CO_2 doubling temperature increase of about 2°C.

Although radiative-convective models cannot be used to predict the horizontal distribution of climate or climate change, these models have provided valuable insights into the problem of climatic change by elucidating the roles of various trace gases. For example, Hansen et al. (1988) used a radiative-convective model to compute and compare the contributions of various greenhouse gases to global warming. Evaluation of scenarios for future emissions by Ramanathan et al. (1985), Wang et al. (1986), and Hansen et al. (1988) suggests that trace gases other than CO_2 could effectively double the climatic effect of CO_2 alone.

The study of the general circulation of the atmosphere might properly be traced to the efforts of Lewis Fry Richardson to use the basic equations of atmospheric motion to develop weather forecasting capability during World War I. An interesting personal account of the early developments of numerical weather prediction has been given by Smagorinsky (1983). The contributions of von Neumann, Charney, Eliassen, Phillips, and others are discussed in some personal detail in that paper. The paper discusses events only until the early 1960s when the application of general circulation models to climate prediction, as opposed to weather prediction, was just beginning.

An account of the first general circulation experiment appeared in the open literature in 1956. Phillips (1956) used the approximate quasi-geostrophic momentum equations in a model with two layers of vertical resolution. The development of larger and more powerful computers quickly led to the use of the more complete primitive equation models (Smagorinski, 1963), with realistic orography (Mintz, 1965), and moist processes (Manabe et al., 1965). Interactions between the atmosphere and the ocean progressed from the "swamp ocean" model of Manabe et al. (1965) to models incorporating mixed layer oceans (Manabe et al., 1980), to those used today that also incorporate complete ocean general circulation models.

The first attempt to study CO_2-induced climate change using a general circulation model of the atmosphere was made by Manabe and Wetherald (1975). The model had a limited computational domain, idealized geography, no seasonal variation in insolation, and employed many other simplifying assumptions. For example, cloudiness was fixed at current values and the ocean was treated as a source of moisture only; that is, it had no heat capacity. Since that time many other groups have developed GCMs. The models have been greatly refined and many realistic processes have been included. Most models now include realistic geography, including variations in the height of land surfaces. Some models still use mixed layer oceans, that is, a layer of finite depth and vertically constant temperature, but many include fully interactive ocean GCMs. Most current models compute cloudiness, albeit with rather crude parameterizations. Hydrology is also rather crudely parameterized although detailed boundary layer and vegetation models have recently been developed. Both seasonal and diurnal cycles are also included in nearly all models. The results of climatic change experiments with many of these models is reported in the IPCC (1990).

A.4 RECENT CHANGES IN THE
TEMPERATURE OF THE EARTH'S SURFACE

Callendar's was not the first attempt to estimate changes in the Earth's temperature. The first may have been Koppen (1873). A number of more recent studies have used improved methodologies and improved data bases. The methods used by early workers as well as the more recent attempts by Mitchell (1961), and the later studies of Vinnikov et al. (1980), Hansen et al. (1981), and Jones et al. (1990) have been reviewed by Ellsaesser et al. (1986).

All of the data sets show the same general features. Each hemisphere exhibits a fairly rapid temperature increase between about 1920 and 1940, followed by a leveling off or cooling from about 1940 until the early 1970s. This event appears to be more pronounced in the Northern Hemisphere. Following the early 1970s, there appears to have been a substantial warming in both hemispheres. Little warming seems to have occurred since about 1982. The difference between the land, air, and sea surface temperature anomalies seems to have decreased substantially between about 1940 and 1970 (IPCC, 1990). Although there has been a general warming on the order of 0.5°C during the past 100 years, the annual, decadal, and longer variability of the temperature record is large enough to cast considerable doubt on whether the expected greenhouse warming signal has yet emerged from the climatic noise. Although the observed global temperature increase is more or less in agreement with that predicted by climate models with moderate sensitivity, climate variability on decadal or longer periods is also large, and its causes remain unresolved, as emphasized by Ellsaesser et al. In Section 1.4.5 we discussed some possible causes of climatic variability.

A.5 ENGINEERING AND ENERGY

Until recently, most economists felt that energy use was strongly coupled to the gross domestic product GDP, and thus to economic growth. Between 1973 and 1986, however, there was almost no growth in energy consumption in OECD countries, while the GDP actually increased by 35%, indicating an effective decoupling of GDP and energy growth. In fact, closer examination shows that the historical long-term evolution of the ratio of primary energy use to GDP in many countries becomes steadily smaller after an initial period when the heavy industrial infrastructure is established. Furthermore, the peak in the ratio is usually smaller for countries that developed their heavy infrastructure later, taking advantage of modern developments in manufacturing and transportation.

A.6 ENGINEERING CLIMATIC CHANGE: GEOENGINEERING

Interest in direct intervention to control the Earth's climate dates at least as far back as 1957 (Hoyle, 1957) and probably much earlier. Rusin and Flit (1962) have discussed many methods of climate control, as has Fletcher (1969). Marchetti (1975) proposed a number of strategies for climate control through reducing the amount of CO_2 emitted to the atmosphere by extracting it from flue gas. This has been elaborated upon by Steinberg et al. (1984). Dyson (1977) and Dyson and Marland (1979) first explored the idea of extracting CO_2 from the atmosphere by storing it in trees through reforestation.

It is interesting that consideration of geoengineering options by early Russian scientists (Rusin and Flit, 1962; Budyko, 1962) was largely inspired by an interest in improving the climate of cold and dry regions. The more recent studies are mainly aimed at exploring possible ways to mitigate anticipated greenhouse gas-induced climatic change.

REFERENCES

Abramopoulos, F., C. Rosenzweig, B. Choudhury, (1977). Improved Ground Hydrology Calculations for Global Climate Models (GCM's): Soil Water Movement and Evapotranspiration, *J. Climate*, 1, 921–941.

Allen, M. R., P. L. Read, L. A. Smith, (1992). Temperature Time Series? *Nature* 355, p. 686.

Angell, J. K., (1990). Variation in Global Tropospheric Temperature After Adjustment for El Niño Influence. *Geophys. Re. Let.* 17:1093–96.

Antonov, J. I., (1992). Recent Climatic Changes of Vertical Thermal Structure of the North Atlantic and North Pacific Oceans, *Meteor. Hydrol.*, No. 2, pp. 66–70.

Antonov, J. I., (1990). Recent Climatic Changes of Vertical Thermal Structure of the North Atlantic and North Pacific Oceans, *Meteorol. i. Gidrol*, Vol. 5, pp. 271–82.

Arnold, J. R., E. C. Anderson, (1957). The Distribution of Carbon-14 in Nature, *Tellus*, Vol. 9, pp. 28–32.

Arrhenius, S., (1896). On the Influence of Carbonic Acid in the Air Upon the Temperature of the Ground, *Phila. Mag.*, Vol. 41, pp. 237–276.

Barnett, T. P., (1989). Global Sea Level Change, NCPO, Climate Variations Over the Past Century and the Greenhouse Effect, National Climate Program Office, NOAA.

Broecker, W. S., (1975). Climatic Change: Are We on the Brink of a Pronounced Global Warming? *Science*, Vol. 189, pp. 460–463.

Budyko, M. I., (1962). Certain Means of Climate Modification, *Meteorol. i Gidrol.*, No. 2, p. 3–8.

Budyko, M. I., (1969). The Effect of Solar Radiation Variations on the Climate of the Earth, *Tellus*, Vol. 21, pp. 611–619.

Brown, C. W., C. D. Keeling, (1965). The Concentration of Carbon Dioxide in Antarctica, *J. Geophys. Res.*, Vol. 70, pp. 6077–85.

Callendar, G. S., (1938). The Artificial Production of Carbon Dioxide and Its Influence on Temperature, *Q. J. R. Meteorol. Soc.*, Vol. 64, pp. 223–37.

Callendar, G. S., (1940). Variations of the Amounts of Carbon Dioxide in Different Air Currents, *Q. J. R. Meteorol. Soc.*, Vol. 66, pp. 395–410.

Callendar, G. S. (1949). Can Carbon Dioxide Influence Climate? *Weather*, Vol. 4, pp. 310–14.

Callendar, G. S., (1958). On the Amount of Carbon Dioxide in the Atmosphere, *Tellus*, Vol. 10, pp. 243–48.

Callendar, G. (1961). Temperature Fluctuation and Trends Over the Earth, *Q. J. R. Meteorol. Soc.*, Vol. 87, pp. 1–12.

Cess, R. D., et al, (1990). Intercomparison and Interpretation of Climate Feedback Processes in 19 Atmospheric General Circulation Models, *J. Geophys. Res.*, 95, p. 19, pp. 16601–16615.

Cess, R. D., (1989). Gauging Water Vapor Feedback, *Nature*, 342, pp. 736–737.

Cess, R. D., et al. (1989). Interpretation of Cloud-Climate Feedback Produced By 14 General Circulation Models, *Science*, 245, pp. 513–516.

Cess, R. D. (1978). Biosphere-Albedo Feedback and Climate Modeling, *J. Atmos. Sci.*, Vol. 35, pp. 1765–68.

Cess, R. D., (1976). Climate Change: An Appraisal of Atmospheric Feedback Mechanisms Employing Zonal Climatology, *J. Atmos. Sci.*, Vol. 33, pp. 1831–1843.

Cess, R.D., G.L. Potter, J.P. Blanchet, G.J. Boer, A.D. Del Genio, M. Deque, V. Dyminikov, V. Galin, W. L. Gates, S.J. Ghan, J.T. Kiehl, A.A. Lacis, H. LeTreut, Z.X. Li, X.Z. Liang, B.J. McAvaney, V.P. Meleshko, J.F.B. Mitchell, J.J. Morcrette, D.A. Randall, L. Rikus, E. Roeckner, J.F. Royer, U. Schlese, D.A. Sheinin, A, Slinge, A.P. Sokolov, K.E. Taylor, W.M. Washington, R.T. Wetherland, I. Yagai, M.H. Zhang, (1990). Intercomparison and Interpretation of Climate Feedback Processes in 19 Atmospheric General Circulation Models, *J. Geophy. Res.*, 95, pp. 16,601–16,615.

Cess, R. D., S. D. Goldenberg, (1981). The Effect of Ocean Heat Capacity upon Global Warming Due to Increasing Atmospheric Carbon Dioxide, *J. Geophys. Res.*, Vol. 86, pp. 498–502.

Chamberlin, R. C., (1899). An Attempt to Frame a Working Hypothesis of the Cause of Glacial Periods on an Atmospheric Basis, *J. of Geol.*, Vol. 7, pp. 575, 667, and 751.

Charlson, R. J., J. E. Lovelock, M. O. Andreae, S. G. Warren, (1987). Oceanic Phytoplankton, Atmospheric Sulfur, Cloud Albedo and Climate, *Nature*, Vol. 326.

Charlson, R. J., J. Langer, H. Rodhe, (1990). Sulfate Aerosol and Climate, *Nature*, Vol. 348.

Charney, J.G., (1975). Dynamics of Deserts and Drought in the Sahel, *Q. J. R. Meteorol. Soc.*, Vol. 101, pp. 193–202.

Charney, J.G., W.J. Quirk, S.H. Chow, J. Kornefield, (1977). A Comparative Study of the Effects of Albedo Change in Drought in Semi-Arid Regions, *J. Atmos. Sci.*, Vol. 34, pp. 1366–1385.

Chen, T., J.J. Tribbia,(1993). An Effect of the Model's Horizontal Resolution on Stationary Eddies Simulated by NCAR CCMI, *J. Climate*, Vol. 6, pp. 1657–1664.

Covey, C., Behavior of an Ocean General Circulation Model at Four Different Horizontal Resolutions, *PCMDI Report No. 4, Program for Climate Model Diagnosis and Intercomparison,* University of California, Lawrence Livermore National Laboratory, Livermore, CA, 1992.

Covey, C., (1991). Chaos in Ocean Heat Transport, *Nature*, Vol. 353.

Cubasch, U., K. Hasselmann, H. Maier-Reimer, E. Mikolajewicz, B. D. Santer, R. Sausen (1992). Time-Dependent Greenhouse Warming Computations with a Coupled Ocean-Atmosphere Model, *Climate Dynamics*, Vol. 8, pp. 55–70.

Dansgaard, W. et al., (1993). *Nature,* Vol. 364, pp. 218–220.

Darwin, C. G., (1953). *The Next Million Years,* Doubleday, New York.

Delworth, T., S. Manabe, R. J. Stouffer, (1993). *J. Clim.*, Vol. 6, pp. 1993–2011.

Dickinson, R., A. Henderson-Sellers, A global climatology of albedo, roughness length and stomatal resistance for atmospheric general circulation models as represented by the Simple Biosphere model (SiB), *Q. J. R. Meteorol. Soc.*, Vol. 114, pp. 439–462, 1988.

Ducoudre, N.I., K. Laval, A. Perrier, (1993). SECHIBA, A New Set of Parameterizations of the Hydrologic Exchanges at the Land-Atmosphere Interface within the LMD Atmospheric General Circulation Model, *J. Climate*, 6, 248–273.

Dyson, F. J., (1977). Can We Control the Carbon Dioxide in the Atmosphere?, *Energy*, Vol. 22, pp. 287–291.

Dyson, F. J., G. Marland, (1979). Technical Fixes for the Climatic Effects of CO_2, in Elliott, W. P., L. Machta (eds.) Workshop on the Global Effects of Carbon Dioxide from Fossil Fuels, Miami Beach, March 7–11, 1979, pp. 111–118, U. S. Department of Energy CONG-770385, UC-11, May, 1979.

Elsner, J. B., R. Vautard, (1991). Interdecadal Oscillations Exist in the Global Temperature Record? *Nature* 353, pp. 551–553.

Ellsaesser, H. W., M. C. MacCracken, J. J. Walton, S. L. Grotch, (1986). Global Climatic Trends as Revealed by the Recorded Data, *Rev. Geophys.*, Vol. 24, pp. 745–792.

Eriksson, R., P. Welander, (1956). On a Mathematical Model of the Carbon Cycle in Nature, *Tellus*, Vol. 8, pp. 155–175.

Field, M. H., B. Huntley, H. Muller, (1994). Eemian Climate Fluctuations Observed in a European Pollen Record, *Nature*, 371, pp. 779–783.

Fletcher, J. O., (1969). Controlling the Planet's Climate, *Impact Sc. Soc.,* Vol. 19, pp. 151–168.

Fonselius, S., F. Koroleff, K. E. Warme, (1956). Carbon Dioxide Variations in the Atmosphere, *Tellus*, Vol. 8, pp. 176–183.

Foukal, P., J. Lean (1990). An Empirical Model of Total Solar Irradiance Variations Between 1874 and 1988. *Science*, 247, pp. 556–558.

Fourier, Jean-Baptiste Joseph, (1824). Memoire sur les Temperatures du Globe Terrestre, *Annales de Chimis est de Physique*, Vol. 27, pp. 136–167.

Garratt, J.R., (1993). Sensitivity of climate simulations to land-surface and atmospheric boundary-layer treatments—A review, *J. Climate*, 6, 419–449.

Garratt, J.R., P.B. Krummel E.A. Kowalczyk, (1993). The surface energy balance at local and regional scales—a comparison of general circulation model results with observations, *J. Climate*, pp. 1090–1109.

Ghil, M., R. Vautard, (1991). Interdecadal Oscillations and the Warming Trend in Global Temperature Time Series. *Nature* 350, pp. 324–327.

Gilliland, R. (1982). Solar, Volcanic, and CO_2 Forcing of Recent Climatic Changes, *Climatic Change*, 4, pp. 111–131.

Gloersen, P., W. J. Campbell, (1988). Variations in the Arctic, Antarctic, and Global Sea Ice Covers During 1979–1987 as Observed With the Nimbus 7 Scanning Multichannel Microwave Radiometer, *J. Geophys. Res.*, Vol. 93, pp. 10,666–10,674.

Gornitz, V. S. Lebedeff, (1987). Global Sea Level Changes During the Past Century, Sea-Level Fluctuation and Coastal Evolution, D. Numeral, O. H. Pilkey, J. D. Howard, (eds.) SEPM Special Publication No. 41.

Grootes, P. M., M. Stuiver, J. W. C. White, S. Johnsen, J. Jouzel, (1993), Comparison of Oxygen Isotope Records from the GISP2 and GRIP Greenland Ice Cores, *Nature,* 366, pp. 552–554.

Grotch, S. L., (1988). Regional Intercomparisons of General Circulation Model Predictions and Historical Climate Data, U.S. Department of Energy, DoE/NBB-0084, Washington, D.C., 291 pp.

Grotch, S. L., M. C. MacCracken, (1991). The Use of General Circulation Models to Predict Regional Climate Change, *J. Climate,* Vol. 4, pp. 286–303.

Haldane, J.B.S., (1923). Daedalus or Science and the Future, K. Paul, Trench, Trubner; London.

Hammond, A. L., (1990). *World Resources*, Oxford University Press, New York, pp. 383.

Hansen, J., D. Johnson, A. Lacis, S. Lededeff, P. Lee, D. Rind, G. Russell, (1981). Climate Impact of Increasing Atmospheric Carbon Dioxide, *Science,* 213, pp. 957–966.

Hansen, J., A. Lacis, D. Rind, G. Russell, P. Stone, I. Fung, R. Ruedy, J. Lerner, (1984). Climate Sensitivity: Analysis of Feedback Mechanisms, Climate Processes and Climate Sensitivity, J. E. Hansen, T. Takahashi, (eds.), (Maurice Ewing Series, No. 5), *Am. Geophys. Union*, pp. 130–163.

Hansen, J., I. Fung, A. Lacis, D. Rind, S. Lebedeff, R. Ruedy, G. Russell, P. Stone, (1988). Global Climate Changes as Forecast by Goddard Institute for Space Studies Three-Dimensional Model, *J. Geophys. Res.*, Vol. 93, pp. 9341–9364.

Hoffert, M. J., A. J. Callegari, C. T. Hsieh, (1980). The Role of Deep Sea Heat Storage in the Secular Response to Climatic Forcing, *J. Geophy. Res.*, Vol. 85, pp. 6667–6679.

Held, I.M., P.J. Phillipps, (1993). Sensitivity of the eddy momentum flux to meridional resolution in atmospheric GCMs, *J. Climate,* Vol. 6, pp. 499–507.

Houghton, R. A., W. H. Schlesinger, S. Brown, J. F. Richards, (1985). Carbon Dioxide Exchange between the Atmosphere and Terrestrial Ecosystems. In *Atmospheric Carbon Dioxide and the Global Carbon Cycle*, (J. R. Trabalka, Ed.), U.S. Department of Energy, DoE/ER-0239, Washington, D. C, pp. 113–140.

Hoyle, F., (1957). *The Black Cloud*, Harper and Brothers, New York, 250 pp.

Hunt, B. G., N. C. Wells, (1979). An Assessment of the Possible Climatic Impact of Carbon Dioxide Increases Based on a Coupled One-Dimensional Atmospheric-Oceanic Model, *J. Geophys. Res.*, 84, pp. 787–781.

Huybrechts, P., J. Oerlemans, (1990). Response of the Antarctic Ice Sheet to Future Greenhouse Warming, *Climate Dynamics,* 5, pp. 93–102.

Huybrechts, P., A. Letreguilly, N. Reeh, (1991). The Greenland Ice Sheet and Greenhouse Warming, *Global and Planetary Change*, 89, pp. 399–412.

IPCC, (1990). Climatic Change: *The IPCC Scientific Assessment.* Report of Working Group 1 of the *Intergovernmental Panel on Climate Changes*, sponsored jointly by the World Meteorological Organization and the United Nations Environment Programme, Meteorological Office, Bracknell, United Kingdom. Cambridge University Press.

Jones, P. D., (1988). The Influence of ENSO on Global Temperatures, *Climate Monitor,* 17(3), pp. 80–89.

Jones, P. D., T. M. L. Wigley, (1990). Global Warming Trends, *Sc. Am.* 263, pp. 84–91.

Jones, P. D., S. C. B. Raper, R. S. Bradley, H. F. Diaz, P. M. Kelly, T. M. L. Wigley, (1986a). Northern Hemisphere Surface Air Temperature Variations: 1851–1984. *J. Climate Appl. Meteorol.* 25(2), pp.161–79.

Jones, P. D., S. C. B. Raper, T. M. L. Wigley, (1986b). Southern Hemisphere Surface Air Temperature Variations: 1851–1984, *J. Climate Appl. Meteorol.* 25(9), pp. 1213–30.

Jones, P. D., T. M. L. Wigley, P. B. Wright, (1986c). Global Temperature Variations Between 1861 and 1984, *Nature* 322, pp. 430–34.

Jones, P. D., T. M. L. Wigley, G. Farmer (1991). Marine and Land Temperature Data Sets: A Comparison and A Look At Recent Trends. *Greenhouse-Gas-Induced Climatic Change: A Critical Appraisal of Simulations and Observations*, M. E. Schlesinger (ed.), Elsevier Science Publishers, Amsterdam, Netherlands, pp. 153–72.

Jones, R. D., S. C. B. Raper, R. S. Bradley, H. F. Diaz, P. M. Kelly, T. M. L. Wigley, (1986). Northern Hemisphere Surface Air Temperature Variations, *J. Clim. Appl. Meteorol.*, Vol. 25, pp. 161–179.

Jones, P.D., T.M. Wigley, K.R. Briffa, (1994). Global and Hemispheric Anomolies—Land and Marine Instrumental Records, pp. 603–608 in T.A. Boden, D.P. Kaiser, R.J. Sepanski, F.W. Stoss (eds.), *Trends '93: A Compendium of Data on Global Change,* ORNL/CDIAC-65. Carbon Dioxide Information Analysis Center, Oak Ridge National Laboratory.

Karl, T. R., (1994). Documenting Observed Climate Variations and Change, in *Global Climate Change: Science, Policy, and Mitigation Strategies,* Air and Waste Management Association International Specialty Conference, Phoenix, Arizona, March 18, 1994.

Keeling, C. D., R. B. Bacastow, A. F. Carter, S. C. Piper, T. P. Whorf, M. Heimann, W. G. Mook, H. Roeloffzen, (1989). A Three Dimensional Model of Atmospheric CO_2 Transport Based on Observed Winds: 1. Analysis of Observed Data, in Aspects of Climate Variability in the Pacific and the Western Americas, D. H. Peterson (ed.), *Geophys. Monog.* 55, AGU, Washingon, pp. 165–236.

Kelly, P. M., T. M. L. Wigley, (1990). The Influence of Solar Forcing Trends on Global Mean Temperature Since 1861, *Nature,* 347, pp. 460–462.

Kessler, E. (1991). Carbon Burning, the Greenhouse Effect and Public Policy, *Bull. Am. Meteorol. Soc.,* 72, pp. 513–514.

Kim, K. Y., Gerald R. North, J. Huang, (1992). On the Transient Response of a Simple Coupled Climate Model, *J. Geophys. Res.,* 97, pp. 10069–10081.

Koppen, W., (1873). Concerning Multiyear Periods of Weather, Especially the 11-year Period of Temperature, Z. Osterr, *Ges. Meteor.,* Vol. 8, pp. 241–248, 257–267.

Lau, N. C., (1992). Climate Variability Simulated in GCMs In *Climate System Modeling*, K. E. Trenberth (ed.) Cambridge University Press, New York.

Lean J., D.A. Warrilow, (1989). Simulations of the regional climatic impact of Amazon deforestation, *Nature*, Vol. 342, pp. 411–413.

Legates, B.R., (1987). A climatology of Global Precipitation, *Publ. Climatol.*, pp. 40, 85.

Legates, D.R., C.J. Willmott, (1990). Mean Seasonal and Spatial Variability in Global Surface Air Temperature, *Theor. Appl. Climatol.*, 41, pp. 11–21.

Levitus, S., (1989). Interpentadal Variation of Temperature and Salinity in the Deep North Atlantic Ocean, 1970 versus 1955–59, *J. Geophys. Revs.*, Vol. 94, pp. 16,125–16,131.

Lorenz, E. N., (1969). The Predictability of a Flow which Possesses Many Scales of Motion, *Tellus*, 21, pp. 289–307.

MacCracken, M. C., H. Moses, (1982). The First Detection of Carbon Dioxide Effects: Workshop Summary, June 8–10, 1981, Harpers Ferry, West Virginia, *Bull. Am. Meteorol. Soc.*, Vol. 63, pp. 1164–1178.

Machta, L., (1972). Mauna Loa and Global Trends in Air Quality, *Bull. Amer. Meteorol. Soc.*, Vol. 53, pp. 402–420.

Mahasenan, N., R.G. Watts, H. Dowlatabadi, (1996). Low-frequency Oscillations in Temperature-proxy Records and Implications for Recent Climate Change, *Geophys. Rev. Let.*, to appear.

Manabe, S., (1964). Climate and Ocean Circulation. 1. The Atmospheric Circulation and the Hydrology of the Earth's Surface, *Mon. Wea. Rev.*, 97, 361–385.

Manabe, S., F. Moller, (1961). On the Radiative Equilibrium and Heat Balance of the Atmosphere, *Mon. Wea. Rev.*, Vol. 89, pp. 503–532.

Manabe, S., J. Smagorinsky, R. F. Strickler, (1965). Simulated Climatology of a General Circulation Model with a Hydrologic Cycle, *Mon. Wea. Rev.*, Vol. 93, pp. 769–798.

Manabe, S., R. T. Wethereld, (1967). Thermal Equilibrium of the Atmosphere with a Given Distribution of Relative Humidity, *J. Atmos. Sci.*, Vol. 24, pp. 241–259.

Manabe, S., R. T. Wethereld, (1975). The Effects of Doubling the CO_2 Concentration on the Climate of a General Circulation Model, *J. Atmos. Sci.*, Vol. 32, pp. 3–15.

Manabe, S., R. J. Stouffer, (1980). Sensitivity of a Global Model to an Increase of CO_2 Concentration in the Atmosphere, *J. Geophys. Res.*, Vol. 85(C10), pp. 5529–5554.

Manabe, S., R. T. Wetherald, (1987). Large Scale Changes of Soil Wetness Induced by an Increase in Atmospheric Carbon Dioxide, *J. Atmos. Sc.*, Vol. 44, pp. 1211–1235.

Manabe, S., M. J. Spelmann, R. J. Stouffer, (1992). Transient Responses of a Coupled Ocean-Atmosphere Model to Gradual changes of Atmospheric CO_2, Part II: Seasonal Response, *J. Climate*, 5, pp. 105–126.

Manabe, S., R. J. Stouffer, M. J. Spelman, K. Bryan, (1991). Transient Responses of a Coupled Ocean-Atmosphere Model to Gradual Changes of Atmospheric CO2, Part I: Annual Mean Response, *J. Climate*, 4, pp. 785–818.

Marchetti, C., (1975). *On Geoengineering and the CO_2 Problem,* International Institute for Applied Systems Analysis, Laxenburg, Austria, 13 pp.

Marchetti, L., (1977). On Engineering the CO_2 Problem, *Climatic Change*, Vol. 1, pp. 59–68.

Marengo, J.A., et al., (1994). Calculations of River-Runoff in the GISS GCM: Impact of a New Land-Surface Parameterization and Runoff Routing Model on the Hydrology of the Amazon River, *Climate Dynamics*, 10, pp. 349–361, 1994.

Marshall Institute (1989). *Scientific Perspectives on the Greenhouse Problem*, F. Seitz (ed.). Marshall Institute, Washington, D.C.

Miller, J., G. Russell, G. Caliri, (1994), Continental Scale River Flow in Climate Models, J. Climate, 7, pp. 914–28.

Mintz, Y., (1965). *Very Long-Term Integration of the Primitive Equations of Atmospheric Motion*, WMO Tech Note No. 66, Geneva, pp. 141–55.

Mitchell, J. F. B., C. A. Wilson, W. M. Cunnington, (1987). On CO_2 Climate Sensitivity and Model Dependence of Results, *Q. J. R. Meteorol. Soc.*, 113, pp. 293–322.

Mitchell, J. F. B., C. A. Senior, W. J. Ingram, (1989). CO_2 and Climate: A Missing Feedback?, *Nature*, Vol. 341, pp. 132–134.

Mitchell, J. F. B., W. J. Ingram, (1992). Carbon Dioxide and Climate: Mechanisms of Changes in Clouds, *J. Climate*, Vol. 5, pp. 5–21.

Mitchell, J. M. Jr.., (1961). Recent Secular Changes of Global Temperature, *Ann. N.Y. Acad. Sc.*, Vol. 95, pp. 245–250.

Moller, F., (1963). On the Influence of Changes in the CO_2 Concentration in Air on the Radiation Balance of the Earth's Surface and On the Climate, *J. Geophys. Res.*, Vol. 68, pp. 3877–86.

Morantine, M. C., R. G. Watts, (1990). Upwelling-Diffusion Climate Models: Analytical Solutions for Radiative and Upwelling Forcing, *J. of Geophy. Res.*, Vol. 95, No. D6, pp. 7563–7571.

Morantine, M, R. G. Watts, (1994). Time Scales in Energy Balance Climate Models: Part 2. The Intermediate Time Solutions, *J. Geophys. Res.*, Vol. 99, No. D2, pp. 3643–3653.

Nobre, C.A., P.J. Sellers, J. Shukla, (1991). Amazonian Deforestation and Regional Climate Change, *J. Climate*, Vol. 4, pp. 957–988.

Pales, J. C., C. D. Keeling, (1965). the Concentration of Atmospheric Carbon Dioxide in Hawaii, *J. Geophys. Res.*, Vol. 70, pp. 6053–76. See History of Mauna Loa in Machta, *Bull. AMS*, Vol. 53, No. 5, May, 1972.

Parrilla, G. et al., (1994). Rising Temperatures in the Subtropical North Atlantic Ocean Over the Past 35 Years, *Nature*, Vol. 369, pp. 48–51.

Phillips, N. A., (1956). The General Circulation of the Atmosphere: A Numerical Experiment, Q. *J. R. Meteorol. Soc.*, Vol. 82, pp. 123–64.

Phillips, T.J., L.C. Corsetti, S.L. Grotch, (1995). The impact of horizontal resolution on moist processes in ECMWF model, *Climate Dynamics*, Vol. 11, pp. 85–102.

Rahmstorf S., (1995). Bifurcations of the Atlantic Thermohaline Circulation in Response to Changes in the Hydrologic Cycle, *Nature*, Vol. 378, pp. 145–149.

Ramanathan, V., (1975). Greenhouse Effect Due to Chlorofluorocarbons: Climatic Implications, *Science*, Vol. 190, pp. 50–52.

Ramanathan, V, (1977). Interaction Between Ice-Albedo Lapse-Rate and Cloud-Top Feedbacks: An Analysis of the Nonlinear Response of a GCM Climate Model, *J. Atmos. Sci.*, 34, pp. 1885–1897.

Ramanathan, V., R. J. Cicerone, H. B. Singh, J. T. Kiehl, (1985). Trace Gas Trends and Their Potential Role in Climate Change, *J. Geophys. Res.*, Vol. 90, pp. 5547–5566.

Randall, D.A. and Co-Authors, (1992). Intercomparison and Interpretation of Surface Energy Fluxes in Atmosphere General Circulation Models, *J. Geophy. Res.*, 97(DA), pp. 3711–3724.

Raval, A., V. Ramanathan, (1989). Observational Determination of the Greenhouse Effect, *Nature*, Vol. 342, pp. 758–761.

Revelle, R., H. E. Seuss, (1957). Carbon Dioxide Exchange Between Atmosphere and Ocean and the Question of and Increase of Atmospheric CO_2 During the Past Decades, *Tellus*, Vol. 9, pp. 18–27.

Rind, D., (1988). The Doubled CO_2 Climate and the Sensitivity of the Modeled Hydrological Cycle. *J. Geophys. Res.*, Vol. 93, pp. 5385–5412.

Rind, D., E. W. Chiose, W. Chu, J. Larson, S. Oltmans, J. Lerner, M. P. McCormick, L. McMaster, (1991). Positive Water Vapor Feedback in Climate Models Confirmed by Satellite Data, *Nature*, 349, pp. 500–503.

Roemmich, D., C. Wunsch, (1984). Apparent Changes in the Climatic State of the Deep North Atlantic, *Nature*, Vol. 307.

Rooth, C., (1982). *Prog. Oceanog.*, Vol. 11.

Rotty, R. M., C. D. Master (1985). Carbon Dioxide from Fossil Fuel Combustion: Trends, Resources and Technical Implication, In *Atmospheric Carbon Dioxide and the Global Carbon Cycle*, Report DoE/ER-0239, Oak Ridge National Laboratories, Oak Ridge, TN.

Rusin, N. P., L. A. Flit, (1962). *Methods of Climate Control*, Moscow.

Sato, N., et al., (1987). Effects of Implementing the Simple Biosphere Model in a General Circulation Model, *J. Atmos. Sci.*, 46, pp. 4058–4097.

Schneider, S. H., S. L. Thompson, (1981). Atmospheric CO2 and Climate: Importance of the Transient Response, *J. Geophys. Res.*, Vol. 86, pp. 3135–47.

Schlesinger, M. E., W. L. Gates, Y. J. Han, (1985). The Role of the Ocean in Carbon-Dioxide-Induced Climate Warming. Preliminary Results from the OSU Coupled Atmosphere-Ocean Model. In *Coupled Ocean-Atmosphere Models*, Elsevier, Amsterdam, pp. 447–478.

Schlesinger, M. E., Z. C. Zhao, (1989). Seasonal Climatic Changes Induced by Doubled CO2 as Simulated by the OSU Atmospheric GCM/Mixed-Layer Ocean Model, *J. Climate*, Vol. 2, pp. 459–495.

Schlesinger, M. E., N. Ramankutty, (1992). Implications for Global Warming of Intercycle Solar Irradiation Variations. *Nature*, 360, pp. 330–333.

Schlesinger, M. E., N. Ramankutty (1994). An Oscillation in the Global Climate System of Period 65–70 Years, *Nature*, 367, pp. 723–726.

Sellers, P.J., Y. Mintz, Y.C. Sud, A. Dalcher, (1986). A Simple Biosphere Model (SiB) for Use Within General Circulation Model, *J. Atmos. Sci.*, 43, pp. 505–531.

Sellers, W. D., (1969). A Global Climate Model on the Energy, Balance of the Earth-Atmosphere System, *J. Appl. Meteorol.*, Vol. 8, pp. 392–400.

Semtner, A.J., Jr., R.M. Chervin, (1988). A simulation of the global ocean circulation with resolved eddies, *J. Geophys. Res.*, Vol. 93, pp. 15502–15522.

Shukla, J.C. Nobre, P. Sellers, (1990). Amazon deforestation and climate change, *Science*, Vol. 247, pp. 1322–1325.

Shuttleworth, W.J., et al., (1984). Eddy Correlation Measurements of Energy Partition for Amazonian Forrest, *Q. J. R. Meteorol. Soc.*, 110, pp. 1143–1163.

Smagorinsky, J., (1963). General Circulation Experiments with the Primitive Equations. I. The Basic Experiment, *Mon. Wea. Rev.*, Vol. 93, pp. 99–164.

Smagorinsky, J., (1983). The Beginnings of Numerical Weather Prediction and General Circulation Modeling: Early Recollections, *Adv. Geophys.*, Vol. 25, pp. 3–36.

SMIC, (1971). *Inadvertent Climate Modification: Report of the Study of Man's Impact on Climate*, MIT Press, Cambridge, MA, 308 pp.

Steinberg, M., H. C. Chen, F. Horn, (1984). A Systems Study for the Removal, Recovery, and Disposal of Carbon Dioxide from Fossil Fuel Power Plants in the U.S., DoE/CH/00016-2, U.S. Department of Energy.

Stouffer, R. J., S. Manabe, K. Bryan, (1989). Interhemispheric Asymmetry in Climate Response to a Gradual Increase of Atmospheric CO_2, *Nature*, 342, pp. 660–662.

Thompson, S. L., S. H. Schneider, (1982). Carbon Dioxide and Climate: The Importance of Realistic Geography in Estimating the Transient Temperature Response, *Science*, 217, pp. 1031–1033.

Thouveny, N., J-L. Beaulieu, E. Bolifay, K. M. Creer, J. Guiot, M. Icole, S. Johnsen, J. Jouzel, M. Reille, T. Williams, D. Williamson, (1994). Climate Variations in Europe Over the Past 140 kyr Deduced from Rock Magnetism, *Nature*, 371, pp. 503–506.

Tolman, C. F., JR., (1899). The Carbon Dioxide of the Ocean and Its Relationships to the Carbon Dioxide of the Atmosphere, *J. Geol.*, Vol. 7, pp. 585–618.

Tyndale, J., (1861). Recognized that slight variations in the Earth's atmospheric composition could lead to climatic change, *Phil. Mag., J. Sc.*, Vol. 22, pp. 169–94, 273–85.

Vautard, R., M. Ghil, (1989). Singular Spectrum Analysis in Nonlinear Dynamics, with Applications to Paleoclimatic Time Series, *Physica,* D 35, pp. 395–424.

Vautard, R., P. Yiou, M. Ghil, (1992). Singular Spectrum Analysis: A Toolkit for Short, Noisy Chaotic Signals. *Physica* D 58, pp. 95–126.

Vellinga, P., R. Swart, (1991). The Greenhouse Marathon: A Proposal for a Global Strategy, *Climatic Change*, Vol. 18, pp. vii–xiii.

Vinnikov, K. Ya., G. V. Gruza, V. F. Zakharov, A. A. Kirillow, N. P. Kovyneva, E. Ya. Ran'kova, (1980). Current Climatic Changes in the Northern Hemisphere (in Russian), *Meteorol. Gidrol.*, Vol. 6, pp. 5–17.

Wang, W. C., Y. L. Yung, A. A. Lacis, T. Mo, J. E. Hansen, (1976). Greenhouse Effects Due to Man-Made Perturbations of Trace Gases, *Science*, Vol. 194, pp. 685–690.

Wang, W. C., D. J. Wuebbles, W. M. Washington, R. G. Isaacs, G. Molnar, (1986). Trace Gases and Other Potential Perturbations to Global Climate, *Rev. Geophys.*, Vol. 24, pp. 110–140.

Washington, W. B., G. A. Meehl, (1989). Climate Sensitivity Due to Increased CO_2: Experiments with a Coupled Atmosphere and Ocean General Circulation Model, *Climate Dynamics*, pp. 1–38.

Washington, W. M., G. A. Meehl, (1983). Seasonal Cycle Experiment on the Climate Sensitivity Due to a Doubling CO_2 with an Atmospheric General Circulation Model Coupled to a Simple Mixed-Layer Ocean Model, *J. Geophys. Res.*, Vol. 89, pp. 9475–9503.

Washington, W. M., G. A. Meehl (1984). Seasonal Cycle Experiment on the Climate Sensitivity Due to a Doubling of CO_2 with an Atmospheric General Circulation Model Coupled to a Simple Mixed-Layer Ocean Model, *J. Geophy. Res.*, 89, pp. 9475–9503.

Watts, R. G., (1985). Global Climate Variation due to Fluctuations in the Rate of Deep Water Formation, *J. Geophys. Res.*, Vol. 90, pp. 8067–70.

Watts, R. G., M. C. Morantine, (1994). Time Scales in Energy Balance Climate Models: Part 1. The Limiting Case Solutions, *J. Geophys. Res.*, Vol. 99.

Watts, R. G., M. C. Morantine, (1991). Is the Greenhouse Gas–Climate Signal Hiding in the Deep Ocean?, *Climatic Change, pp.* iii–vi.

Weaver, A. J., E. S. Sarachik, J. Marotze, (1991). Freshwater Flux Forcing of Decadal and Interdecadal Oceanic Variability, *Nature*, Vol. 353, pp. 836–839.

Weinberg, A. M. (1966). Can Technology Replace Social Engineering?, *Bull. Atomic Sc.*, 22, pp. 4–8.

Wetherald, R. T., S. Manabe, (1986). An Investigation of Cloud Cover Change in Response to Thermal Forcing, *Clim. Change*, 8, pp. 5–23.

Wigley, T. M. L., S. C. B. Raper, (1990). Climatic Change Due to Solar Irradiance Changes, *Geophys. Res. Lett.*, 17, pp. 2169–2172.

Wigley, T. M. L., S. C. B. Raper, (1990). Natural Variability of the Climate System and Detection of the Greenhouse Effect, *Nature*, Vol. 344, pp. 324–27.

Willmott, C.J., D.R. Legates, (1993). A Comparison of GCM-Simulated and Observed Mean January and July Surface Air Temperature, *J. Climate*, 6, pp. 274–291.

Wood, F. B., (1988). Global Alpine Glacier Trends, 1960s to 1980s, *Arct. Alp. Res.*, Vol. 20, pp. 404–413.

Wood, E.F., V. Lakshmi, Scaling Eater and Energy Fluxes in Climate Systems: Three Land-Atmospheric Modeling Experiments, *J. Climate*, 6, pp. 839–857, 1993.

Chapter 2

EMISSIONS AND BUDGETS OF RADIATIVELY IMPORTANT ATMOSPHERIC CONSTITUENTS

Author: Don Wuebbles
Co-Authors: Jae Edmonds, Jane Dignon, William Emanuel,
Donald Fisher, Richard Gammon, Robert Hangebrauck,
Robert Harris, M.A.K. Khalil, John Spence, Thayne M. Thompson

2.1 INTRODUCTION

Carbon dioxide (CO_2), as the single largest contributor, has received much of the attention in the concerns about climate change and the greenhouse effect. However, climate models (e.g., Ramanathan et al., 1985; Hansen et al., 1988; IPCC, 1990, 1992)

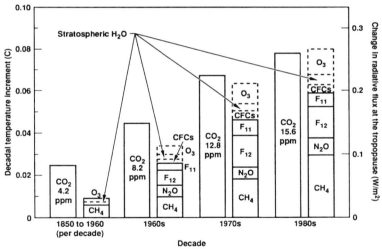

Figure 2.1 Decadal additions to global-mean greenhouse radiative forcing of the climate system based on studies using a one-dimensional radiative-convective model by Hansen et al. (1988). Shown is the computed commitment to future surface temperature change (°C) once equilibrium is reestablished. The change in temperature is estimated for decadal increases in trace gas abundances, without including the amplification that would be induced by climate feedbacks. The change in net radiative forcing of the troposphere–surface system is indicated on the right ordinate.

Table 2.1   Greenhouse gases and other inportant gases that affect climate.

Trace Constituent	Common Name	Importance For Climate
CO_2	Carbon dioxide	Absorbs infrared (IR) radiation; affects stratospheric O_3
CH_4	Methane	Absorbs IR radiation; affects tropospheric O_3 and OH; affects stratospheric O_3 and H_2O; produces CO_2
CO	Carbon monoxide	Affects tropospheric O_3 and OH cycles; produces CO_2
N_2O	Nitrous Oxide	Absorbs infrared radiation; affects stratospheric O_3
NOx	Nitrogen oxides $(= NO + NO_2)$	Affects O_3 and OH cycles; precursor of acidic nitrates
$CFCl_3$	CFC-11	Absorbs infrared radiation; affects stratospheric O_3
CF_2Cl_2	CFC-12	Absorbs infrared radiation; affects stratospheric O_3
$C_2F_3Cl_3$	CFC-113	Absorbs infrared radiation; affects stratospheric O_3
C_2F_5Cl	CFC-115	Absorbs infrared radiation; affects stratospheric O_3
CHF_2Cl	HCFC-22	Absorbs infrared radiation; affects stratospheric O_3
CCl_4	Carbon tetrachloride	Absorbs infrared radiation; affects stratospheric O_3
CH_3CCl_3	Methyl chloroform	Absorbs infrared radiation; affects stratospheric O_3
CF_2ClBr	Ha-1211	Absorbs infrared radiation; affects stratospheric O_3
CF_3Br	Ha-1301	Absorbs infrared radiation; affects stratospheric O_3
SO_2	Sulfur dioxide	Forms aerosols, which scatter solar radiation and affect cloud properties
COS	Carbonyl sulfide	Forms aerosol in stratosphere which alters albedo
$(CH_3)2S$	Dimethyl sulfide (DMS)	Produces cloud condensation nuclei, affecting cloudiness and albedo
C_2H_4, etc.	NMHC	Absorbs infrared radiation; affects tropospheric O_3 and OH
O_3	Ozone	Absorbs ultraviolet (UV), visible, and IR radiation
OH	Hydroxyl	Scavenger for many atmospheric pollutants, including CH_4, CO, CH_3CCl_3, and CHF_2Cl
H_2O	Water vapor	Absorbs near-IR and IR radiation

indicate that the sum of radiative effects from the growing atmospheric concentrations of other greenhouse gases, along with the induced effects on the distribution of ozone and other atmospheric constituents, could be comparable to the radiative effects projected for CO_2 alone (see Figure 2.1). In addition, increases in concentrations of aerosols may counteract some of the expected climatic warming. It is essential to understand the changing atmospheric concentrations of these gases and aerosols in order to evaluate the effects they may have on future climate.

The chemical structures of many of the gases important to climate change are shown in Table 2.1. Also shown in this table are the known direct and indirect ways in which these gases can influence climate forcing. Most of the gases are themselves absorbers of longwave terrestrial radiation, i.e., they are greenhouse gases. Several of the gases, such as OH and CO, do not have a direct influence in climate, but because of their importance in atmospheric chemical processes, these gases still have a strong influence on the rate of climate change. Other gases react in the atmosphere to form aerosols that can affect climate (tending to cool) through scattering of solar radiation and through effects on cloud properties.

For each of the gases in Table 2.1, the current atmospheric concentration (in the troposphere for those with long atmospheric lifetimes; the range of tropospheric and stratospheric concentrations for the others), the trend in concentration, and the atmospheric lifetime are shown. As the table indicates, the atmospheric concentrations of a number of globally important trace gases are increasing. Likewise, there is ample evidence suggesting surface emissions, largely from anthropogenic sources, are primarily responsible for the increase in atmospheric concentrations of such gases as CO_2, CH_4, CO, N_2O, and the chlorofluorocarbons (CFCs) (and several other halocarbons).

The concentrations of the most important infrared absorbing gases, namely carbon dioxide and tropospheric water vapor, are much larger than the concentrations of other greenhouse gases. Nonetheless, a number of these other gases are important to climate change. As indicated in Figure 2.2, there are a number of wavelengths where CO_2 and H_2O absorb the radiation emitted at the surface before it can be lost to space. However, there is a region from about 7 to 13 μm, referred to as the window region, where absorption by CO_2 and H_2O is weak. Most of the non-CO_2 gases with the potential to influence climate change absorb in this region, including CH_4, N_2O, O_3, and the CFCs.

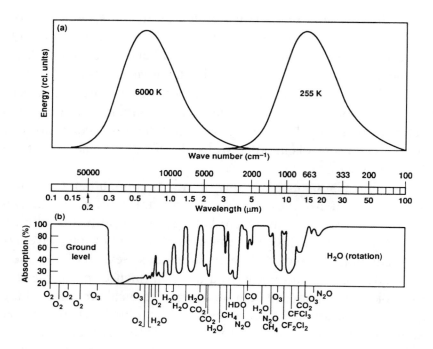

Figure 2.2   (a) Blackbody curves showing the variation of emitted energy with wavelength for temperatures typical of the sun and the Earth, respectively, and (b) percentage of atmospheric absorption for radiation passing from the top of the atmosphere to the surface. Note the comparatively weak absorption of the solar spectrum and the region of weak absorption from 7 to 13 μm in the long wavelength spectrum, referred to as the "window region." (Based on MacCracken and Luther, 1985.)

Atmospheric chemistry is important to determining the concentrations of most of the important greenhouse gases. With the exception of carbon dioxide, atmospheric chemical processes largely determine the rate of removal of greenhouse gases from the atmosphere. In addition, there are indirect effects on climatic forcing that result from chemical interactions on other greenhouse gases such as ozone or water vapor.

The primary goals of this chapter are: (1) establishing what we know about radiatively important atmospheric constituents; (2) establishing how well we know emissions and budgets for these constituents; and (3) determining how we can use this knowledge and the recognized uncertainties to determine credible engineering responses. This chapter examines the emissions and budgets of the radiatively important trace gases that may be having a major influence on climate. We also examine some of the important factors affecting potential attempts to reduce the future emissions of these gases and, thus, reduce their impact on climate.

2.2 BUDGETS OF RADIATIVELY IMPORTANT CONSTITUENTS

2.2.1 Carbon Dioxide
2.2.1.1 Trends and Emissions

Accurate measurements of atmospheric CO_2 concentration began in 1958 at the Mauna Loa Observatory in Hawaii (Keeling et al., 1976; Keeling, 1986). Figure 2.3 shows that the average annual concentration of CO_2 in the atmosphere has risen from 316.1 ppmv in 1959 to 352.6 ppmv in 1989. The average annual rate of increase is about 1.2 ppmv or 0.4% per year.

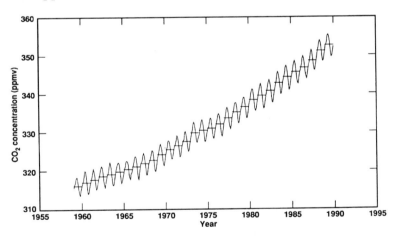

Figure 2.3   Atmospheric CO_2 concentration at Mauna Loa Observatory, Hawaii (Keeling, 1986). Lines connect monthly average values. Horizontal bars indicate annual average concentrations. (From Keeling, 1986.)

Seasonal variations in plant primary productivity and in the decomposition of dead organic matter on land cause an annual cycle in the Mauna Loa data (Keeling et al., 1989). Carbon dioxide concentrations increase during the fall and winter in the northern hemisphere and decline during spring and summer. This cycle is reversed (relative to the season in the northern hemisphere) and of smaller amplitude in the southern hemisphere. The smaller amplitude in the southern hemisphere reflects its smaller land mass, covered by ecosystems that are less seasonal in cycling carbon. A weaker three-to-four year cycle in atmospheric CO_2 levels is associated with the occurrence of El Niño/Southern Oscillation events (Bacastow et al., 1980).

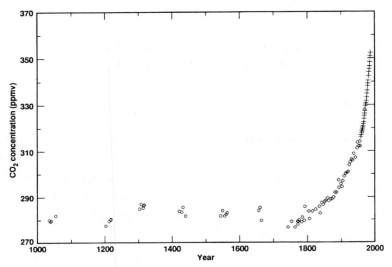

Figure 2.4   Concentration of CO_2 in air bubbles trapped in an ice core extracted at the South Pole (o) (Siegenthaler et al. 1988), at Siple Station Antarctica (◊) (Friedli et al., 1986; Neftel et al., 1985), and as recorded at Mauna Loa, Hawaii since 1959 (+) (Keeling, 1986).

Measurements of CO_2 concentration in air trapped in ice cores indicate that the pre-industrial concentration of CO_2 was approximately 280 ppmv (Neftel et al., 1985; Friedli et al., 1986; Siegenthaler et al., 1988). Figure 2.4 displays these measurements along with annual average concentrations at Mauna Loa.

Two types of human activities release CO_2 into the atmosphere: fossil fuel use and land use (including deforestation, biomass burning, etc.). Fossil fuel use released about 6.0 Pg of carbon into the atmosphere in 1990—of which cement manufacturing was responsible for about 0.15 Pg (Marland et al., 1989). Land use may have contributed about 1.6 Pg in addition to that from fossil fuels (Houghton and Skole, 1991). Approximate errors in estimates of annual fossil fuel emission of plus or minus 10% are frequently cited (Marland and Rotty, 1984). Even if these error estimates are overly optimistic by a factor of two, they are considerably narrower than the confidence on the emissions of other greenhouse gases, and fossil fuel CO_2 emissions are known with much greater certainty than are those due to land use.

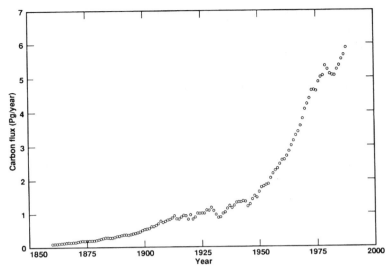

Figure 2.5   Past releases of carbon into the atmosphere from fossil fuels. These estimates are derived from United Nations energy use data. (From Marland et al., 1989.)

Since 1860 fossil fuel emissions have increased from less than 0.1 PgC/year to about 6.2 Pg in 1991 (see Figure 2.5). Emissions increased steadily at about 4.5% per year from 1945 through 1979. They declined from 1979 until 1983, but have increased since. The U.S., U.S.S.R., and the Peoples Republic of China are responsible for about half of the world's fossil fuel CO_2 emissions. U.S. emissions accounted for more than 40% of 1950 global emissions; this share declined steadily to less than 25% in 1988. U.S. CO_2 emissions peaked in 1973 at 1.27 PgC/year and again in 1979 at 1.3 PgC/year.

Prior to impacts by human activities, carbon in vegetation approached 1000 Pg (Olson, 1974). Plants now hold about 560 PgC (Olson et al., 1983). Carbon storage on the contemporary landscape is certainly less than under natural conditions, but the magnitude of releases since the onset of significant fossil fuel use and their timing are very uncertain (Dale et al., 1991).

Houghton et al. (1983) report estimates of carbon releases from vegetation and soils since 1800. Their analysis tracks the area, age, and carbon content of disturbed regions, using response curves to specify changes in carbon stocks in different ecosystem types. They explicitly account for the oxidation rates of fuel wood and wood products. Recent estimates derived by this approach (Houghton and Skole, 1991) indicate that land use decreased carbon storage in vegetation and soil by about 170 Pg since 1800. Estimates of the net flux of carbon into the atmosphere in 1980 due to land use range from 0.4 PgC/year to 2.5 PgC/year (Detwiler and Hall, 1988; Houghton et al., 1987; IPCC, 1990, 1992).

2.2.1.2 Carbon Dioxide and the Global Carbon Cycle

Fossil fuel CO_2 emissions force carbon to redistribute within its global cycle, and the dynamics of this redistribution determine the corresponding rise in CO_2 concentration (Post et al., 1990; Bolin, 1986). The oceans, which contain roughly 50–60 times more carbon than does the atmosphere, are the most important reservoir. Vegetation and soil on land also sequester carbon, but the turnover is more rapid than for the oceans, and in many regions land use changes, particularly forest clearing, cause net CO_2 releases in addition to those from fossil fuels. Figure 2.6 summarizes the major reservoirs and exchanges that affect the response of atmospheric CO_2 concentration to fossil fuel emissions.

The oceans contain about 38,000 Pg of carbon. Three sets of processes determine the rate of CO_2 uptake by the oceans: (1) CO_2 exchange across the air–sea boundary; (2) the incorporation of surface water carbon into compounds other than CO_2; and (3) mixing and circulation, which transport carbon from surface water into deeper layers where it is sequestered from atmospheric exchange (Broecker et al., 1979).

The difference between the partial pressures of CO_2 in the atmosphere and in surface waters governs the exchange of CO_2 across the atmosphere–ocean interface, which is generally assumed to be by molecular diffusion through a thin film and is controlled by temperature, wind stress, and processes such as turbulent mixing. The partial pressure of CO_2 in the atmosphere is proportional to the ratio of the masses of CO_2 and dry air, while the partial pressure of CO_2 in surface waters depends on the chemical equilibria between carbon compounds and transport into deeper water. As CO_2 is added to surface waters, chemical equilibration is rapid compared with transport and mixing processes. The reactions of primary concern are:

$$CO_2(aq) + H_2O \leftrightarrows H_2CO_3$$
$$H_2CO_3 \leftrightarrows HCO_3^- + H^+$$
$$HCO_3^- \leftrightarrows CO_3^{--} + H^+$$

(Eq. 2.1)

so that while the concentration of total carbon is

$$[\Sigma C] = [CO_2(aq)] + [H_2CO_3] + [HCO^-] + [CO^{--}]$$

(Eq. 2.2)

the carbon flux from surface waters back into the atmosphere is proportional only to $[CO_2(aq)]$ (Keeling, 1973).

Mixing and circulation remove carbon from surface waters into deeper layers where it may be sequestered for some time. Analyzing

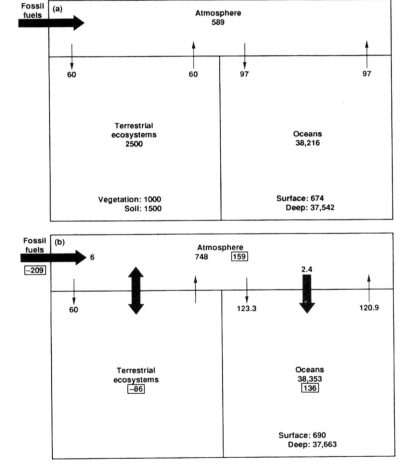

Figure 2.6   Reservoirs and fluxes in the global carbon cycle that control its response to fossil fuel CO_2 emissions. (a) Terrestrial ecosystems and the oceans each in equilibrium with the atmospheric pool in the middle of the 18th century. (b) 1989 fluxes and reservoir contents. Boxed values are changes in reservoir contents over the equilibrium levels in (a). Reservoir carbon contents are in Pg of carbon, and fluxes are in Pg/year of carbon.

[14]C measurements collected by the Geochemical Ocean Sections Study, Stuiver et al. (1983) find that the average residence time of carbon in the deep ocean out of contact with the atmosphere is about 500 years. While some water is isolated from the surface for 1700 years, deep water resides in the Atlantic for only about 275 years.

Broecker et al. (1980) summarize estimates of the annual invasion of carbon into the surface waters of the oceans. Their best estimate is 82.3 Pg/year, with values ranging from 69.3 to 104 Pg/year. The invasion flux estimate in Figure 2.6 is derived from a diffusive model of carbon turnover in the atmosphere and oceans that is calibrated to natural ^{14}C observations (Killough and Emanuel, 1981).

Land plants contain about 560 Pg of carbon (Olson et al., 1983), and soils, about 1500 Pg in actively turning over dead organic matter (Post et al., 1982). Plants assimilate atmospheric CO_2 by photosynthesis. Decomposition and fire release CO_2 from dead organic matter. For reasonably undisturbed ecosystems and over sufficiently long time periods, uptake and loss fluxes are frequently assumed to balance. The balance certainly shifts on shorter time scales with changes in nutrient availability, climate, and sporadic natural disturbances.

Net primary production, the difference between gross assimilation and the return of CO_2 to the atmosphere by plant respiration, is responsible for longer term accumulation of carbon in vegetation and ultimately in dead organic matter. The rapid variations in gross photosynthesis and autotrophic respiration that occur within hours and days do not affect the steadily increasing trend in atmospheric CO_2 concentration. But changes in the relative magnitudes of cumulative net production and losses of organic matter from litter and soil pools due to decomposition and fire can be significant.

Land plants assimilate about 60 PgC/year from the atmosphere as net primary production (Olson et al., 1983). While the oceans apparently act as a net sink for carbon from the atmosphere in any given year, it seems likely that imbalances between terrestrial income and loss fluxes might cause either net increases or decreases in vegetation and soil pools in a given year, depending on environmental conditions, nutrient availability, and sporadic disturbances. A widely distributed 0.5 Pg change in terrestrial carbon pools within a year is likely undetectable.

Disturbance to terrestrial ecosystems by land use change alters the standing levels of carbon and the dynamics of cycling. With land use change such as forest clearing, carbon is immediately released into the atmosphere by burning or, after a delay, by decomposition. In many instances, ground vegetation (e.g., annual crops, pasture, or other herbaceous types) rather than trees is established after clearing. This altered land cover with lower carbon storage than in the original ecosystems may be maintained indefinitely. Following clearing, net productivity generally exceeds losses, and disturbed ecosystems may act as sinks for some time. Abandoned areas recover toward natural conditions and, during recovery, may also serve as carbon sinks. Human management of the landscape has created a complex pattern of sources and sinks in different regions which has changed through time.

Increasing atmospheric CO_2 concentrations affect photosynthesis and perhaps decomposition as well (Eamus and Jarvis, 1989; Bazzaz, 1990). Plants respond to atmospheric CO_2 variations by rapidly changing enzyme activities and stomatal aperture. With increases in atmospheric CO_2, average stomatal conductance can probably be decreased, thus reducing transpiration and increasing water-use efficiency (the ratio of assimilated CO_2 to transpired water).

If fossil fuel use continues to increase at about 2% per year, then the relative contribution of future terrestrial releases to atmospheric CO_2 increase will likely be small compared to that from fossil fuels. The current standing stock of carbon in vegetation is too small compared with the likely fossil fuel source. However, terrestrial sources may be important if substantially lower rates of fossil fuel use occur.

In addition to the major carbon exchanges between the atmosphere and the oceans, and between the atmosphere, vegetation, and soil that appear to dominate the response of atmospheric CO_2 concentration to fossil fuel releases, numerous other carbon fluxes may also be significant. Furthermore, atmospheric exchanges and the turnover of carbon in the oceanic and terrestrial components of the carbon cycle depend on climate—climatic change caused by increases in greenhouse gas concentrations may very well feed back on the controls of atmospheric composition in the global carbon cycle.

Numerous terrestrial and oceanic processes influence changes in atmospheric CO_2 concentration over decades and centuries. As a result, the lifetime of CO_2 in the atmosphere cannot be characterized accurately in terms of a single characteristic response time as is commonly done for the other greenhouse gases (see Shine et al., in IPCC,

1990). Maier-Reimer and Hasselmann (1987) describe the impulse response of the atmospheric CO_2 level in an oceanic general circulation model by the sum of four exponential functions; IPCC (1990, 1992) determined a similar fit using the carbon cycle model of Siegenthaler and Oeschger (1987).

2.2.1.3 Simulating Past and Projecting Future CO_2 Concentrations

An accounting of the uptake of carbon from the atmosphere by the oceans has generally been assumed as being the most important requirement in explaining observed past changes in CO_2 concentration or in estimating future CO_2 increases that would be expected to result from different levels of fossil fuel use. But historic changes in terrestrial carbon storage, whether due to natural phenomena or because of human activities, cannot be ignored nor can future changes, unless the magnitude of fossil fuel releases increases substantially.

Oeschger et al. (1975) proposed a model of carbon turnover in the atmosphere and oceans that is very widely applied. It has become a benchmark against which other models are referenced and compared (Emanuel et al., 1985; Kratz, 1985). In the Oeschger model, a diffusion equation with constant diffusivity represents the vertical transport of carbon in a globally averaged water column. The diffusivity parameter and the invasion flux of carbon from the atmosphere are set for agreement between the equilibrium distributions of carbon and radiocarbon implied by the model and idealized profiles derived from observations.

The effects of surface water chemistry on oceanic CO_2 uptake can be accounted for by solving the equilibrium conditions for the significant reactions (Keeling, 1973). In this approach the partial pressure of dissolved CO_2 in surface waters is a nonlinear function of their total carbon content. Many models use a linear approximation of this relationship.

Interactions between the terrestrial and oceanic reservoirs are weak except through the atmosphere, and if the exchanges between the atmosphere and terrestrial systems are assumed to be independent of atmospheric CO_2 concentration—this assumption is made in many modeling studies that concentrate on land use releases from vegetation and soil—then models of the atmosphere–ocean system can be decoupled from those that account for changes in terrestrial storage. Terrestrial sources are introduced as net fluxes into the atmospheric compartment of atmosphere–ocean models in a manner similar to the fossil fuel CO_2 input. Stand-alone models can be used to simulate the time course of the net terrestrial source, given the intensity of forest harvest and changes in the areal extents of different ecosystem types, such as conversion from forest to cropland (Houghton et al., 1983).

If the Siple ice core and Mauna Loa CO_2 records accurately describe past changes in atmospheric CO_2 concentration, then estimates of historical terrestrial carbon releases are inconsistent with estimates of oceanic uptake derived from models of carbon turnover in the atmosphere and oceans (Siegenthaler and Oeschger, 1987; Enting and Mansbridge, 1987). Until such inconsistencies are resolved, one recourse is to match the history of atmospheric CO_2 concentration to the Siple ice core and Mauna Loa records by calculating the net residual flux into or out of the atmosphere required to balance simulated oceanic uptake. This procedure takes the model solution from initial equilibrium conditions to the present in a way that is consistent with our best understanding of past changes in atmospheric CO_2 concentration. Future changes can then be projected with assumed inputs of CO_2 into the atmosphere from fossil fuels and because of land use change.

Figure 2.6(b) summarizes changes in the major carbon reservoirs from 1745 through 1988 if the net residual flux is assumed to be from terrestrial ecosystems into the atmosphere and if a one-dimensional, discrete-diffusion model (Killough and Emanuel, 1981) similar to Oeschger's box-diffusion model, describes carbon turnover in the oceans. The increase in the carbon content of the atmosphere is 76% of the 209 Pg of carbon released from fossil fuels and 54% of the total release from fossil fuels and from the residual flux assumed to be from vegetation and soils. Again the 86 Pg residual release required to match simulated oceanic uptake and observed changes in the atmosphere is substantially less than the decreases in terrestrial carbon storage estimated from land use data.

Inconsistencies in our understanding of the relationship between observed increases in atmospheric CO_2 and past fossil fuel emissions force skepticism regarding CO_2 projections. However, all evidence indicates that fossil fuel use can raise CO_2 levels to twice preindustrial concentrations over the next 50 years. Drastic emissions reductions are required in order to hold CO_2 levels constant (Emanuel et al., 1985; Shine et al., 1990). As climatic change occurs, it will affect the carbon cycle and may make detailed prognosis for CO_2 increases over the next several decades much different than current models imply.

2.2.2 Methane

Although its atmospheric abundance is less than 0.5% that of CO_2, methane (CH_4) is one of the most important greenhouse gases based on the rapid rate of increase in its atmospheric concentrations. On a per molecule basis, an additional methane molecule in the current atmosphere is about 21 times more effective at affecting climate than an additional molecule of CO_2 (IPCC, 1990). Since the 1700s, the growing concentrations of atmospheric methane have been responsible for about 23% of the change in radiative forcing on climate (IPCC, 1990).

The globally averaged atmospheric concentration of methane is about 1.72 ppmv, with slightly higher concentrations in the Northern Hemisphere (1.76 ppmv) compared to the Southern Hemisphere (1.68 ppmv). Continuous monitoring of methane trends in ambient air (1979–1989) indicate that concentrations have been increasing at an average of about 16 ppbv/yr (1%/yr). The data show that in the late 1980s the trends have decreased to about 10 ppbv/yr (Khalil and Rasmussen, 1993; Steele et al., 1992). The change in growth rate of atmospheric methane may reflect changes in source emissions over the last few decades and possible changes in the removal rates (OH and soils).

Methane concentrations in the atmosphere exhibit seasonal and interannual variations (Khalil and Rasmussen, 1985, 1987; SWMO, 1985, 1989, 1991). The magnitude of the seasonal variability varies with latitude, being controlled by a combination of temporal variability in the atmospheric oxidizing capacity and in the source emission strengths. The relatively short continuous records for atmospheric methane concentrations do not permit a detailed analysis of possible causes of interannual variations. An understanding of controls on the seasonal and interannual variations in atmospheric methane will provide the basis for a potentially sensitive early warning indicator of changes in regional sources of methane and in the atmospheric oxidizing capacity. A carefully chosen network of monitoring sites is needed with highest priority given to sites downwind of major continental source areas. Figure 2.7 shows the recent trend, through 1992, in surface concentrations of methane based on measurements since 1980. Ice core data going back as far as 160,000 years indicate that the concentration of CH_4 in the pre-industrial atmosphere was about 0.7 ppmv (0.35 ppmv during glacial periods), which is less than half of the current concentration (Pearman et al., 1986; Khalil and Rasmussen, 1987; Raynaud et al., 1988; Chappellaz, 1990). The trend in CH_4 over the last 1000 years is shown in Figure 2.8.

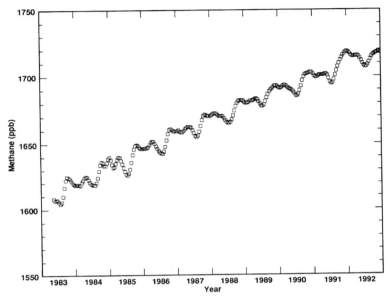

Figure 2.7   Globally averaged, biweekly CH_4 mixing ratios for the marine boundary layer determined from 37 sites in the NOAA CMDL cooperative air sampling network (see Steele et al., 1992, for details).

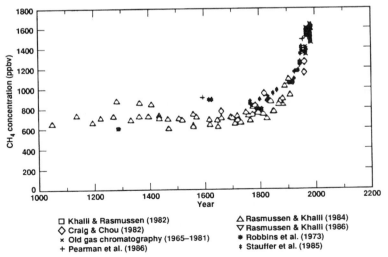

Figure 2.8   Atmospheric concentration of CH_4 over the last 1000 years, most of which is determined from ice core data. (From Khalil, M.A.K. and Rasmussen, *Atmos. Environ.*, 21, 2447, 1987. With permission.)

Table 2.2　　Estimated sources and sinks of methane (Tg CH_4/yr). Primarily based on IPCC (1990, 1992, 1994) and references therein; also see Khalil and Shearer (1993).

Sources	Best Estimate and Range[a]	
Natural		
Enteric fermentation (wild)	4	(1–7)
Wetlands (swamps, etc.)	115	(55–155)
Lakes	5	(1–25)
Tundra	4	(2–7)
Oceans	10	(5–50)
Termites and other insects	20	(10–50)
Methane hydrates	5	(0–100)
Other	40	(0–80)
Energy		
Natural gas losses	40	(25–50)
Coal mining	30	(15–45)
Petroleum industry	15	(5–30)
Biomass burning (e.g., fuel wood)	15	(5–30)
Landfills	40	(20–70)
Animal waste	25	(20–30)
Sewage treatment	25	(15–80)
Agriculture/non-energy		
Enteric fermentation (domesticated)	81	(65–100)
Rice paddies	60	(20–100)
Biomass burning	40	(20–80)
Sinks		
Reaction with tropospheric OH	445	(330–560)
Removal in stratosphere	40	(25–55)
Microorganisms uptake by soils	30	(15–45)
Accumulation	30	(30–35)

[a] Lack of balance between total sources and sinks reflects uncertainties.

Estimated sources and sinks of methane are shown in Table 2.2. Data on the photochemical sink for methane in the lower atmosphere (troposphere) have recently been revised (Vaghjiani and Ravishankara, 1991), and are thought to be one of the most accurate terms in the budget. Using current estimates of global OH concentrations (see Section 2.2.9), the total quantity of methane that reacts with OH must be about 445 Tg/yr (IPCC, 1994). Soil uptake of atmospheric methane as a result of microbial oxidation is estimated to remove an additional 30 ±15 Tg/yr. A recent experiment suggests that the soil sink for methane may be sensitive to acid deposition.

Measurements at a few sites in temperate forest soils demonstrated that methane uptake is reduced with increasing nitrogen deposition (Steudler et al., 1990). A relatively small sink of methane results from an approximately 10 ± 5 Tg/yr loss due to stratospheric reactions with $O(^1D)$ and Cl. These reactions are important to the understanding of stratospheric O_3 and water budgets (see Sections 2.2.10 and 2.2.11).

The total atmospheric sink for methane is approximately 510 Tg/yr. The total methane source should be equal to the total sinks plus the observed annual increase in the atmospheric burden (about 30 Tg/yr), which adds up to 540 ± 105 Tg/yr. Isotopic measurements indicate that about 20% of the methane emitted each year does not contain radioactive ^{14}C and must be from fossil sources like natural gas use, fossil fuel combustion, and decomposition of "old" organic matter (Wahlen et al., 1989).

Using the information discussed above and the limited emissions data available for specific methane sources, numerous attempts have been made to construct global source estimates (Khalil and Rasmussen 1990a; Fung et al., 1991; WMO 1985, 1989, 1991; IPCC 1990, 1992, 1994). Recent budget analyses have been designed to be consistent with the total set of observed measurements and variables including the ice core data.

For considerations of possible engineering responses to increasing global atmospheric methane, the sources related to human activities are most pertinent. These sources include emissions from food production systems (e.g., domestic ruminants, rice paddies, etc.), energy systems, leakage from natural gas production and use, coal mining and anaerobic waste management systems (landfills, sewage treatment plants). Several of these sources offer interesting opportunities for cost effective and rapid mitigation of methane emissions. According to the IPCC (1990) report a 15–20% reduction in emissions would be required to stabilize atmospheric concentrations at present levels.

Quantitative local-to-regional scale source inventories will be useful for developing specific engineering approaches to mitigate methane emissions. For example, from an examination of the geographical distribution of methane emissions from coal mining in the U.S., it is obvious that a U.S. component of a global strategy would focus on emissions from deep coal mines in such states as Kentucky, West Virginia, Pennsylvania, and Illinois. An existing technology for commercial extraction of coalbed methane prior to mining is a cost-effective method for reducing methane degassing during coal mining operations.

In summary, the global sources and sinks of methane are sufficiently well known to suggest some target sources for emissions controls. Collection of methane from animal waste treatment lagoons, large landfills, and anaerobic sewage treatment can reduce emissions and provide a potential energy source for local applications. A reduction of leakage losses from natural gas transmission and distribution systems is feasible and in most cases, cost effective. However, it is important to emphasize that to be effective in reducing atmospheric methane these efforts must be global in scale.

2.2.3 Chlorofluorocarbons, Halons, Methyl Bromide and Their Replacements

Halocarbons are of environmental concern since they have both the potential to affect stratospheric ozone as well as contribute to climate change. Those halocarbons containing chlorine and/or bromine are of concern to stratospheric ozone due to the effectiveness of the catalytic ozone destruction cycles based on bromine and chlorine. The same group plus the halocarbons containing fluorine also have the potential to affect climate change since these species characteristically have strong infrared absorption features in the radiation "window" region of the atmosphere. The most potent halocarbons in the current atmosphere are the chlorofluorocarbons $CFCl_3$ (CFC-11) and CF_2Cl_2 (CFC-12) due to their relatively high concentrations. Molecules of CFC-11 and CFC-12 in the atmosphere are about 12,400 and 15,800 times more effective, respectively, at affecting climate than an additional molecule of CO_2 (IPCC, 1990).

With the exception of the naturally occurring emissions of CH_3Cl and CH_3Br, essentially all the halocarbons in the atmosphere are man-made. CFC-11 and CFC-12 have the largest atmospheric concentrations, 0.28 and 0.51 ppbv, respectively. The tropospheric concentrations of both of these gases were increasing at about 4% per year in the early 1990s, as shown in Figure 2.9, but have now slowed to 1% or less for CFC-11 and 2% or less for CFC-12 (Elkins et al., 1993). Table 2.3 gives the product and use statistics for important halocarbons. CFC-11 is used primarily as a blowing agent for plastic foams and, to a lessening amount, as a propellant in aerosol cans; CFC-12 is

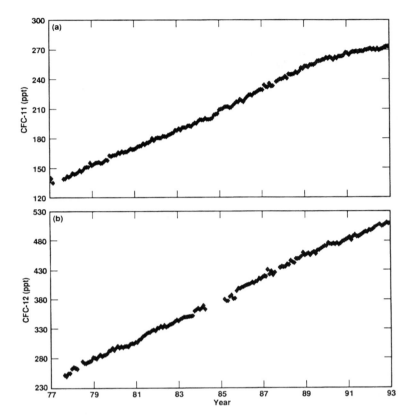

Figure 2.9   Globally averaged monthly mean (a) CFC-11 and (b) CFC-12 mixing ratios from NOAA/CMDL sites and cooperative stations (From Elkins et al., 1993).

used primarily as a refrigerant and as a propellant. The atmospheric concentrations of several other halocarbons have until recently been growing at even a faster rate than CFC-11 and CFC-12. For example, the concentration of CFC-113 ($C_2F_3Cl_3$), was increasing about 10% per year in the early 1990s but has also slowed greatly, with a current concentration of about 0.08 ppbv. Abundances of CH_3CCl_3 and HCFC-22 (CHF_2Cl) are increasing at about 2 and 10% per year from their current concentrations of 0.16 and 0.10 ppbv, respectively (Montzka et al., 1993; IPCC, 1994). Whereas CFC-113 and CH_3CCl_3 are both used primarily as solvents, HCFC-22 is used in air conditioning and refrigeration.

All of the fully halogenated chlorofluorocarbons have long atmospheric lifetimes. The lifetime of CFC-11 is about 50 years, while the lifetime of CFC-12 is about 102 years. The long atmospheric lifetimes

Table 2.3   Estimated global sources of major halocarbons. Production has fallen substantially in the early 1990s.

Gas	Production in Mid-to Late-1980s (Tg/yr)	Uses	
CFC-11 (CFCl3)	~0.33	Refrigeration, air conditioning	~8%
		Closed-cell foams	~36%
		Open-cell foams	~19%
		Aerosol propellants	~31%
		Other uses	6%
CFC-12 (CF_2Cl_2)	~0.44	Refrigeration, air conditioning	~49%
		Closed-cell foams	~8%
		Open-cell foams	~5%
		Aerosol propellants	~32%
		Other uses	6%
CFC-113 ($CFCl_2CF_2Cl$)	~0.19	Aerosol propellants; cleaning agents	~98%
		Closed cell foams; refrigerants; heat transfer fluid	~2%
$CH_3 CCl_3$	~0.70	Cleaning operations	100%
HCFC-22 (CF_2HCl)	~0.20	Blowing agents, aerosol propellant	~15%
		Refrigeration, air conditioning	~85%

contribute to the sustained increasing concentrations of these gases. The CFCs are primarily destroyed by photolysis (and, to a lesser extent, by reaction with excited oxygen atoms, $O(^1D)$) in the stratosphere, resulting in the release of their chlorine at altitudes where it can destroy ozone (see later discussion on ozone trends). Fifteen percent of the increase in radiative forcing on climate since 1900 is due to the CFCs, while they contribute almost one fourth of the increased climatic forcing during the 1980s (IPCC, 1990); the cooling effect due to halocarbon-induced decreases in stratospheric ozone over the last two decades would reduce the net radiative warming effect from CFCs and other halocarbons.

The atmospheric lifetimes of HCFCs, HFCs, and other halocarbons containing hydrogen tend to be much shorter than the CFCs. These gases can react with OH in the troposphere. The lifetime of CH_3CCl_3 is 4 to 5 years, while the lifetime of HCFC-22 is about 14 years. Because of these shorter lifetimes, fewer of these compounds reach the stratosphere and they have less effect on ozone than the CFCs.

Bromine-containing halons, most notably CF_3Br (Ha-1301) and CF_2ClBr (Ha-1211), are used in fire extinguishing. These compounds have small atmospheric concentrations, about 2 pptv for both compounds (Butler et al., 1992). Because bromine is more effective at destroying ozone than chlorine, these compounds have caused concern, especially in light of their early rapid increases of atmospheric concentration, but now seem to be leveling off. Primary destruction of these compounds occurs through photolysis, resulting in long atmospheric lifetimes (110 years for Ha-1301; 25 years for Ha-1211). At their present atmospheric levels, their contribution to radiation change is considered minimal.

Recent modifications to the Montreal Protocol by the United Nations Environment Programme at the June 1990 London Agreement and at the November 1992 Copenhagen Agreement essentially call for the elimination of the production of CFCs by 1996, and a phase out for the production of CH_3CCl_3, CCl_4, and the halons. The chemical industry is currently developing a number of possible replacement compounds. The suggested replacements for the CFCs are generally halogenated hydrocarbons, some containing chlorine (HCFCs), and others not (e.g., HFCs). All of the suggested replacements have shorter atmospheric lifetimes, generally less than 20 years, than the compounds they would replace. As a result, their concentrations should always be well below the levels reached for CFCs. The potential for these compounds to destroy ozone is less than 10% of the CFCs they would replace (Fisher et al., 1990a; WMO, 1989, 1991). While the replacement compounds are greenhouse gases, their potential for affecting climate is also smaller than that of the CFCs (Fisher et al., 1990b; WMO, 1989; IPCC, 1990). Nonetheless, the replacement compounds could affect ozone and climate if emissions and concentrations grow enough (WMO, 1989, 1991; IPCC, 1990, 1992, 1994).

Although production of CFCs will be essentially eliminated by the turn of the century, their atmospheric concentration is expected to grow for a limited time, as material presently held in use is emitted to the atmosphere. Additionally, if countries which have not signed the Montreal Protocol continue to produce and use CFCs, atmospheric levels could be sustained for an additional period of time.

2.2.4 Nitrous Oxide

Nitrous oxide (N_2O) is a greenhouse gas for which one molecule is 200 times more efficient than a CO_2 molecule as an infrared absorber.

Also, it is the primary source (through reaction with excited oxygen atoms) of the nitrogen oxides that account for a significant fraction of the natural destruction of ozone in the stratosphere.

The mean atmospheric concentration of N_2O in 1990 was about 311 ppb (IPCC, 1994). Its concentration is increasing at a rate of 0.2–0.3% per year (see Figure 2.10; Khalil and Rasmussen, 1992). The mean northern hemispheric mixing ratio is about 1 ppbv higher

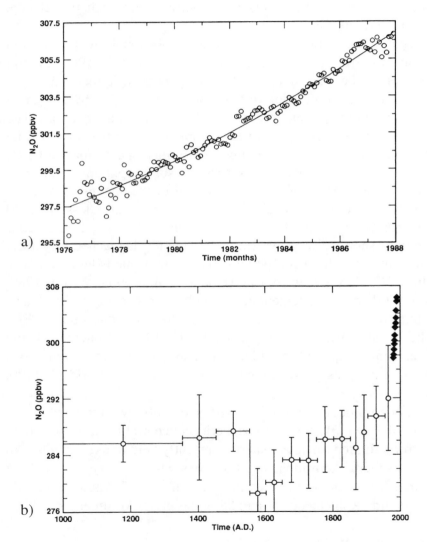

Figure 2.10   Recent and long-term trends in N_2O concentrations based on available atmospheric and ice core data. (From Khalil, M.A.K. and Rasmussen, *J. Geophys. Res.*, 97, 14653, 1992.)

than the mean southern hemispheric value. Ice core data indicate that the concentration of N_2O in the atmosphere until about 150 years ago was about 285 ppb (Pearman et al., 1986; Khalil and Rasmussen, 1988a; Zardini et al., 1989). The atmospheric lifetime of N_2O is about 120 years.

The cause of the increase in N_2O is highly uncertain. The various natural and anthropogenic sources are not well quantified. Until recently, combustion, biomass burning, and use of fertilizers were thought to explain the increase. However, current estimates of the production from biomass burning of less than 0.2 TgN per year are much smaller than prior estimates (IPCC, 1990; Crutzen, 1989; Elkins et al., 1990; Griffith et al., 1990). Recent papers (Muzio and Kramlich, 1988; Linak et al., 1990) also suggest that earlier measurements of the combustion source of N_2O were an artifact of the measurement technique, and that the actual combustion source may be as small as a few tenths of a teragram of nitrogen per year. The source from nitrification and denitrification of nitrogen from use of industrially produced fertilizers is difficult to quantify, but is likely less than 2.2 TgN per year (IPCC, 1990). A new source of N_2O was recently identified in the manufacturing industry (Thiemens and Trogler, 1991). Nitric acid oxidation to form adipic acid results in N_2O as a by-product. The upper estimate of this source is 0.4 TgN per year. There may be other industrial processes not yet accounted for. The total source needs to be 10–17.5 TgN/yr (assuming a removal rate primarily by photolysis in the stratosphere to be 7–13 TgN per year and its atmospheric accumulation to be 3–4 TgN per year). The natural sources from oceans, soils, and aquifers can explain 4–14 TgN per year. Thus, there is some indication that a source of N_2O is missing in the current budget analyses. An estimated budget, reflecting the many uncertainties, for N_2O is shown in Table 2.4.

The IPCC (1990) estimates that a 70–80% reduction in nitrous oxide emissions would be necessary to stabilize concentrations at present day levels.

It is very difficult to project future emissions of N_2O when large uncertainties are associated with numerous small sources making up the budget. In particular, there are few measurements of the biogenic sources where the processes and the factors controlling them are not well understood. As countries develop, and agriculture employs fertilizer, global emissions could grow. As mitigation efforts are employed

Table 2.4   Estimated sources and sinks of nitrous oxide. Based primarily on IPCC (1994) and Khalil and Rasmussen (1992). Budget is unbalanced due to uncertainties.

Sources of NO_2	Estimated Value and Range (TgN/yr)	
Natural		
Oceans and estuaries	3	(1–5)
Natural soils	5	(3–8)
Aquifers	0.8	(0.8–2)
Energy		
Fossil fuel combustion	0.5	(0–3)
Agriculture/non-energy		
Biomass burning	0.5	(0.2–1.0)
Cultivated soils	1.5	(0.03–3)
Industry processes	1.3	(0.7–1.8)
Cattle and feedlots	0.4	(0.2–0.5)
Sinks		
Stratospheric photolysis	12	(9–17)
Removal by soils	small	
Accumulation	3.9	(3.1–4.7)

to reduce one kind of emission, another can be produced. For example, new combustion methods used to reduce acid rain precursors employing fluidized beds can produce N_2O; automotive catalytic converters can also produce N_2O (Knapp, 1990).

2.2.5 Non-Methane Hydrocarbons

A number of non-methane hydrocarbons (NMHC) (many of these are also referred to as volatile organic compounds, VOCs) are potentially important greenhouse gases if their atmospheric concentrations become large enough. However, due to their reactivity with hydroxyl, these gases tend to have short atmospheric lifetimes. As a result, it is unlikely that their concentrations, in the foreseeable future will be large enough to directly affect climate.

The indirect affect of NMHCs on climate lies with their role in tropospheric chemistry. NMHCs in the presence of NO_x contribute to the formation of tropospheric ozone and stratospheric H_2. They can also react with nitrogen species to produce PAN which is a long-lived reservoir of reactive nitrogen. Oxidation of NMHCs is an additional source of CO and ultimately CO_2.

NMHCs have many anthropogenic and natural sources into the atmosphere. Anthropogenic sources vary from all aspects of human activity including chemical manufacturing, vehicle exhaust, food processing, refuse disposal, biomass burning, and energy production. Global emissions from anthropogenic activities are estimated at roughly 100 Tg by WMO (1991). Emissions from the natural sources are less well known. The observed gradients in concentration of NMHCs from northern to southern hemisphere, ocean to land, and lower to upper troposphere (WMO, 1991) suggest that the distribution of all of the sources is highly variable. Because little information has been published about these gases, a data base is required to understand the geographical variabilities, seasonal cycles, and chemical mechanisms involved. Only then will we be able to quantify their cumulative affect on other radiatively important gases including O_3, CO, and hydroxyl.

Table 2.5   Estimates of global carbon monoxide (CO) sources and sinks. Budget is unbalanced due to uncertainties.

Sources of CO	Estimated value and Range (TgC/yr)	
Natural		
Plant emissions	75	(60–160) [a]
Oxidation of natural hydrocarbon	250	(50–500) [a]
Oceans	20	(10–100) [a]
Oxidation of methane	400	(260–500) [a]
Energy		
Energy use	400	(300–550) [b]
Agriculture/non-energy		
Agriculture	110	(40–170) [a]
Biomass burning	350	(300–700) [a]
Oxidation of man-made hydrocarbons	40	(0–80) [a]
Sinks		
Reaction with OH	900	(500–1200) [a]
Soil uptake	250	(100–390) [a]
Accumulation	~2	

[a] Wuebbles and Edmonds (1991); IPCC (1992, 1994).
[b] Wuebbles and Edmonds (1991); residential/commercial (52%), industrial (1%), transport (47%), electric utilities (0%); IPCC (1992, 1994).

2.2.6 Carbon Monoxide

Although carbon monoxide (CO) is not important for its direct impact on climate, it is important to climate change through its reactivity in the atmosphere. The reaction of carbon monoxide with OH is the primary sink for atmospheric OH. This reaction also yields an additional source of the greenhouse gas CO_2. In addition to its global role, CO is also a local air pollutant.

The global emissions of CO are still poorly understood. Sources of carbon monoxide include incomplete combustion processes (complete combustion yields CO_2 rather than CO). Emissions of CO from fossil fuel combustion peak between 30 and 60 degrees north latitude. More than 70% of the biomass burning source is emitted in tropical regions (Dignon, 1994). Chemical decomposition of methane and other hydrocarbons are also an important source. Total sources are about 1100 TgC per year with a large uncertainty range (IPCC, 1992). As a result of its short lifetime (2–3 months) and the high spatial variability of its sources, the atmospheric concentration of CO exhibits significant spatial and temporal inhomogeniety. Annually averaged concentrations of CO peak at about 0.2 ppmv at high northern latitudes; minimum concentrations of about 0.06 ppmv occur throughout most of the Southern Hemisphere. Long-term trends suggest that CO concentrations have been increasing in the Northern Hemisphere at about 1% per year, with little evidence of an increase in the Southern Hemisphere (Khalil and Rasmussen. 1988; Cicerone, 1988; WMO, 1991).

2.2.7 Hydroxyl

Hydroxyl (OH) does not have a direct radiative impact on climate. The effect of OH on climate change comes entirely through the importance of its chemical reactivity, particularly in the troposphere. Reaction with OH is the primary chemical sink for a number of radiatively important gases, including CH_4, CO, CH_3CCl_3, CH_3Cl, CH_3Br, SO_2, DMS, and other hydrocarbons and hydrogen-containing halocarbons. Most of the compounds being considered as replacement compounds in the uses of CFCs have reaction with OH as their primary sink. Not only is it important as a scavenger of greenhouse gases, but OH chemistry is also important in determining the concentrations of ozone (and other gases) in the troposphere and stratosphere. Reactions with OH in the troposphere also limit the amount of

CH_4 and other gases reaching the stratosphere, where these species can affect the ozone distribution.

The hydroxyl radical is formed primarily through the reaction of an excited oxygen atom with a water vapor molecule. This reaction is the dominant source in the production of odd-hydrogen or hydrogen oxides, HO_x (=OH, HO_2, H_2O_2, etc.) in the atmosphere. The excited oxygen atom is generated by photolysis of ozone at wavelengths less than about 300 nm. Regions with high concentrations of hydrocarbons can add to the HO_x formation. The primary sinks for the removal of OH are the reactions of OH with CO and CH_4.

The global distribution of OH concentrations in the troposphere is poorly understood, primarily as a result of the difficulty of measuring the small concentrations (105–106 molecules per cubic centimeter). Measurement techniques to determine the global tropospheric distribution of OH are badly needed. The high chemical reactivity of OH contributes to it having an extremely short atmospheric lifetime in the troposphere. While it is possible to use measured trends in CH_3CCl_3 and other species to estimate the globally-averaged tropospheric OH concentration (Prinn et al., 1987; WMO, 1989), the paucity of direct measurements and possible systematic errors in indirect methods (Butler et al., 1991) make it difficult to determine the trend in tropospheric OH concentrations.

Increases in the concentrations of CO and CH_4 can lead to decreased concentrations of OH, with a subsequent positive feedback on the atmospheric lifetimes (and consequently the concentrations) of CO, CH_4 and other molecules scavenged by OH. These interactions provide an important nonlinear process that affects climate change. If, as seems likely, CO and CH_4 emissions and concentrations continue to increase, the average OH abundance could decrease. This would further enhance the concentrations of CH_4 and enhance the radiative effects of methane on climate. Other greenhouse gases would be similarly affected.

Significant uncertainties are also encountered in investigating the implied trends of OH in the recent past. Studies of CH_4 measured trends have suggested that as much as 20 to 50% of the increase in methane may be attributable to decreasing OH concentrations (Khalil and Rasmussen, 1985; Thompson and Cicerone, 1986; Levine et al., 1985). However, uncertainties in the budgets and concentration trends of CO, NO_x, and NMHC contribute to a wide range of possible past trends in OH. For the troposphere as a whole, it is thought that the current atmosphere is NO_x-poor with respect to net production of odd-hydrogen from oxidation of CO, CH_4, and NMHC. Projected increases in these compounds may thus lead to continued decrease of tropospheric OH, depending on how much NOx emissions increase. However, Isaksen and Hov (1987) found in their model calculations that concurrent increases in concentrations of NO_x, CO, CH_4 and NMHC produced a slight increase in OH concentration.

Climate-driven increases in tropospheric water vapor would tend to enhance the concentrations of OH. Global increases in temperature are expected to increase the water holding capacity of the atmosphere (IPCC, 1990; MacCracken and Luther, 1985; Ramanathan et al., 1987). Model calculations indicate that a 2°C increase in temperature could be associated with an increase in water vapor concentration of 10–30%, implying a few percent increase in tropospheric OH concentrations (Thompson et al., 1989). Increases in tropospheric concentrations of O_3 and NO_x can lead to increases in OH concentrations by enhancing cycling reactions that convert HO_2 to OH. Increasing biogenic emissions of NMHC from global warming may decrease the amount of tropospheric OH (Trainer et al., 1987; Penner et al., 1989).

2.2.8 Nitrogen Oxides

Emissions of nitrogen oxides have historically been considered important due to their role as primary pollutants in photochemical smog and their contribution to acid wet and dry deposition. Nitrogen oxides are also important because of their indirect effect on climate through their role in the apparent increase in tropospheric ozone concentrations. Although NO_x species are relatively short lived, they can react chemically with NMHCs to produce PAN. PAN has a strongly temperature-dependent lifetime. This provides a reservoir for nitrogen oxides which can be transported long ranges to affect ozone chemistry well downstream from the sources.

The largest sources of reactive nitrogen in the troposphere are fossil fuel combustion, biomass burning, lightning discharges, microbial activity in soils, and transport from the stratosphere. An estimated budget of the sources and sinks of nitrogen oxides is given in Table 2.6. Emissions of nitrogen oxides from combustion of fossil fuel has increased globally at 1–2% per decade during this century (Dignon and Hameed, 1989; Hameed and Dignon, 1988) resulting in increased

Table 2.6   Recent estimates for the global sources and sinks of reactive nitrogen oxides. Based on Wuebbles and Edmonds (1991) and IPCC (1990, 1992, 1994).

Sources	Estimated value and Range (TgN/yr)	
Natural		
Stratospheric oxidation of N_2O	1.0	(0.5–1.5)
Lightning	5	(2–80)
Soil microbial activity	12	(1–20)
Oceans	0.15	(<1)
Energy		
Fossil fuel combustion	24	(20–25)
Agriculture/non-energy		
Biomass burning	8	(3–13)
Jet aircraft	0.4	(0.2–0.5)
Sinks		
Wet and dry deposition		(25–85)

tropospheric concentrations particularly over continents in the boundary layer, and in the flight corridors used by commercial aircraft. One of the nitrogen oxides, namely NO_2, is an important absorber of visible solar radiation, and could affect climate directly if tropospheric and/or stratospheric concentrations continue to increase.

Control strategies for NO_x emissions may be included into a "no regrets" policy scenario because of the benefits from reduction in photochemical smog and acid rain. For these reasons, many developed nations have joined the "30 Percent Club," agreeing to reduce NOx emissions by 30% of their mid-1980s values (Rosencranz, 1986).

2.2.9 Hydrogen

Molecular hydrogen exists at about 500 ppbv in the earth's atmosphere and it is increasing at about 5 ppbv/yr (Khalil and Rasmussen, 1990b) or about 1% per year. The anthropogenic sources of H_2 are similar to sources of CO — the important ones being oxidation of methane and other hydrocarbons, biomass burning, and automobiles. The major natural source is the oceans. The major sink is removal by the soils and lesser amounts are removed by reactions with OH. The increasing trends are attributed to increasing contributions from human activities (see Khalil and Rasmussen, 1990b).

In the present context H_2 has two notable effects. It produces water vapor in the stratosphere, where increasing concentrations may contribute to climate change, and secondly, suggestions to convert from carbon-based fuels to H_2 may cause significant perturbations to the stratospheric water vapor concentrations. If all carbon-based fuels are replaced by H_2 even a 1% leakage would release as much extra H_2 as all present sources, thus doubling the H_2 concentration and its contribution to stratospheric water vapor. This may cause a serious aggravation of the stratospheric water vapor budget, unknown and unquantified effects on greenhouse forcing, and stratospheric ozone.

2.2.10 Water Vapor

Concentrations of water vapor in the atmosphere vary from as much as a few percent of air density near the surface in the tropics to as low as 3 ppmv in the lower stratosphere. The spatial distribution of water vapor in the troposphere is primarily determined by evapotranspiration, condensation, and transport processes. Human activities currently have little impact on tropospheric water vapor concentrations. Increased water vapor concentrations as a result of global warming is a well recognized climatic feedback process; increasing temperatures allow more water vapor to remain in the atmosphere, but, since water vapor is one of the most important greenhouse gases, the added water vapor further enhances the greenhouse radiative forcing. Resulting changes in cloudiness also have important effects on climate.

Very little of the tropospheric water vapor penetrates into the stratosphere. The mechanism limiting the transport of tropospheric water vapor into the stratosphere is still not well understood. As a consequence, it is not known how water vapor concentrations in the lower stratosphere will respond to climate change effects on tropospheric water vapor concentrations. Concentrations of water vapor increase with altitude in the stratosphere, from 3 ppmv in the lower stratosphere to about 6 ppmv in the upper stratosphere. This increase in concentration with altitude occurs primarily as a result of the oxidation of methane. The methane oxidation reactions produce about two water molecules for each CH_4 molecule that is destroyed. Stratospheric water vapor concentrations should increase as concentrations of methane increase. Since methane concentrations have increased from about 0.7 ppmv in the pre-industrial atmosphere to the current concentration of 1.7 ppmv; this implies that upper stratospheric water vapor concentrations have increased by about 2 ppmv over this time period. Such an increase in stratospheric water vapor concentrations further enhances the greenhouse effect, increasing the radiative forcing from the added methane by about an additional 5%. As mentioned earlier, increasing concentrations of hydrogen also would contribute to increasing concentrations of stratospheric water vapor. Another potential source of lower stratospheric water vapor could result from emissions of high flying commercial aircraft.

Another climatic feedback process may result from the increasing stratospheric water vapor concentrations. Water vapor produced in the middle and upper stratosphere will be transported poleward and downward in the winter. This water vapor can contribute to the formation of stratospheric clouds in the polar lower stratosphere, and to increased effectiveness of the heterogeneous processes determining the destruction of ozone presently occurring in the Antarctic (the springtime "ozone hole") and, to a lesser degree, in the Arctic (Blake and Rowland, 1988; Ramanathan, 1988; WMO, 1989, 1991). Cooler stratospheric temperatures as a result of increasing concentrations of CO_2 and other greenhouse gases could also enhance the formation of polar stratospheric clouds (Blanchet, 1989; Shine, 1988). The implications of these processes are currently unknown. Although speculative, increasing levels of CH_4 and H_2 may be partially responsible for the apparent increase in the frequency of polar stratospheric clouds during the last decade.

2.2.11 Ozone

Ozone has several important effects on climate. Although the direct radiative effect of CO_2 and other trace gases considered above largely depend on their concentration in the troposphere, the climatic effect of ozone depends on its distribution throughout the troposphere and stratosphere. Ozone is the primary absorber of ultraviolet and visible radiation in the atmosphere, and its concentrations determine the amount of ultraviolet radiation reaching the Earth's surface. It is the absorption of solar radiation by ozone that explains the increase in temperature with altitude in the stratosphere. However, ozone is also a greenhouse gas, with a strong infrared absorption band at 9.6 μm.

It is the balance between these radiative processes that determines the net effect of ozone on climate (Lacis et al., 1990). Increases in ozone above about 30 km tend to decrease the surface temperature as a result of the increased absorption of solar radiation, effectively decreasing the solar energy that would otherwise warm the Earth's surface. Below 30 km, increases in ozone tend to increase the surface temperature. The infrared greenhouse effect dominates in this region. Changes in ozone in the upper troposphere and lower stratosphere are particularly effective in affecting climate forcing. This is because the greenhouse effect produced by ozone is directly proportional to the temperature contrast between the radiation absorbed and the radiation reemitted. This contrast is greatest near the tropopause (at about 13 km altitude at midlatitudes) where temperatures are at a minimum compared to the surface temperature.

Although ozone is a major component of photochemical smog in urban areas, this ozone contributes very little to the global ozone budget. Ozone is produced in the atmosphere by the rapid reaction of an oxygen atom with an oxygen molecule in the presence of any third molecule. However, the means by which the oxygen atom is primarily generated differs greatly between the troposphere and the stratosphere.

Approximately 90% of the ozone in the atmosphere is contained in the stratosphere. In the stratosphere, the production of ozone begins with the photodissociation of O_2 at ultraviolet wavelengths less than 242 μm. This reaction produces two ground-state oxygen atoms that can react with O_2 to produce ozone. Since an oxygen atom is essentially the same as having an ozone, it is common to refer to the sum of the concentrations of O_3, O and $O(^1D)$ as odd-oxygen. The primary destruction of odd-oxygen in the stratosphere comes from catalytic mechanisms involving various free radical species. Nitrogen oxides, chlorine oxides, and hydrogen oxides participate in catalytic reactions that destroy odd-oxygen. The odd-nitrogen (or nitrogen oxides) cycle above is believed to be responsible for about 70% of the total odd-oxygen destruction. While most of the nitrogen oxides in the stratosphere are naturally occurring, the increasing concentration of N_2O is leading to increased amounts of nitrogen oxides and increased effectiveness of the catalytic reactions that destroy stratospheric ozone.

The chlorine catalytic mechanism is particularly efficient. Because of the growing levels of reactive chlorine in the stratosphere resulting from emissions of $CFCl_3$, CF_2Cl_2, and other halocarbons, this mechanism has been the subject of much concern due to the potential effects on concentrations of stratospheric ozone. The chlorine catalytic cycle can occur thousands of time before the catalyst is converted to a less reactive form. Because of this cycling, relatively small concentrations of reactive chlorine can have a significant impact on the amount and distribution of ozone in the stratosphere.

Methane has several effects on stratospheric ozone. Its reaction with reactive chlorine results in the less reactive HCl. Hydrogen oxides produced from the dissociation of methane can react catalytically with ozone, particularly in the upper stratosphere. In the lower stratosphere, the primary effect of these hydrogen oxides is to react with nitrogen oxides, reducing the effectiveness of the nitrogen oxide catalytic cycle. Current models of the processes controlling the stratosphere calculate that increasing methane concentrations result in a net ozone production in the troposphere and lower stratosphere, and net ozone destruction in the upper stratosphere.

Carbon dioxide is not chemically reactive in the troposphere or stratosphere. However, the radiative cooling of the stratosphere resulting from increasing CO_2 concentrations results in decreased effectiveness of the catalytic ozone destruction mechanisms, whose reactions tend to be temperature-dependent. The net result is that the effect of increasing CO_2 concentrations is to increase ozone, particularly in the upper stratosphere.

The downward transport of ozone from the stratosphere has traditionally been thought to be the major source of tropospheric ozone. It is now generally regarded, however, that the net tropospheric photochemical production of ozone is of similar magnitude to the downward transport source (e.g., Fishman, 1985; WMO, 1985). The formation of tropospheric ozone occurs primarily through the so-called smog formation-type mechanisms where nitrogen oxide, NO, reacts with HO_2, CH_3O_2 (produced from methane or other methyl-containing molecules), or RO_2 (where RO_2 represents a variety of complex organic peroxy radicals) to form NO_2. In the troposphere, NO_2 usually photolyzes to produce an oxygen atom, which then forms ozone. Ozone is removed from the troposphere by surface deposition and by photochemical loss processes. Surface deposition occurs when ozone is taken up by contact with a surface, such as leaves or soils. Tropospheric photochemical ozone destruction occurs through a variety of processes. The reactions of OH and HO_2 directly with ozone are particularly important. However, the effectiveness of these reactions depends on the amount of NO_x. The rate of loss of ozone is almost independent of NO_x for concentrations below about 200 pptv (WMO, 1985).

The rate for production of ozone is roughly proportional to the concentration of NO, while, as mentioned above, the rate of loss is almost independent of NO_x. Over oceans or regions of the world characterized by low NO concentrations, there is likely a net photochemical sink of odd-oxygen (Liu et al., 1983; WMO, 1985). Conversely, high concentrations of NOx over the continental boundary layer are likely a net source of ozone. In the presence of adequate NO_x, oxidation of CO, CH_4, and hydrocarbons in the troposphere generally leads to net production of ozone. However, the magnitude of the odd-oxygen production is ultimately limited by the supply of carbon monoxide, methane, and hydrocarbons. The oxidation of one CO molecule can lead to one ozone molecule. In contrast, the complete oxidation of CH_4 can produce 3 to 4 molecules of ozone. Generally, higher hydrocarbon complexity is associated with higher ozone forming potential.

Measurements of ozone from ground-based stations and from satellites indicate that concentrations of ozone in the upper stratosphere are decreasing. Ozone at 3 mbars or 40 km altitude is decreasing globally by 3–4% per decade, in good agreement with the model calculations of the expected effects from CFCs and other trace gas emissions (WMO, 1989, 1991; DeLuisi et al., 1989; Wuebbles et al., 1991). Surface measurements indicate that the total ozone column at midlatitudes in the Northern Hemisphere has decreased since 1969 (WMO, 1989). Since 1969, the annually-averaged total ozone column between 30° and 64°N has decreased by approximately 2%, with larger ozone decreases found in winter than in summer (WMO, 1989, 1991). These data sets also indicate that ozone in the lower stratosphere is decreasing at a faster rate than can be explained by current theory; part of this decrease appears attributable to the dilution of the Antarctic ozone hole (to be discussed below) after its late springtime breakup, but the rest of the lower stratospheric ozone decrease is still not understood.

Beginning in the late 1970s, a special phenomenon began to occur in the springtime over Antarctica, referred to as the Antarctic ozone "hole" (Solomon, 1988). A large decrease in total ozone occurs over Antarctica beginning in early spring. Decreases in total ozone column of more than 50% as compared to historical values have been observed by both ground-based and satellite techniques. Measurements made in 1987 indicated that more than 95% of the ozone over Antarctica at altitudes from 15 to 20 km had disappeared during September and October (WMO, 1989). The Antarctic ozone hole was smaller in 1988 than in 1987. In general, the odd years appear to have larger ozone decreases than the alternate years; dynamical effects resulting from the quasi-biennial oscillation seem to explain these variations (Garcia and Solomon, 1987). The Antarctica ozone holes after 1988 have been progressively larger, but more similar in magnitude to that in 1987 (IPCC, 1991).

Measurements also indicate that the unique meteorology during the winter and spring over Antarctica sets up special conditions producing a relatively isolated air mass (the polar vortex). Polar stratospheric clouds form if the temperatures are cold enough in the lower stratosphere, a situation which often occurs within the vortex over Antarctica. Heterogeneous reactions can occur between atmospheric gases and the particles composing these clouds. Measurements indicate that reactions of HCl and $ClONO_2$ on these particles can release reactive chlorine once the sun appears in early spring. Thus, the reactions on the cloud particles allow chlorine to be in a very reactive state with respect to ozone. The ozone hole ends in late spring with the breakup of the vortex. Scientists have generally concluded that the weight of scientific evidence strongly indicates that man-made chlorinated (produced from CFCs) and brominated chemicals are primarily responsible for the substantial decreases of stratospheric ozone over Antarctica in springtime (WMO, 1989).

Ozonesonde balloon measurements indicate that northern midlatitude concentrations of ozone in the troposphere have been increasing for at least 15–20 years (WMO, 1985, 1989; Logan, 1985; Lacis et al., 1990). The ozone increase from these measurements is largest in the lower troposphere although the exact amount on a global basis is still poorly understood. Liu et al. (1987) propose that the statistically significant winter and spring increasing trends at Northern Hemisphere stations are the result of increasing anthropogenic fluxes of NO_x and hydrocarbons directly affecting atmospheric ozone production. Ozone increases in the upper troposphere may be related to NO_x emissions from commercial aircraft (Wuebbles and Kinnison, 1990) or perhaps due to the effects of convective processes bringing surface emissions of NO_x and other pollutants to the upper troposphere. There are large uncertainties in the global extent of changes in tropospheric ozone concentrations; satellite measurement capabilities are needed to determine trends in tropospheric ozone. Fishman et al. (1990) have had some success with determining tropospheric ozone column densities from existing satellite measurements.

2.2.12 Particulates and Aerosols

Emissions of sulfur dioxide and other gases can result in the formation of aerosols that can influence climate. Aerosols affect climate directly by absorption and scattering of solar radiation and indirectly by acting as cloud condensation nuclei (CCN). The effects of aerosols on climate depend directly on atmospheric chemistry through the transformation of gas phase species to the aerosol species. In addition, heterogeneous chemistry on aerosols can influence concentrations of important constituents like ozone. Recent studies (Charlson et al., 1990) suggest that anthropogenic emissions of sulfur, and the resulting increased sulfuric acid concentrations in the troposphere, may be cooling the Northern Hemisphere sufficiently to compensate for much of the warming expected from greenhouse gases. Emissions of carbonaceous aerosols from biomass burning may also be having a substantial effect on climate forcing (Penner et al., 1992; Penner, 1990). Volcanic emissions can influence climate for short periods (1 to 3 years) through emissions of sulfur dioxide into the lower stratosphere.

Over half of the sulfur dioxide, SO_2, emitted into the atmosphere comes from anthropogenic sources, mainly from the combustion of coal and other fossil fuels (see Table 2.7; Crutzen and Graedel, 1986; Andreae, 1989; Hameed and Dignon, 1988). Other SO_2 sources come from biomass burning, from volcanic eruptions, and from the oxidation of di-methyl sulfide (DMS) and hydrogen sulfide (H_2S) in the atmosphere (Bates et al., 1991). Atmospheric SO_2 has a lifetime of less than a week due its rapid reaction with OH, leading to formation of sulfuric acid and eventually sulfate aerosol particles. Gas-to-particle conversion can also occur in cloud droplets; when precipitation does not soon occur, the evaporation of such droplets can then leave sulfate aerosols in the atmosphere. Recent studies suggest that sulfate aerosol concentrations may have increased substantially over North America and Europe during the last century (Langner and Rodhe, 1990; IPCC, 1990).

Table 2.7   Estimated global sources and sinks of sulfur dioxide (SO_2). Based on IPCC (1990), Difnon (1991) and Wuebbles and Edmonds (1991).

Sources	Estimated Value and Range (TgS/yr)	
Natural		
Volcanoes	10	(5–30)
Oxidation of H_2S, (CH3)2S	60	(15–100)
Energy		
Fossil fuel combustion	65	(40–100)
Other industrial processes	15	(10-20)
Agriculture/non-energy		
Biomass burning	7	(3–10)
Smelting of ores	13	(10–16)
Sinks		
Reaction with OH	~30%	
Aqueous conversion/depositor	~60%	
Dry deposition	~10%	

Other aerosols may also be having an influence on climate, but these effects are also poorly understood. As an example, changes in oceanic emissions of DMS as a result of global warming could provide an important feedback on climate (Charlson et al., 1987; Penner, 1990). Chemical conversion of DMS through reaction with OH to form aerosols is thought to be the major source of CCN over the oceans.

Volcanic eruptions can inject large amounts of SO_2, dust, and other materials directly into the stratosphere. The SO_2 is rapidly converted to sulfuric acid aerosol. As a result of their long stratospheric lifetime, concentrations of stratospheric aerosols may be greatly enhanced over a large area of the globe for a few years following a major volcanic eruption. The subsequent aerosol interactions with solar and infrared radiation warm the stratosphere, as verified by measurements following the 1982 eruption of El Chichon. These aerosols may also cause a cooling of the troposphere and Earth's surface. Various studies have estimated that a global mean cooling of at least 0.1–0.2 K may result for a one to two year period following a major eruption (IPCC, 1990). The 1991 eruption of Mt. Pinatubo is estimated to have caused about a 0.4 K cooling in 1992 and 1993 (Hansen et al., 1993). However, attempts to correlate past surface temperature changes with volcanic activity have largely been inconclusive and remain controversial.

During periods of low volcanic activity, carbonyl sulfide (COS) is thought to be responsible for the maintenance of the sulfuric aerosol layer found in the lower stratosphere. Natural emissions explain most of the COS in the present atmosphere, while the relatively long atmospheric lifetime (about 2 years) of COS explains why much of it reaches the stratosphere before its conversion to sulfuric acid aerosol. However, if coal combustion and other anthropogenic sources of COS (or its precursor, CS_2) were to increase dramatically, the background aerosol layer concentration would increase along with resulting climatic implications (Wang et al., 1986; Wuebbles et al., 1989).

2.3 REDUCING THE UNCERTAINTIES IN GREENHOUSE GAS BUDGETS

Two key questions relate to uncertainties of trace gases. The first is, How well do we need to know the emission rate to address a particular scientific or engineering issue? The second is: How well can we know the uncertainties given the limitations of existing technology, available financial resources and also the nature of the sources? In some sense there is no trace gas, including CO_2, for which we know the sources and sinks sufficiently well to be able to answer most of the important questions regarding environmental influences of human activities.

The annual emissions can be described as a matrix in which the rows for instance may represent different sources and the columns may represent different regions. Khalil and Rasmussen (1990c) have developed statistical methods to evaluate the uncertainties in the global budgets expressed in this way. In this method it becomes clear that the individual elements of this matrix are the most uncertain component. Adding the emissions from each region to estimate the total emissions from a given source produces uncertainties that are smaller than would be obtained by adding the lower and upper limits of emissions from each region. Similarly, the total emissions or the total anthropogenic or natural emissions of a trace gas are less uncertain than the uncertainties implicit in each source. In brief, the less resolution, either in space, or time or in the source characterization that is demanded the less the uncertainties. The total global emissions of a trace gas can therefore be much better known than global emissions from individual sources or regions.

It is also apparent that there are fundamental limitations to how well the emissions can be known. Some of these limitations are from the inability to measure the fluxes accurately, but limitations also arise from the fundamental natural variability which differs among gases. As time goes on, reducing uncertainties becomes more and more costly and each increment is a smaller reduction of uncertainties (see Khalil and Rasmussen, 1990c).

In spite of these philosophical and practical limitations there are some important examples of trace gas sources that can provide significant new knowledge and have not yet reached close to the limitations mentioned earlier. Methane emissions from waste disposal, coal mining, landfills and septic systems need to be measured. Programs are already underway to determine global fluxes of methane from rice fields and domestic ruminants. It is likely that further research will be justified after the present studies are concluded. The present sources of N_2O are in considerable disarray. Therefore any new flux measurements or calculations are of considerable interest. The same is true for non-methane hydrocarbons. For CO_2, much can still be learned from experiments to determine the role of temperate forests. Expanded use of isotopic measurements may help to establish the extent of net sources or sinks from the oceans and terrestrial biosphere. On the sink side, improvements are needed in quantifying the removal of N_2O and soil sinks of CH_4.

If we consider uncertainties in predictions of future concentrations under any scenario of interest, new difficulties arise in reducing uncertainties. These problems include the changing nature of sources and sinks, changing technological and agricultural processes, production and demands, and the possibility of legislative action of precursors of related pollutant emissions.

2.4 STABILIZING THE ATMOSPHERE

In terms of engineering response, we offer the following challenge:

- Stabilizing the atmosphere such that by the year 2030 radiative forcing does not exceed current levels. If we believe that the environment of the Earth must be preserved in its present form, then we must meet such a challenge. The challenge is to accomplish the goal with our present knowledge and our present abilities through global cooperation. It is clear to us that the goal cannot be accomplished unless we design integrated programs in which all the important sources of climatically important trace constituents are managed simultaneously and globally. Controlling one or two sources of one gas or another in one country or another will simply not be enough.

Other objectives included within the chapter:

- Develop the necessary technology to quantify budgets, both natural and anthropogenic, for radiatively important atmospheric constituents (both long-lived and short-lived species). Where possible, analyses need to consider constraints from observations, such as those from ice core and isotopic data.
- Develop technology (if possible) to enable development of worldwide database (inventory) for emissions of radiatively important atmospheric constituents (RIAC) from individual countries to within an uncertainty range of ±10% for individual sources.
- Develop monitoring techniques needed to determine global distributions of key constituents such as hydroxyl, sulfuric aerosols, and a number of other tropospheric trace species.
- Develop criteria to judge the desirability and feasibility of proposed engineering and scientific solutions.

In order to assess the consequences of alternative technologies on emissions, a set of tools is necessary. At the global level of analysis a set of coupled models linking emissions of all radiatively important gases from all sources, including both anthropogenic and natural, atmospheric stocks of radiatively important gases, and radiative forcing provide a useful mechanism for assessment. For regions below the level of global aggregation, and for individual technology assessment, another, more disaggregated set of tools is appropriate. The Global Warming Potential Coefficients (GWP) were developed to provide such guidance.

As we have already discussed, global emissions of radiatively important gases, atmospheric stocks, and radiative forcing are all related but different. The relationship between global emissions of

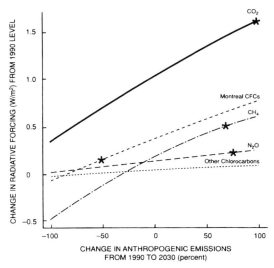

Figure 2.11 Radiative forcing of principal greenhouse gases from 1990 to 2030 for different emission rates. (From NAS, *Policy Implications of Greenhouse Warming,* National Academy Press, Washington, D.C., 1991, 13. With permission.)

radiatively important gases and stocks is governed by important global natural processes including the carbon cycle, nitrogen cycle, and atmospheric chemistry. While uncertainty exists concerning the exact relationship between a fixed rate of emissions and radiative forcing, models can be used to provide some guidance with regard to this relationship. Figure 2.11 shows this relationship between change in radiative forcing (W/m^2) from 1990 levels and a change in the annual rate of anthropogenic emissions from 1990 to 2030 (NAS, 1991). Note the increase in radiative forcing relative to 1990 levels that occurs even when emissions are held constant.

2.4.1 Assessing the Relative Importance of Gases, Human Activities, and Technologies

For the assessment of regional, national, and sectoral emissions a different set of tools is necessary: global warming potential coefficients. Since different gases have different lifetimes, different direct effects on the radiative balance of the Earth, and different effects on the overall composition of the atmosphere, some measure of relative importance is needed in order to compare the release of different amounts of each gas into the atmosphere. There is no unambiguously correct method for approaching this problem. Research directed at the development of alternative metrics for assessing releases of gases into the atmosphere began in about 1989.

By 1990, formal papers were beginning to appear in the literature. These include: Lashof and Ahuja, 1990; Rodhe, 1990; Derwent, 1990; Nordhaus, 1990; and IPCC, 1990. All of these papers have adopted a change in radiative forcing as its common unit of impact. Models of the atmosphere and energy are used to simulate the effect on radiative forcing of the release of a kilogram of CO_2 in each year, over a period of time into the future, relative to a standard reference scenario. The change in radiative forcing is then added over a specific period of time, e.g., 20 years, 100 years, 500 years. The experiment is then run for other gases. The cumulative effect of each gas on radiative forcing is then compared to the cumulative effect of CO_2 on radiative forcing. While all of the above cited formulations are similar in construct, each varies somewhat in the details of its implementation. Differences include the way different gases are treated with regard to their residence time in the atmosphere after emission, their solar and infrared radiative absorption properties, and the indirect climatic effects resulting from chemical interactions with other greenhouse gases. In addition, different studies treat the time profile of the change in radiative forcing differently. Some studies simply add changes in radiative forcing on an equal basis over different numbers of years. Others apply discount rates to each year's change in radiative forcing before summing. From the perspective of its natural science, the construct developed by Derwent, Rodhe, and Wuebbles for IPCC (1990) is the most complete to date.

We have adopted the IPCC (1990, 1992) conventions for GWP coefficients as a standard set of weights for evaluating emissions of radiatively important gases. These appear in Table 2.8. In the IPCC reports, the GWP is defined as the time-integrated commitment to climatic radiative forcing from the instantaneous per unit mass release of a species relative to the climatic radiative forcing from the per unit mass release of CO_2. These GWPs were evaluated for time-integration periods of 20, 100, and 500 years, reflecting the concern by policy makers that global warming effects should be considered over a range of time periods.

Note that GWP is defined relative to CO_2, the greenhouse gas of current primary concern to climate change, and thus carries the units CO_2 equivalent emissions.

Table 2.8   Global warming potentials following the instantaneous injection of 1 kg of each trace gas, relative to carbon dioxide. (From IPCC, 1990, 1992). Note that lifetimes are different in some cases from more recent analyses presented in text (GWPs should scale with the change in lifetime).

Trace Gas	Estimated Lifetime (years)	Global Warming Potential Integration Time Horizon (years)		
		20	100	500
Carbon Dioxide	*	1	1	1
Methane	10	63	24	9
(including indirect effects**)				
Nitrous Oxide	150	270	290	190
CFC-11	60	4500	3500	1500
CFC-12	130	7100	7300	4500
HCFC-22	15	4100	1500	510
CFC-113	90	4500	4200	2100
CFC-114	200	6000	6900	5500
CFC-115	400	5500	6900	7400
HCFC-123	1.6	310	85	29
HCFC-124	6.6	1500	430	150
HFC-125	28	4700	2500	860
HFC-134a	16	3200	1200	420
HCFC-141b	8	1500	440	150
HCFC-142b	19	3700	1600	540
HFC-143a	41	4500	2900	1000
HFC-152a	7.1	510	140	47
CCl_4	50	1900	1300	460
$CH_3 CCl_3$	6	350	100	34
CF_3Br	110	5800	5800	3200

CFCs and other gases do not include effect through depletion of stratospheric ozone. Changes in lifetime and variations of radiative forcing with concentration are neglected. The effects of N_2O forcing due to changes in CH_4 (because of overlapping absorption), and vice versa, are neglected.

* The persistence of carbon dioxide has been estimated by explicitly integrating the box-diffusion model of Siegenthaler (1983); an approximate lifetime is 120 years.

** Methane includes direct radiative effects and indirect effects due to chemical effects on ozone and water vapor. Error in IPCC (1990) direct GWP is corrected based on IPCC (1992).

While natural scientists continue to refine the construct of GWP coefficients for different gaseous emissions, social scientists have also explored the concept. Reilly (1990) points out that the whole point in constructing a GWP index in the first place is to obtain a relative measure of damage. Implicit in the standard formulation of a GWP is the assumption that the effect of a one-degree change in radiative forcing at any two points in time is the same. This need not be the case. Marginal damages are likely to vary over time. Reilly (1990; p.2) also points out that "gases have non-climate-related economic effects that differ amongst gases and these should be counted as credits (e.g., direct CO_2 fertilization of crops) or debits (e.g., CFCs as contributors to ozone depletion)." While the introduction of a calculation of full damage and benefit from a pulse emission of a gas adds considerable complexity to the problem, Reilly (1990) shows that the successful introduction of such a concept could significantly change GWP measures. Reilly (1990) developed coefficients using alternative damage functions for the effects of a change in radiative forcing on a steady-state economy and alternative allowances for a CO_2 fertilization effect on agriculture. He found that the application of these varying assumptions in conjunction with a simple model of the atmosphere, yielded a range of values for coefficients at least as large as the range of values associated with range of integration times investigated by the IPCC.

A direct implication of Reilly's work is that GWP measures are not directly comparable from year to year. That is, the base atmosphere will change over time as a result of the accumulation of greenhouse gases from previous emissions. This will not only change the chemical and radiative interactions of the gases, but in addition, the damage associated with each degree change in radiative forcing will be affected by the scale of previously inflicted damage.

It is also important to note that GWP measures are marginal. That is, they measure the consequences of a one-time, one-kilogram release of an individual gas into a standard atmosphere. The consequence of teragram and pentagram scale releases of gases into the atmosphere are not taken into account, neither are the consequences of simultaneous releases of gases. As these gases interact both chemically and radiatively, scale effects are potentially non-trivial.

2.4.1.1 Using GWP Coefficients as Weights on Emissions

We have taken estimated emissions rates for the year 1988 and multiplied emissions rates by GWP values for three different integration periods. The relative contributions of the different gases and human activities associated with emissions are then calculated. It is important to note that these calculations apply a concept developed for small, marginal changes in the release of a gas independent of the release of other gases, to large releases of multiple gases simultaneously. The computation is therefore different from the computation of direct radiative forcing.

The role of different gases in contributing to total GWP weighted emissions varies directly with the integration time of the GWP coefficients. Short integration times yield a high relative share for CO, CH_4, and NO_x. Longer integration times show CO_2 and the CFCs to be more important. The relative contributions of these gases is displayed graphically in Figure 2.12.

The role of energy varies from approximately 52% of GWP weighted total emissions when the 20-year coefficients are used to approximately 67% when the 500-year coefficients are used. The role of agriculture and land use change declines from a maximum of 36% when 20-year coefficients are applied to 22% when 500-year coefficients are used. The role of other activities, principally the manufacture of CFCs and CFC substitutes, also increases as the integration time for the GWP coefficient increases, rising from approximately 11 to 15% as the integration increases from 20 to 500 years. The shifting relative importance of human activities in emissions can be directly traced to the nature of the dominant emissions associated with the activity. Agriculture and land use change are known to have relatively higher rates of emission of CH_4 and CO than other activities. Methane is a potent radiative absorber, but it is short lived. Similarly, CO is more potent than CO_2 through its indirect effects, but is short lived. Thus, agriculture and land use appear relatively less important as a source of GWP weighted emissions when longer integration times are used to calculate the GWP coefficients. The relative roles of human activities for different integration periods of GWP coefficients are shown in Figure 2.13. Recent considerations suggest that the NO_x indirect effects on tropospheric ozone are overstated in the GWPs developed by IPCC (1990).

2.4.1.2 Atmospheric Composition and GWP Weighted Emissions

The usefulness of GWP coefficients is limited. The GWP-weighted emission calculation provides only a mechanism for determining relative ranking. It is limited by virtue of the fact that the weights would be expected to change as the composition of the atmosphere changed. Furthermore, the calculated total yields only indirect information about the rate of change of the atmosphere. It is a relative measure of atmospheric loading. To derive information about the actual

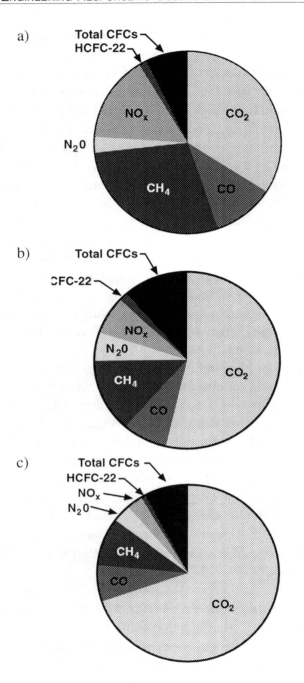

Figure 2.12 Relative contributions of each greenhouse gas determined using the IPCC global warming potential for 1988 emissions and following integration times: a) 20 years, b) 100 years, and c) 500 years.

a)

b)

c)

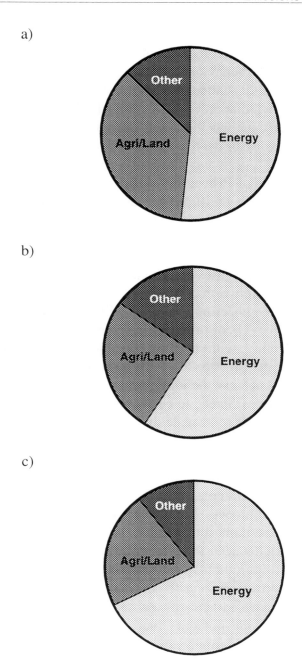

Figure 2.13 Relative roles of human activities for the calculations of global warming potentials as show in Figure 2.12.

composition of the atmosphere requires that a set of linked models be employed, including a full carbon cycle model (including both ocean and terrestrial ecosystem behavior), and an atmospheric chemistry module. To derive a time-dependent model of radiative forcing would require the addition of a radiation code. Without all of these elements, it is impossible to determine whether the underlying radiative forcing of the climate is increasing or decreasing over time.

2.4.2 Important Aspects of Prevention and Control

The chapters which follow will provide detailed analysis of engineering options and a wide array of engineering responses to climate change. Nevertheless, having given the engineering community the admittedly challenging goal of stabilizing the atmospheric concentration of radiatively important gases at 1990 levels by the year 2030, a number of observations are worth making.

The development of engineering responses will doubtlessly require the development of new ways of thinking. Holistic approaches should be considered in dealing with the use of domestic and global resources. Small changes in the "business as usual" evolution of human systems will not achieve the long-term goal. A "pollution prevention pays" philosophy will need to be adopted. This means that by restructuring processes so as to provide the same or greater final services, it may be possible to simultaneously reduce emissions and to increase profitability. This can occur because the search for process improvements in any one area (pollution emissions, energy efficiency, or labor productivity, for example) lead to changes that are general improvements in the total process. Methods for dealing with the problem require looking at the domestic and global resource systems, recognizing that such systems need to be adjusted and optimized to (1) eliminate misplaced resources; (2) rely on non-destructive use of ancient resources via recycling; and (3) rely increasingly on renewable resources. In the development of new technologies (where we take technologies to include not only the physical device, but the human and environmental system in which they operate) a new design criterion is needed. Technologies must be designed not only to perform their traditional functions, but in addition, minimize releases to the environment of radiatively important gases and other criteria and toxic pollutants.

What is the potential for mitigation? Projecting realistic application rates of mitigation and resulting emissions cannot be taken lightly.

Radiative forcing needs to be stabilized, not just emissions. Some of the barriers include:

* Population and economic growth;
* Necessary massive reductions. The rate of application of mitigation measures has to first exceed the rates of emission growth before any temperature reduction can take place;
* The possibility that warming has already triggered terrestrial and ocean feedback sources which will also have to be overcome;
* Difficulties in changing major industry infrastructures;
* Limited potential for stopping the onslaught of deforestation and limited impact of improved forest and soil management measures;
* Concerns about the economic effects of mitigation;
* Lack of appreciation of the cost-effectiveness of conservation and lack of quantification of the environmental benefits of conservation;
* Fear and cost of nuclear power;
* Lack of R&D aimed at cost-effective solutions, massive expenditures required for necessary R&D, and fear that if such solutions were available there would be pressure to implement them;
* Doubt by many that there really is a warming problem and the distant-in-time prospect of impact; and
* Perception by those in colder climates that global warming is a change for the better despite other manifestations such as a rising sea level and warnings of unpredictable or run-away changes.

Prevention, innovation, and cost-effectiveness: We live in a world used to dealing with environmental problems largely through fixing problems once they have occurred. An alternative approach is averting the problem. There is plenty of evidence that a "pollution prevention pays" approach is effective. In industrial situations emissions have been avoided from industrial processes through optimization or changing the process or product—source reduction and recycling. We now need to apply this general approach on a broader scale to our domestic and global "processes" (systems) for example, petroleum (petrochemicals and energy), coal (energy), timber (pulp and solid wood products).

Optimization in the use of these ancient and renewable resources should be possible given a clear statement of the objectives and ability to undertake gradual change to redirect misplaced resources, eliminate or minimize destructive uses/applications of ancient resources, and build up energy applications of renewable resources.

Effective assessment and mitigation requires dealing with all gases. It is unlikely that adequate understanding of the global climate problem or solutions to the global climate problem can be arrived at without attacking all important components of the problem.

While the chapters that follow will provide more specific analysis of alternative technologies for reducing the impact on radiative forcing of human activities, we briefly discuss the various pathways through which emissions can be reduced. Each radiatively important species and each human activity associated with greenhouse related emissions, presents unique challenges and opportunities.

2.4.3 Reducing Future Fossil Fuel CO_2

More attention has been focused on the question of CO_2 emissions reductions than on emissions reductions for any other gas or human activity, with the exception of CFC applications. The two principal sources of CO_2 emissions are fossil fuel use and deforestation/land use changes.

CO_2 is an unavoidable byproduct of the combustion of carbonaceous fuels. The level of emission depends upon the amount of carbon per unit energy contained in the fuel. For fossil fuels, the hydrogen to carbon ratio is critical. Emissions can be reduced in six different ways:

1. Energy Conservation: Reducing the energy required to provide any energy service, for example, space heating, goods transport, electric power generation. Energy conservation does not mean doing without.
2. Fuel Substitution: Changing the fuel mix to increase the proportion of low or non-CO_2 emitting fuels relative to high CO_2 emitting fuels, for example from coal to natural gas or from natural gas to nuclear or renewable energy.
3. Scale: Changing the scale of the overall human and energy systems, for example by changing population, or the level of overall economic activity or simply doing without and thereby having less of the fruits of energy use.

4. Activity Mix: Changing the composition of energy using activities, for example moving away from the production of goods that use ferrous metals to those which use plastics or moving away from goods and toward services.
5. Carbon Removal and Disposal or Recycling: (1) Flue-gas sequestration consists of removal and or concentration of CO_2 in flue gas followed by disposing of the carbon in a permanent repository; or (2) atmospheric fixation consists of a variety of techniques for terrestrial and marine carbon fixation.
6. Reduce non-energy use of fossil hydrocarbons: such uses include solvents, surface coatings, lubricating oils, asphalt, and chemicals.

All emissions reductions options apply one or more of the above principles. The nuclear and renewable power options are fuel switching options, while lifestyle changes would exert an influence through several avenues including energy conservation, fuel switching, and activity mix.

2.4.3.1 Energy Conservation

We begin by noting that energy conservation is achieved by applying technologies that produce the same level of energy service while using less energy. We use the term technology in the broadest possible context to describe how societies organize to produce the goods and services they desire. Thus, the application of technology can mean either the introduction of a new machine, or simply a new way to organize the present array of humans and machines. It is important to emphasize that conservation means technological improvement, not doing without.

One of the earliest studies to focus on the potential for fossil fuel CO_2 emissions reductions was Rose et al. (1983). This study concluded that one of the most important factors governing the rate of future fossil fuel CO_2 emissions was the rate of introduction of new energy conserving technologies. Edmonds et al. (1986) concluded that this was one of the three critical factors governing future fossil fuel CO_2 emissions. Both of these studies were "top down" studies which focused on the macro level of the energy system, and did not provide details regarding individual technologies that might be applied.

More recently a series of studies have been published that have applied a "bottom up" analysis of the role energy conserving technologies might play in CO_2 emissions reductions. These studies include Cheng et al. (1985), Goldemberg et al. (1987), Chandler et al. (1988), and NAS (1991). These studies suggest substantial unutilized technological potential for energy efficiency improvements, ranging from improved lighting technologies, electric motors, internal combustion vehicles, gas turbines, and building shells. These studies also indicate that these technologies are likely to penetrate the market because they are financially attractive under present economic conditions. Furthermore, the accelerated introduction of these technologies would further reduce energy and in particular fossil fuel demands and thereby reduce CO_2 emissions.

The following areas were outlined in NAS (1991): Residential and commercial energy management includes: (1) electric efficiency measures: (white surfaces/vegetation, lighting, water heating, cooking, refrigeration, appliances, domestic space heating, commercial and industrial space heating, and commercial ventilation); (2) oil and gas efficiency; and (3) fuel switching. Industrial energy management includes: co-generation, electricity efficiency, fuel switching, efficiency, and new process technology. Transportation energy management includes: vehicle efficiency (light trucks, heavy trucks, aircraft); alternative fuels; and transportation demand management. Electricity and fuel supply includes: heat rate improvements, advanced coal technologies, and natural gas use.

Several studies, for example, Goldemberg et al. (1987), Chandler et al. (1988), and NAS (1991) advocate policies such as CAFE standards, energy taxes, and/or building code changes, to accelerate the introduction of these technologies. These studies have generally been microeconomic in nature and have not assessed feedback consequences of policies within a market equilibrium setting. Key issues remaining to be explored include the effect of accelerated investments on capital markets, economic growth, competitiveness, and energy prices. For example, will the adoption of new energy conserving technologies lower the cost of energy services sufficiently that additional quantity demanded of the service increases, and reduces the overall effectiveness of the measure? Will the introduction of new, more efficient technologies improve both energy and labor productivity simultaneously, and since labor productivity improvements increase GNP, and thereby the scale of energy demand, result in lesser CO_2 reductions than anticipated by the "bottom up" studies?

Quantifying the full emission reduction benefits (reduction of emissions other than CO_2 as well as CO_2) from conservation is a vital task and remains to be done. It should be noted that conservation developments also help the renewable option.

2.4.3.2 Fuel Substitution

There are two types of fuel substitution: the substitution of natural gas for coal in the short-run and substitution of non-fossil fuels for fossil fuels in the long-run. The former option has been looked at by numerous researchers including Mintzer (1987), Chandler and Nicholls (1990), and Edmonds et al. (1989). These studies find that the substitution of natural gas for coal can reduce near-term emissions; however, the strategy is limited both in the amount that it can contribute in the near-term and in its ability to contribute at all in the long-term. There are two problems. First the difference between the carbon-energy ratio between coal and natural gas is approximately a factor of two. But the use of natural gas has associated direct releases into the atmosphere of methane, an extremely potent greenhouse gas in its own right. The overall reduction in greenhouse warming potential is not as great then as, for example between non-fossil fuels and coal. The second problem is that the resource base of natural gas is limited. Therefore, the substitution of natural gas for coal today means that there will be less easily accessible natural gas available in the future. Unless the present estimates of the resource base of natural gas are found to be substantially in error, which is always possible, or technologies are developed which can access gas which is presently highly uneconomic to produce, the substitution of natural gas will remain a relatively limited mechanism for reducing greenhouse warming potential.

The substitution of non-fossil fuels for fossil fuels is a very different matter. Non-fossil fuels have either low to modest releases associated with their direct application. Non-fossil fuel options include such technologies as hydroelectric power, nuclear power including both fission and fusion, solar energy in its various forms such as for example, biomass including urban waste, rural waste, and biomass cultivation, wind, ocean thermal energy conversion (OTEC), PV arrays, and solar thermal conversion and geothermal. (Note that passive solar design is included here as a form of energy conservation.) Most studies conclude that in the long-run that fossil fuel greenhouse gas emissions reductions can be accomplished at minimum cost only if conservation technologies are augmented by increases in the supply of energy by non-fossil technologies. See for example Edmonds and Barns (1990), Mintzer (1987), Manne and Richels (1990a, b, c, d). Conservation in turn reduces the required renewable capacity base.

Biomass energy production raises a number of interesting issues. While the net carbon emission of biomass is approximately zero over the full lifecycle, uptake and release occur at different points in time. Uptake occurs slowly over the period prior to harvest while carbon is returned to the atmosphere as a pulse. For short rotation crops, and annual crop residues, this issue is minor, but for longer-lived biomass feedstocks, the issue is non-trivial and can lead to either significant reductions in actual annual net emissions due to an excess of biomass growth over harvest, or an increase in actual annual net emissions as with deforestation. Borgwardt (1992) has shown that the use of biomass as a feedstock for hydrogen with carbon recapture could provide a mechanism for negative net annual emissions from a biomass energy.

Other issues that arise with regard to the use of biomass energy include the competition for land with other potential land users, especially afforestation and agriculture. Afforestation and bioengineering approaches to carbon sequestration are discussed in a later section of this chapter, as well as in later chapters of this book.

Ocean biomass farming would result in little additional land use, but is a technology whose cost, effectiveness, and environmental properties have not yet been fully developed.

2.4.3.3 The Relative Scale of Emissions

At base, it is the scale of human activity that raises concerns regarding the emission of CO_2 and other radiatively important gases. These gases, released at smaller scales relative to global natural systems, would be of no concern. The main features affecting the scale of global releases of greenhouse gases is the level of human population and the attendant levels of economic activity associated with achieving desired standards of living. Since the achievement of higher standards of living is generally considered to be the objective of development, reducing the level of per capita economic growth, particularly in developing nations, is generally not taken seriously as an option for reducing the rate of growth of greenhouse gas emissions. This is not necessarily the case for developed nations. The powerful influence the scale of economic activity has on the level of emissions is sometimes cited as a reason for giving up a "profligate lifestyle." In general, even in developed nations, this concern regarding lifestyle is translated into a need for developing alternative means for assuring a good standard of living without environmental degradation.

The development of population control policies is taken more seriously. The proposal, especially when coming from developed nations, that developing nations control their population to prevent climate change is an extremely delicate subject. Also, Edmonds et al. (1986) found population growth to be a lesser factor in determining the potential variability in future "business as usual" emissions. It should also be noted that due to the fact that people do not enter the labor force for at least 15 years, the principal effects of changes in population growth are delayed.

2.4.3.4 Activity Mix

Because the energy and carbon intensity of production in various economic activities varies greatly, shifting the mix of final products consumed can have an important effect on total greenhouse gas emissions. Similarly, a shift in the mix of technologies used to produce goods and services can directly affect greenhouse gas emissions. The importance of the shift in materials is emphasized by such researchers as Williams and Larson (1987), Goldemberg et al. (1987), and Williams (1990) who argue that two things are occurring in developed nations, saturation in the demand for energy-intensive goods, and a shift in the technologies toward relatively lower energy-intensive final products. The shift toward less energy-intensive final products is subtle, because on the surface, the energy intensity of per unit mass of "new" materials has been rising relative to the energy intensity of the traditional materials they replace. But the mass requirements of "new" materials per unit final product are so much smaller than those of traditional materials that the overall energy intensity of products has been falling and will likely continue to fall.

2.4.3.5 Carbon Removal and Recovery

Carbon removal and disposal or recycling approaches can be categorized as: (1) precombustion removal of carbon from fuels; (2)

flue-gas or fuel-gas sequestration, which consists of removal and or concentration of CO_2 followed by disposing of the carbon in a permanent repository or recycling; or (3) atmospheric fixation, which consists of a variety of techniques for terrestrial and marine carbon fixation, followed in some cases by utilization. Combinations of these are also possible.

It is possible to use coal as a feedstock for hydrogen, removing the carbon for return to the mine (Steinberg and Grohse, 1989). The feasibility, economics and environmental consequences of such technologies have not been addressed.

Flue gas CO_2 sequestration and ocean/land CO_2 disposal approaches appear to be costly and in need of environmental assessment. Separation might be done using flue gas scrubbing on power plant stack gases or fuel-gas scrubbing in gasification combined cycles systems. If carbon is removed from the exhaust stream in the form of gaseous CO_2, the problem of disposition emerges. Several options have been discussed. These include disposal in the deep ocean, depleted oil/gas wells, aquifers, and salt domes, or utilization for enhanced oil recovery. Among the many factors to be assessed is the rate at which the CO_2 will reenter the atmosphere, including the possibility of a new source of CO_2 not dependent on the rate of fossil fuel combustion.

Carbon recycling, on the other hand, may be a way around disposing of CO_2 as a waste. Although current commercial markets require small amounts of CO_2 relative to the scale of global emissions, recycling could possibly be done through (1) the use of biomass as fuels; (2) innovative co-siting of CO_2 users with CO_2 generators (e.g., ammonia plants or ethanol plants co-sited with methanol plants; (3) use of biomass and/or CO_2 as feedbacks along with renewable-derived hydrogen to make hydrocarbons; and (4) microalgae flue as capture concepts.

Several studies have begun to look at the issue of reducing net CO_2 emissions by reforesting. Three notable examples of such studies are Marland (1988), Sedjo and Solomon (1989), and Moulton and Richards (1990). Both Marland (1988) and Sedjo and Solomon (1989) look at the issue from the global perspective. Marland (1988) concludes that the scale of the effort required to remove an additional 5 PgC/yr from the atmosphere is approximately equal to doubling the net annual yield of all the world's closed forests or planting new fast growing forests over an area equivalent to the total of global forest clearing to date.

Sedjo and Solomon (1989) conclude that it would require approximately 465 million hectares of newly planted, fast-growing forest to recover 2.9 PgC/yr. The economic cost would be approximately $372 billion if planted in the temperate zone and $186 billion if planted in tropics. The average cost of carbon emissions reductions in temperate zones would be approximately $128/TC/yr with initial costs perhaps half the cost of the final tonne of carbon. If the trees were planted in the tropics the average cost would be halved, or $64/TC/yr.

Moulton and Richards (1990) investigate tree planting in the U.S. only. They undertake a detailed analysis of tree planting in which they conclude that approximately 0.71 PgC/yr could be removed annually for 40 years or more at a total cost of $20 billion per year on approximately 142 million acres. The marginal cost of carbon emissions reductions range from approximately $5/TC/yr to $50/TC/yr.

In principle, it should similarly be possible to use ocean biomass as a store of carbon. Because soil carbon comprises a reservoir of approximately 1200 PgC, an increase in the stock of soil carbon of 1% per year would remove more than 15% of annual anthropogenic carbon emissions to the atmosphere. Increasing soil carbon uptake rates could be a powerful geoengineering response to climate change. It is difficult to imagine any technological response which would enhance soil carbon uptake on the scale of a pentagram of carbon or more which would not have major environmental externalities.

2.4.4 Reducing CFCs, HCFCs, and Halons

As discussed in the section considering the prospects for future emissions, the process of reducing future CFC emissions is already well underway. The initial timetable for emissions reduction is laid down in the Montreal Protocol as well as a process by which this timetable can be altered. Since the Montreal Protocol focuses on reducing the risk to ozone depletion, reductions in greenhouse gas emissions are a byproduct. Under the November 1992 Copenhagen Agreement, production and new uses of CFCs and halons will end by 1996 and 1994, respectively. Provisions to end production of other chlorine-containing solvents were added, as were non-binding provisions to gradually phase out production of HCFCs. Compliance with the Montreal Protocol, Copenhagen Agreement, and possible future modifications and extensions, results in the technical problem of finding replacements for the current suite of ozone-depleting substances. Replacement compounds being suggested are mostly chlorocarbons containing hydrogen, which will react with OH in the troposphere. The suggested replacement compounds, such as HCFC-123, HCFC-141b, and HFC-134a, are generally halogenated hydrocarbons, some containing chlorine (the HCFCs) and others not (the HFCs). All of the suggested replacements have shorter lifetimes than CFCs, generally less than 20 years, and their concentrations will, as a result, be lower than the concentrations of the compounds they are replacing. Nonetheless, most of these compounds are greenhouse gases and could affect the climate if concentrations become large.

Substitutes, alternatives, and different ways of doing things comprise the major means for reducing emissions of halogenated organic materials having important global warming potentials. NAS (1991) lists the following approaches for dealing with halocarbons:

- Modify or replace equipment to use non-CFC materials;
- Conservation and recycling of the halocarbons;
- HCFC/HFC substitutes for cleaning, blowing agents, and aerosols;
- HFC-chiller retrofits and replacements to use alternatives;
- HFC-auto air conditioning replacement to use substitutes;
- HFC-appliance replacements for domestic refrigerators;
- HCFC-other refrigeration replacements, e.g., supermarkets or transportation; and
- HCFC/HFC-appliance insulation replaced with substitutes.

Other measures include alternatives for halons used for fire and explosion protection.

It is important to note that "third generation" chlorofluorocarbon replacement compounds, which could perform the same services as present CFCs, HCFCs, and HFCs with similar desirable properties and without undesirable properties, are in an early stage of evaluation. The creation of such a new set of compounds is essential if the production of all chlorocarbon compounds are to be suspended early in the 21st century, and must be considered an important component of an overall engineering response to global environmental change.

2.4.5 Reducing Methane

Anthropogenic emissions of methane are presently thought to be associated with agricultural activities such as rice cultivation, ruminant livestock herds, animal wastes, and slash and burn agricultural practices, with the deep mine production of coal, with oil production, with natural gas production, transmission, and distribution and with organic decomposition (such as at landfills and waste treatment sites).

Engineering responses to reduce anthropogenic emissions could involve, but are not necessarily limited to, the development of new wetland rice strains or rice cultivation practices, reengineering or changing the diet of the cow, transforming waste to energy, and developing relevant replacement practices for areas which presently practice slash and burn agriculture.

The energy sector presents engineers with the challenge of reducing natural gas everywhere between the field and the burner tip. They must find ways to maximize gas capture in coal mining and minimize the cost of capture and use, and similarly for landfill methane. There are currently a number of projects worldwide evaluating the conversion of landfill gas to energy.

In light of the fact that the methane budget has not been satisfactorily balanced, it is possible that new sources will be identified in the future.

2.4.6 Limiting Human Effects on Tropospheric Ozone

Reducing CH_4, other volatile organic emissions, CO, and NO_x is an important component of an overall strategy for stabilizing atmospheric composition. Methane is a key ozone precursor. Means for reducing emissions are covered in Section 2.4.5. Potential approaches for NMHC, CO, and NO_x are listed here, but it should be noted that conservation and renewable resource mitigation approaches undertaken for CO_2 will also in many cases reduce these criteria pollutants. Not only are many of the conservation approaches cost-effective, but costs for controlling the criteria pollutants such as NO_x are covered to some extent by existing or future regulatory requirements.

Subsonic and supersonic aircraft emissions are a special case because they inject emissions directly into the free troposphere and stratosphere. Although these emissions are known to be important, there has been little focus on reducing emissions at cruise altitudes for jet aircraft. Most concerns on emission have been related to pollution resulting from take-off, landing, and ground operations. Other issues affecting engine design and operation include efficiency enhancement and noise reduction. It is likely that new emissions sources will be identified.

The reduction of NMHCs, and CO and NO_x is a difficult and diverse engineering challenge. Emissions of these gases are frequently associated with a geographically dispersed set of heterogeneous processes including: industrial processes, solvent usages, consumer products (e.g., lighter fluids and aerosols), pesticides, gasoline, waste, low temperature biomass burning, power plants, industrial combustion, and aircraft emissions. Prevention-oriented approaches, such as taking the volatiles out of gasoline and improving combustion, are the most attractive because they may even reduce costs. Use of conservation and renewables are also prevention oriented approaches, which, if applied, may result in cost-effective multipollutant control.

2.4.7 Reducing N_2O Emissions

N_2O sources overall have not been well quantified. Therefore, less is known about means for prevention and control. The NAS (1991) report suggests the option of reduction of the use of nitrogenous fertilizer. Other options which may need research are alternatives for direct N_2O use, such as modifying or eliminating emissions from certain industrial processes, and assuring that emissions are low from auto exhaust catalysts.

2.4.8 Geoengineering Responses

In addition to engineering responses which affect the emission of radiatively important gases, there are an array of engineering responses which could affect the radiation balance of the planet directly. These options include things such as mirrors orbited about the Earth to reduce incoming solar radiation, or injecting aerosol particulates directly into the stratosphere. These options are discussed in Chapter 8. The full cost, efficiency, and environmental consequences of these options have not been adequately assessed. At present, most must be viewed as potential backstops, which might be used to mitigate or reverse climate change if other strategies fail and climate change becomes disastrous.

REFERENCES

Andreae, M.O., 1989: The global biogeochemical sulphur cycle, 1989: a review. In *Trace Gases and the Biosphere*, B. Moore, (Ed.), University of Arizona Press, Tucson, Arizona.

Bacastow, R.B., J.A. Adams, C.D. Keeling, D.J. Moss, T.P. Whorf, C.S. Wong, 1980: Atmospheric carbon dioxide, the Southern Oscillation, and the weak 1975 El Niño. *Science*, 210, 66–68.

Bazzaz, F.A., 1990: The response of natural ecosystems to the rising global CO_2 levels. *Ann. Rev. Eco. Systematics*, 21, 167–196.

Blake, D.R., and F.S. Rowland, 1988: Continuing worldwide increases in tropospheric methane, 1978–1987. *Science*, 239, 1129–1131.

Blanchet, J.P., 1989: The response of polar stratospheric clouds to increasing carbon dioxide. *Proceedings of the International Radiation Symposium*, Lille, France, August, 1988, A. Deepak Publishing, Hampton, VA.

Bolin, B., 1986: How much CO_2 will remain in the atmosphere? In *The Greenhouse Effect, Climatic Change and Ecosystems, SCOPE 29*, B. Bolin et al., (Eds.), 93–155, John Wiley & Sons, New York.

Borgwardt, R.H., 1992: A technology for reduction of CO_2 emissions from the transportation sector. *Energy Conser. Manage.*, 33, 443–449.

Broecker, W.S., T. Takahashi, H.J. Simpson, T.-H. Peng, 1979: Fate of fossil fuel carbon dioxide and the global carbon budget. *Science*, 206, 409–417.

Broecker, W.S., T.-H. Peng, R. Engh, 1980: Modeling the carbon system. *Radiocarbon*, 22, 565–598.

Butler, J.H., J.W. Elkins, T.M. Thompson, B.D. Hall, T.H. Swanson, V. Koropalov, 1991: Oceanic consumption of CH_3CCl_3: implications for tropospheric OH. *J. Geophys. Res.*, 96, 22347–22355.

Butler, J.H., J.W. Elkins, B.D. Hall, S.O. Cummings, S.A. Montzka, 1992: A decrease in the growth rates of atmospheric halon concentrations. *Nature*, 359, 403–405.

Chandler, W.U., and A.K. Nicholls, 1990: *Assessing Carbon Emissions Control Strategies: A Carbon Tax or a Gasoline Tax?* The American Council for an Energy-Efficient Economy, ACEEE Policy Paper No. 3.

Chandler, W.U., H.S. Geller, M.R. Ledbetter, 1988: *Energy Efficiency: A New Agenda.* The American Council for an Energy-Efficient Economy, 1001 Connecticut Avenue, N.W., Washington, DC 20036. (July).

Chappellaz, J., J.M. Barnola, D. Raynaud, Y.S. Korotkevich, C. Lorius, 1990: Ice-core record of atmospheric methane over the past 160,000 years. *Nature*, 345, 127–131.

Charlson, R.J., J.E. Lovelock, M.O. Andreae, S.G. Warren, 1987: Oceanic phytoplankton, atmospheric sulphur, cloud albedo and climate. *Nature*, 326, 655–661.

Charlson, R.J., J. Langner, H. Rodhe, 1990: Sulfate aerosol and climate. *Nature*, 348, 22–26.

Cheng, H.C., M. Steinberg, M. Beller, 1985: Effects of Energy Technology on Global CO_2 Emissions. U.S. Department of Energy report DOE/NBB-0076. National Technical Information Service, U.S. Department of Commerce, Springfield VA 22161.

Cicerone, R.J., 1988: How has the atmospheric concentration of CO changed? In *The Changing Atmosphere*, F.S. Rowland, I.S.A. Isaksen, (Eds.), John Wiley & Sons, New York.

Cicerone, R.J., and R.S. Oremland, 1988: Biogeochemical aspects of atmospheric methane. *Global Biol. Cycles*, 2, 299–327.

Crutzen, P.J., 1989: Emissions of CO_2 and other trace gases to the atmosphere from fires in the tropics. 28th Liege International Astrophysical Colloquium, University of Liege, Belgium, June 26–30.

Crutzen, P.J., and T.E. Graedel, 1986: The role of atmospheric chemistry in environment-development interactions. *In Sustainable Development of the Biosphere*, W.C. Clark and R.E. Munn, (Eds.), Cambridge University Press, New York.

Crutzen, P.J., I. Aselmann, W.S. Seiler, 1986: Methane production by domestic animals, wild ruminants, other herbivorous fauna, and humans. *Tellus*, 38B, 271–284.

Dale, V.H., R.A. Houghton, C.A.S. Hall, 1991: Estimating the effects of land-use change on global atmospheric CO_2 concentrations. *Can. J. Forest Res.*, 21, 87–90.

DeLuisi, J.J., D.U. Longenecker, C.L. Mateer, D.J. Wuebbles, 1989: An analysis of northern mid-latitude Umkehr measurements corrected for stratospheric aerosols for 1979–1986. *J. Geophys. Res.*, 94, 9837–9845.

Derwent, R.G., 1990: *Trace gases and their relative contribution to the greenhouse effect*. Atomic Energy Research Establishment, Harwell, Oxon, Report AERE-R13716.

Detwiler, R.P., and C.A.S. Hall, 1988: Tropical forests and the global carbon cycle. *Science*, 239, 42–47.

Dignon, J., and S. Hameed, 1989: Global emissions of nitrogen and sulfur dioxides from 1860–1980. *JAPCA*, 39, 180–186.

Dignon, J., 1994: Impact of biomass burning on the atmosphere. *Proceedings of NATO Advanced Workshop on Ice Core Studies of Global Biochemical Cycles*, March 26–31, 1993, Annecy, France.

DOE Multi-Laboratory Climate Change Committee, 1990: *Energy and Climate Change*. Lewis Publishers, Chelsea, MI.

Eamus, D., and P.G. Jarvis, 1989: The direct effects of increase in the global atmospheric CO_2 concentration on natural and commercial temperate trees and forests. *Adv. Ecol. Res.*, 19, 1–55.

Edmonds, J., and D.W. Barns, 1990: Estimating the Marginal Cost of Reducing Global Fossil Fuel CO_2 Emissions. Pacific Northwest Laboratory, Washington, DC.

Edmonds, J., B. Ashton, H. Cheng, M. Steinberg, 1989: An Analysis of U.S. CO_2 Emissions Reduction Potential in the Period to 2010. U.S. Department of Energy report DOE/NBB-0085 Dist. Category UC-11, National Technical Information Service, U.S. Department of Commerce, Springfield, VA 22161.

Edmonds, J.A., J.M. Reilly, R.H. Gardner, A. Brenkert, 1986: Uncertainty in Future Global Energy Use and Fossil Fuel CO_2 Emissions 1975 to 2075. U.S. Department of Energy report DOE/NBB-0081 Dist. Category UC-11, National Technical Information Service, U.S. Department of Commerce, Springfield VA 22161.

Elkins, J.W., B.D. Hall, J.H. Butler, 1990: Laboratory and field investigations of the emissions of nitrous oxide from biomass burning. Chapman Conference on Global Biomass Burning: *Atmospheric, Climatic, and Biospheric Implications*, J. S. Levine, Ed., Williamsburg, VA.

Elkins, J.W., T.M. Thompson, T.H. Swanson, J.H. Butler, B.D. Hall, S.O. Cummings, D.A. Fisher, A.G. Raffo, 1993: Decrease in the growth rates of atmospheric chlorofluorocarbons 11 and 12. *Nature*, 364, 780–783.

Emanuel, W.R., Y.-S. Fung, G.G. Killough, B. Moore, T.-H. Peng, 1985: Modeling the global carbon cycle and changes in the atmospheric carbon dioxide levels. pp. 141–173. In *Atmospheric Carbon Dioxide and the Global Carbon Cycle*, J.R. Trabalka (Ed.) DOE/ER-0239. Carbon Dioxide Research Division, U.S. Department of Energy, Washington, DC.

Enting, I.G., and J.V. Mansbridge, 1987: The incompatibility of ice-core CO_2 data with reconstructions of biotic CO_2 sources. *Tellus*, 39B, 318–325.

Fisher, D. A., C.H. Hales, D.L. Filkin, M.K.W. Ko, N.D. Sze, P.S. Connell, D.J. Wuebbles, I.S.A. Isaksen, F. Stordal, 1990a: Model calculations of the relative effects of CFCS and their replacements on stratospheric ozone. *Nature*, 344, 508–512.

Fisher, D.A., C.H. Hales, W-C. Wang, M.K.W. Ko, N.D. Sze, 1990b: Model calculations of CFCs and their replacements on global warming. *Nature*, 344, 513–516.

Fishman, J., 1985: Ozone in the troposphere. In *Ozone in the Free Troposphere*. R.C. Whitten and S.S. Prasad, (Eds.), Van Nostrand Reinhold, New York.

Fishman, J., C.E. Watson, J.C. Larsen, J.A. Logan, 1990: Distribution of tropospheric ozone determined from satellite data. *J. Geophys. Res.*, 95, 3599–3617.

Friedli, H., H. Lötscher, H. Oeschger, U. Siegenthaler, B. Stauffer, 1986: Ice core record of $^{13}C/^{12}C$ ratio of atmospheric carbon dioxide in the past two centuries. *Nature*, 324, 237–238.

Fung, I., J. John, J. Lerner, E. Matthews, M. Prather, L.P. Steele, P.J. Fraser, 1991: Three-dimensional model synthesis of the global methane cycle. *J. Geophys. Res.*, 96, 13,033–13,065.

Garcia, R.R., and S. Solomon, 1987: Interannual variability in antarctic ozone and the quasi-biennial oscillation. *Geophys. Res. Lett.*, 14, 848–851.

Gerbers, D., T. Kram, P. Lako, P.A. Okken, J.R. Ybema, 1990: Opportunities for New Energy Technologies to Reduce CO_2-Emissions in the Netherlands Energy System Up to 2030. Draft, prepared for ETSAP-workshop, Geneva 8–12 October 1990. Netherlands Energy Research Foundation, Energy Studies, P.O. Box 1, 1755 ZG PETTEN, The Netherlands.

Goldemberg, J., T.B. Johansson, A.K.N. Reddy, R.H. Williams, 1987: *Energy for a Sustainable World*. Wiley-Easton, New Delhi, India.

Griffith, D.W.T., W.G. Mankin, M.T. Coffey, D.E. Ward, A. Riebau, 1990: FTIR remote sensing of biomass burning emissions of CO_2, CO, CH_4, CH_2O, NO, NO_2, NH_3 and N_2O. Chapman Conference on Global Biomass Burning: *Atmospheric, Climatic, and Biospheric Implications*, J.S. Levine, (Ed.), Williamsburg, VA.

Hameed, S., and J. Dignon, 1988: Changes in the geographical distributions of global emissions by NO_x and SO_x from fossil fuel combustion between 1966 and 1980. *Atmos. Environ.*, 22, 441–449.

Hansen, J., A. Lacis, M. Prather, 1989: Greenhouse effect of chlorofluorocarbons and other trace gases. *J. Geophys. Res.*, 94, 16417–16421.

Hansen, J., I. Fung, A. Lacis, D. Rind, S. Lebedeff, R. Ruedy, G. Russell, P. Stone, 1988: Global climate changes as forecast by Goddard Institute for Space Studies three-dimensional model. *J. Geophys. Res.*, 93, 9341–9364.

Hansen, J.E., A. Lacis, R. Ruedy, M. Sato, H. Wilson, 1993: How sensitive is the world's climate? *Nat. Geograph. Res. Explor.*, 9, 142–158.

Houghton, R.A., J.E. Hobbie, J.M. Melillo, B. Moore, B.J. Peterson, G.R. Shaver, G.M. Woodwell, 1983: Changes in the carbon content of terrestrial biota and soils between 1860 and 1980: A net release of CO_2 to the atmosphere. *Ecological Monographs*, 53, 235–262.

Houghton, R.A., R.D. Boone, J.R. Fruci, J.E. Hobbie, J.M. Melillo, C.A. Palm, B.J. Peterson, G.R. Shaver, G.M. Woodwell, B. Moore, D.L. Skole, N. Myers, 1987: The flux of carbon from terrestrial ecosystems to the atmosphere in 1980 due to changes in land use. *Tellus*, 39B, 122–139.

Houghton, R.A., and D.L. Skole, 1991: Carbon. pp. 393–408. In *The Earth as Transformed by Human Action.*, R.L. Turner, W.C. Clark, R.W. Kates, J.F. Richards, J.T. Mathews, W.B. Meyer (Eds.), Cambridge University Press, Cambridge, U.K.

IPCC (Intergovernmental Panel on Climate Change), 1990: *Scientific Assessment of Climate Change*. Cambridge University Press, Cambridge, U.K.

IPCC (Intergovernmental Panel on Climate Change), 1992: *Climate Change: 1992*. Cambridge University Press, Cambridge, U.K.

IPCC (Intergovernmental Panel on Climate Change), 1994: *Radiative Forcing of Climate Change: 1994*. Cambridge University Press, Cambridge, U.K.

Isaksen, I.S.A., and O. Hov, 1987: Calculation of trends in the tropospheric concentration of O_3, OH, CO, CH_4 and NO_x, *Tellus*, 39B, 271–285.

Keeling C.D., 1973: The carbon dioxide cycle: Reservoir models to depict the exchange of atmospheric carbon dioxide with the oceans and land plants. pp. 251–328. In *Chemistry of the Lower Atmosphere*. S.I. Rasool (Ed.), Plenum, New York.

Keeling, C.D., 1986: Atmospheric CO_2 Concentrations—Mauna Loa Observatory, Hawaii 1958–1986. NDP-001/R1. Carbon Dioxide Information Center, Oak Ridge National Laboratory, Oak Ridge, TN.

Keeling, C.D., R.B. Bacastow, A.E. Bainbridge, C.A. Ekdahl, P.R. Guenther, L.S. Waterman, J.F. Chin, 1976: Atmospheric carbon dioxide variations at Mauna Loa Observatory, Hawaii. *Tellus*, 28, 538–551.

Keeling, C.D., R.B. Bacastow, A.F. Carter, S.C. Piper, T.P. Whorf, M. Heimann, W.G. Mook, H. Roeloffzen, 1989: A three-dimensional model of atmospheric CO2 transport based on observed winds: 1. Analysis of observational data. pp. 165–236. In *Aspects of Climate Variability in the Pacific and the Western Americas.*, D.H. Peterson (Ed.), American Geophysical Union, Washington, DC.

Khalil, M.A.K., and R.A. Rasmussen, 1985: Causes of increasing atmospheric methane: Depletion of hydroxyl radicals and the rise of emissions. *Atmos. Environ.*, 19, 397–407.

Khalil, M.A.K., and R.A. Rasmussen, 1987: Atmospheric methane: Trends over the last 10,000 years. *Atmos. Environ.*, 21, 2445–2452.

Khalil, M.A.K., and R.A. Rasmussen, 1988: Carbon monoxide in the Earth's atmosphere: Indications of a global increase. *Nature*, 332, 242–245.

Khalil, M.A.K., and R.A. Rasmussen, 1990a: Atmospheric methane: recent global trends. *Env. Sci. Tech.*, 24, 549–553.

Khalil, M.A.K., and R.A. Rasmussen, 1990b: Global Increase of Atmospheric Molecular Hydrogen. *Nature*, 347, 743–745.

Khalil, M.A.K., and R.A. Rasmussen, 1990c: Constraints on the global sources of methane and an analysis of recent budgets. *Tellus*, 42B, 229–236.

Khalil, M.A.K., and R.A. Rasmussen, 1992: The Global Sources of Nitrous Oxide. *J. Geophys. Res.* 97, 14651–14660.

Khalil, M.A.K., and M.J. Shearer, 1993: Sources of Methane: an Overview. In *Atmospheric Methane: Sources, Sinks, and Role in Global Change*, M.A.K. Khalil (Ed.), Springer-Verlag, Berlin.

Killough, G.G., and W.R. Emanuel, 1981: A comparison of several models of carbon turnover in the ocean with respect to their distributions of transit time and age and response to atmospheric CO$_2$ and ^{14}C. *Tellus*, 33, 274–290.

Knapp, K.T., 1990: Continuous FTIR measurements of mobile source emissions. Report of the Proceedings of the European Workshop on N$_2$O Emissions. Departmento de Energias Convencionais, Lisboa, Portugal, June 1990.

Kratz, G., 1985: Modelling the global carbon cycle. pp. 29–81. In *The Handbook of Environmental Chemistry, The Natural Environment and the Biogeochemical Cycles.*, O. Hutzinger, (Ed.), Springer-Verlag, Berlin.

Lacis, A.A., D.J. Wuebbles, J.A. Logan, 1990: Radiative forcing of climate by changes in the vertical distribution of ozone. *J. Geophys. Res.*, 95, 9971–9981.

Langner, J., and H. Rodhe, 1990: Anthropogenic impact on the global distribution of atmospheric sulphate. *Proceedings of First International Conference on Global and Regional Atmospheric Chemistry*, Beijing, China, 3–10 May, 1989, L. Newman et al., (Eds.), U.S. Department of Energy report.

Lashof, D.A., and D.R. Ahuja, 1990: Relative global warming potentials of greenhouse gas emissions. *Nature*, 344, 529–531.

Levine, J.S., C.P. Rinsland, G.M. Tennille, 1985: The photochemistry of methane and carbon monoxide in the troposphere in 1950 and 1985. *Nature*, 318, 254–257.

Linak, W. P., J.A. McSorley, R.E. Hall, J.V. Ryan, R.K. Srivastava, J.O.L. Wendt and J.B. Mereb, 1990: Nitrous oxide emissions from fossil fuel combustion. J. *Geophys Res.*, 95, 7533–7541.

Liu, S.C., M. McFarland, D. Kley, O. Zafiriou, B.J. Huebert, 1983: Tropospheric NO$_x$ and O$_3$ budgets in the equatorial Pacific. *J. Geophys. Res.*, 88, 1360–1368.

Liu, S.C., M. Trainer, F.C. Fehsenfeld, D.D. Parrish, E.J. Williams, D.W. Fahey, G. Hubler, P.C. Murphy, 1987: Ozone production in the rural troposphere and the implications for regional and global ozone distributions. *J. Geophys. Res.*, 92, 4191–4207.

Logan, J.A., 1985: Tropospheric ozone: Seasonal behavior, trends, and anthropogenic influence. *J. Geophys. Res.*, 90, 10463–10482.

MacCracken, M.C., and F.M. Luther, 1985: *Projecting the Climatic Effects of Increasing Carbon Dioxide*. Report DoE/ER-0237, U.S. Dept. of Energy, Washington, DC.

Maier-Reimer, E., and K. Hasselmann, 1987: Transport and storage of CO_2 in the ocean—an inorganic ocean-circulation carbon cycle model. *Climate Dynamics*, 2, 63–90.

Manne, A.S., and R.G. Richels, 1990a: CO_2 Emission Limits: An Economic Cost Analysis for the USA. *Energy J.*, 11, 51–75.

Manne, A.S., and R.G. Richels, 1990b: *Global CO_2 Emissions Reductions—the Impacts of Rising Energy Costs*. Electric Power Research Institute, Palo Alto, CA.

Manne, A.S., and R.G. Richels, 1990c: Buying Greenhouse Insurance. Electric Power Research Institute, Palo Alto, CA.

Manne, A.S., and R.G. Richels, 1990d: *The Costs of Reducing U.S. CO_2 Emissions—Further Sensitivity Analysis*. Electric Power Research Institute, Palo Alto, CA.

Marland G., and R.M. Rotty, 1984: Carbon dioxide emissions from fossil fuels: A procedure for estimation and results for 1950–1982. *Tellus*, 36B, 232–261.

Marland G., T.A. Boden, R.C. Griffin, S.F. Huang, P. Kanciruk, T.R. Nelson, 1989: Estimates of CO_2 Emissions from Fossil Fuel Burning and Cement Manufacturing, Based on the United Nations Energy Statistics and the U.S. Bureau of Mines Cement Manufacturing Data, ORNL/ CDIAC-25. Oak Ridge National Laboratory, Oak Ridge, TN.

Marland, G., 1988: The Prospect of Solving the CO_2 Problem through Global Reforestation. U.S. Department of Energy report DOE/NBB-0082; TR039; National Technical Information Service, U.S. Department of Commerce, Springfield VA 22161.

Mintzer, I., 1987: *A Matter of Degrees: The Potential for Controlling the Greenhouse Effect*, Research Report #5, World Resources Institute, Washington, DC.

Montzka, S.A., R.C. Myers, J.H. Butler, J.W. Elkins, 1993: Global tropospheric distribution and calibration scale of HCFC-22. *Geophys. Res. Lett.*, 20, 703–706.

Moulton and Richards, 1990: *Costs of Sequestering Carbon Through Tree Planting and Forest Management in the United States*. U.S. Department of Agriculture, Washington, DC.

Muzio, L.J., and J.C. Kramlich, 1988: An artifact in the measurement of N_2O from combustion sources. *Geophys. Res. Lett.*, 15, 1369–1372.

National Academy of Sciences, *1991: Policy Implications of Greenhouse Warming*. National Academy Press, Washington, DC.

Neftel, A., E. Moor, H. Oeschger, B. Stauffer, 1985: Evidence from polar ice cores for the increase in atmospheric CO_2 in the past two centuries. *Nature*, 315, 45–47.

Nordhaus, W.D., 1990: *Contribution of Different Greenhouse Gases to Global Warming: A New Technique for Measuring Impact.* Department of Economics, Yale University, New Haven, CT.

Oeschger, H., U. Siegenthaler, A. Gugelman, 1975: A box diffusion model to study the carbon dioxide exchange in nature. *Tellus*, 27, 168–192.

Olson, J.S., 1974: Terrestrial ecosystem. pp. 144–149. In *Encyclopedia Britannica*, 15th edition, Helen Hemingway Benton, Chicago, Illinois.

Olson, J.S., J.A. Watts, L.J. Allison, 1983: *Carbon in Live Vegetation of Major World Ecosystems.* ORNL-5862. Oak Ridge National Laboratory, Oak Ridge, TN.

Pearman, G.I., D. Etheridge, F. de Silva, P.J. Fraser, 1986: Evidence of changing concentrations of atmospheric CO_2, N_2O, and CH_4 from air bubbles in Antarctic ice. *Nature*, 320, 248–250.

Penner, J. E., 1990: Cloud albedo, greenhouse effect, atmospheric chemistry, and climate change. *J. Air Poll. Control Assoc.*, 40, 456–461.

Penner, J.E., and R.E. Dickinson, C.A. O'Neill, 1992: Effects of aerosol from biomass burning on the global radiation budget. *Science*, 256, 1432–1434.

Penner, J.E., P.S. Connell, D.J. Wuebbles, C.C. Covey, 1989: *Climate Change and its Interactions with Air Chemistry: Perspectives and Research Needs.* Lawrence Livermore National Laboratory report UCRL-21111; also in The Potential Effects of Global Climate Change on the United States, J.B. Smith and D.A. Tirpak, (Eds)., U.S. Environmental Protection Agency report EPA-230-05-89-056.

Post, W.M., W.R. Emanuel, P.J. Zinke, A.G. Stangenberger, 1982: Soil carbon pools and world life zones. *Nature*, 298, 156–159.

Post, W.M., T.-H. Peng, W.R. Emanuel, A.W. King, V.H. Dale, D.L. DeAngelis, 1990: The global carbon cycle. *Am. Sc.*, 78, 310–326.

Prinn, R., D. Cunnold, R. Rasmussen, P. Simmonds, F. Alyea, A. Crawford, P. Fraser, R. Rosen, 1987: Atmospheric trends in methylchloroform and the global average for the hydroxyl radical. *Science*, 238, 945–950.

Ramanathan, V., 1988: The greenhouse theory of climate change: A test by an inadvertent global experiment. *Science*, 240, 293–299.

Ramanathan, V., L. Callis, R. Cess, J. Hansen, I. Isaksen, W. Kuhn, A. Lacis, F. Luther, J. Mahlman, R. Reck, M. Schlesinger, 1987: Chemical-climate interactions and effects of changing atmospheric trace gases. *Rev. Geophys.*, 25, 1441–1482.

Ramanathan, V., R.J. Cicerone, H.B. Singh and J.T. Kiehl, 1985: Trace gas trends and their potential role in climate change. *J. Geophys. Res.*, 90, 5547–5557.

Raynaud, D., J. Chappellaz, J.M. Barnola, Y.S. Korotkeviich, C. Lorius, 1988: Climatic and CH_4 cycle implications of glacial-interglacial CH_4 change in the Vostok ice core. *Nature*, 333, 655–657.

Reilly, J.M., 1990: *Climate Change Damage and the Trace Gas Index Issues.* U.S. Department of Agriculture, Economic Research Service, Room 528, Washington, DC 20005-4788.

Rodhe, H., 1990: A comparison of the contributions of various gases to the greenhouse effect. *Science*, 248, 1217–1219.

Rose, D.J., M.M. Miller, C. Agnew, 1983: *Global Energy Futures and CO_2-Induced Climate Change.* MITEL 83–015, MIT Energy Laboratory also as MITNE-259, MIT Department of Nuclear Engineering, Cambridge MA.

Rosencranz, A., 1986: The acid rain controversy in Europe and North America. *AMBIO*, 15, 47–50.

Sedjo, R.A., and, A.M. Solomon, 1989: Climate and forests. In *Greenhouse Warming: Abatement and Adaptation.*, N.J. Rosenberg, W.E. Easterling III, P.R. Crosson, J. Darmstadter, (Eds.), Resources for the Future, Washington, DC.

Shine, K.P., 1988: Comment on "southern hemisphere temperature trends: A possible greenhouse effect." *Geophys. Res. Lett.*, 15, 843–848.

Shine, K.P., R.G. Derwent, D.J. Wuebbles, J.-J. Morcrette, 1990: Radiative forcing of climate. pp. 41–69. In *Climate Change*, J.T. Houghton, G.J. Jenkins, J.J. Ephraums, (Eds.), Cambridge University Press, Cambridge, U.K.

Siegenthaler, U., and H. Oeschger, 1987: Biospheric CO_2 emissions during the past 200 years reconstructed by deconvolution of ice core data. *Tellus*, 39B, 140–154.

Siegenthaler, U., H. Friedli, H. Loetscher, E. Moor, A. Neftel, H. Oeschger, B. Stauffer, 1988: Stable-isotope ratios and concentrations of CO_2 in air from polar ice cores. *Ann. Glaciol.*, 10, 1–6.

Solomon, S., 1988: The mystery of the Antarctic ozone hole. *Rev. Geophys.*, 26, 131–148.

Steinberg M., and E.W. Grohse, 1989: *The Hydrocarb Process for Environmentally Acceptable and Economically Competitive Coal-Derived Fuel for the Utility and Heat Engine Market.* BNL-43554, Brookhaven National Laboratory, Upton, NY.

Steele, L.P., E.J. Dlugokencky, P.M. Lang, P.P. Tans, R.C, Martin, K.A. Masarie, 1992: Slowing down of the global accumulation of atmospheric methane during the 1980s. *Nature*, 358, 313–316.

Steudler, P.A., R.D. Bowden, J.M. Melillo, J.D. Aber, 1989: Influence of nitrogen fertilization on methane uptake in temperate forest soils. *Nature*, 341, 314–316.

Stuiver, M., P.D. Quay, H.G. Ostlund, 1983: Abyssal water carbon-14 distribution and the age of the world oceans. *Science*, 219, 849–851.

Thiemens, M.H., and W.C. Trogler, 1991: Nylon production: an unknown source of atmospheric nitrous oxide. *Science*, 251, 932–934.

Thompson, A.M., and R.J. Cicerone, 1986: Possible perturbations to atmospheric CO, CH_4, and OH. *J. Geophys. Res.*, 89, 10853–10864.

Thompson, A.M., R.W. Stewart, M.A. Owens, J.A. Herwehe, 1989: Sensitivity of tropospheric oxidants to global chemical and climate change. *Atmos. Environ.*, 23, 519–532.

Trainer, M., E.Y. Hsie, S.A. McKeen, R. Tallamraju, D.D. Parrish, F.C. Fehsenfeld, S.C. Liu, 1987: Impact of natural hydrocarbons on hydroxyl and peroxy radicals at a remote site. *J. Geophys. Res.*, 92, 11879–11894.

Vaghjiani, G.L., and A.R. Ravishankara, 1991: Rate coefficient for the reaction of OH with CH4: implications to the atmospheric lifetime and budget of methane. *Nature*, 350, 406–409.

Wahlen, M., N. Tanaka, R. Henry, B. Deck, J. Zeglen, J.S. Vogel, J. Southon, A. Shemish, R. Fairbanks, W. Broecker, 1989: Carbon-14 in methane sources and in atmospheric methane: the contribution from fossil carbon. *Science*, 245, 286-290.

Wang, W.-C., D.J. Wuebbles, W.M. Washington, R.G. Isaacs, G. Molnar, 1986: Trace gases and other potential perturbations to global climate. *Rev. Geophys.*, 24, 110–140.

Williams, R.H. and E.D. Larson, 1987: Materials, affluence, and industrial energy use. *Ann. Rev. of Energy 1987*, 12, 99–144.

Williams, R.H., 1990: *Will Constraining Fossil Fuel Carbon Dioxide Emissions Really Cost So Much?* Center for Energy and Environmental Studies, Princeton University, Princeton, NJ.

World Meteorological Organization, 1985: Stratospheric Ozone 1985: *Assessment of our understanding of the processes controlling its present distribution and change*. Global Ozone Research and Monitoring Project—Report No. 16.

World Meteorological Organization, 1989: *Scientific Assessment of Stratospheric Ozone: 1989*. Global Ozone Research and Monitoring Project—Report No. 20.

World Meteorological Organization, 1991: Scientific Assessment of Ozone Depletion: 1991. Global Ozone Research and Monitoring Project—Report No. 25.

Wuebbles, D.J., and D.E. Kinnison, 1990: Sensitivity of stratospheric ozone to present and possible future aircraft emissions. In *Air Traffic and the Environment: Background Tendencies and Potential Global Atmospheric Effects*, U. Schumann, (Ed.), Springer-Verlag, Berlin.

Wuebbles, D.J., and J. Edmonds, 1991: *A Primer on Greenhouse Gases*, Lewis Publishers, Chelsea, MI.

Wuebbles, D.J., K.E. Grant, P.S. Connell, J.E. Penner, 1989: The role of atmospheric chemistry in climate change. *J. Air Poll. Control Assoc.*, 39, 22–28.

Zardini, D., D. Raynaud, D. Scharffe, W. Seiler, 1989: N_2O measurements of air extracted from Antarctic ice cores: implications on atmospheric N_2O back to the last glacial-interglacial transition. *J. Atmos. Chem.*, 8, 189–201.

Chapter 3

ENERGY DEMAND REDUCTION

Authors: Arthur Rosenfeld, Barbara Atkinson,
Lynn Price, Bob Ciliano, J.I. Mills, Kenneth Friedman
Co-Authors: Ed Flynn, Mary Hopkins,
Henry Shaw, John Wilson, Francis Wood
Contributors: Ruth Reck, Eric Larson

3.1 INTRODUCTION

In this chapter, we discuss only CO_2, the greenhouse gas most directly involved in energy utilization through the combustion of fossil fuels (see Chapter 4). There are three reasons for this focus. First, CO_2 represents more than half of the total "global warming potential" (NAS, 1992). Second, because of our worldwide reliance on the production of energy through combustion of fossil fuels, reduction of CO_2 presents a great engineering challenge. Finally, CO_2 is the only greenhouse gas that can be potentially reduced at a net economic benefit. Gross economic potential savings for the U. S. are about 40 % of the nation's $450 billion annual energy bill, and net economic savings (after paying for improved energy efficiency) are about half of the gross savings (see NAS, 1992).

3.1.1 Recent Energy, GNP, and CO_2 Emissions Trends

In Figure 3.1 we see U.S. energy and electricity consumption (E) and GNP trends since 1960. For the U.S., we define end use efficiency as energy (primary) consumed per dollar of GNP, i.e.:

$$\text{end use efficiency} = \text{E/GNP.} \qquad \text{(Eq. 3.1)}$$

As a nation industrializes, E/GNP first rises as inefficient technology is used to build an infrastructure, then falls as more of a service economy prevails and as end use efficiency improves with time

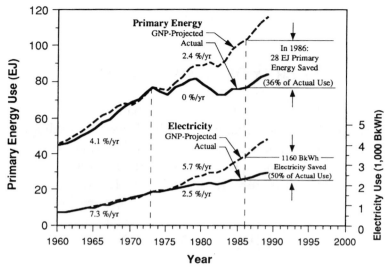

Figure 3.1   Total U.S. primary energy and electricity use: actual vs. GNP projected (1960-1989). Before the 1973 oil embargo, total primary
energy use was growing at the same rate as GNP. Between 1973 and
1986, growth in energy consumption was halted by high oil prices
and progressive energy policies. In 1986, projected primary energy
use was 36% (28 exajoules) higher than actual use. Electricity use
followed a similar pattern. Between 1973 and 1986, electricity use
decreased, growing only 2.5% per year, or 3.2% per year less than
projected by pre-1973 trends. (GNP-projected energy values are
based on 1973 efficiency and GNP. The electricity projections
include an additional 3% per year to account for increasing electrification). In 1986, projected electricity use was 50% higher than
actual energy use, indicating a savings of 1160 BkWh.

Electricity use is given in terms of total equivalent primary energy
input (exajoules—left-hand scale), and net consumption (1,000
BkWh—right-hand scale). (From EIA, 1990.)

(Figure 3.2). Between 1960 and 1973, energy was inexpensive, and no
distinction was made between providing energy and providing the
energy services of space heating, lighting, motor shaft power, etc.
Accordingly, little attention was paid to improving efficiency, which
remained "frozen." Primary energy use and U.S. GNP were thus
linked, and each climbed at about 4% per year. In 1973, the
Organization of Petroleum Exporting Countries (OPEC) oil embargo
introduced a powerful incentive to improve energy efficiency. During
the 13 years of high oil prices and progressive energy policies from
1973 to 1986, national energy use stayed constant, while U.S. GNP
grew by a total of 35%, i.e., by 2.4% per year. Of the total savings of

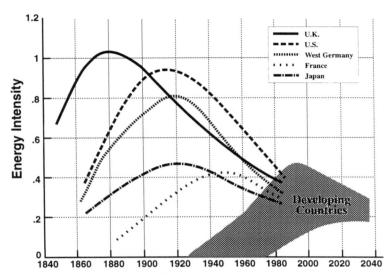

Figure 3.2   In industrialized countries the energy intensity (ratio of energy consumption to gross national product) rose, then fell. Because of improvements in materials science and energy efficiency, the maxima reached by countries during industrialization have progressively decreased over time. Developing nations can avoid repeating the history of the industrialized world by using energy efficiently. (From A.K. Reddy and J. Goldemberg, Energy for the Developing World, *Sc. Am.* Special Issue: Energy for Planet Earth, Sept. 1990. With permission.)

28 exajoules (the difference between GNP-projected and actual consumption during this period), one third is attributed to structural changes in the economy and the remaining two thirds is attributed to improved energy efficiency. Efficiency measures implemented during this period avoided a sharp increase in coal use, and hence avoided a rise of 50% in U.S. greenhouse gas emissions.

Even more impressive than the past reductions in primary energy use is the electricity conservation also shown in Figure 3.1. Until 1973, total electricity use was growing at a rate of 7.3% per year (3.2% faster than GNP). Between 1973 and 1986, electricity use grew only as fast as GNP, for an annual savings of 3.2%, or a total of 50% in the 13-year period. This 50% savings, 1160 billion kilowatt hours (kWh) per year, is equivalent to the annual output of 230 baseload (1000 megawatt) power plants.

In late 1985, when OPEC's oil prices collapsed, gains in energy efficiency nearly stopped. Primary energy consumption is now climbing again at a rate of about 2.5% per year, compared to 3% per year for GNP, directly contributing to increased emissions of CO_2.

Following the National Academy of Sciences study (see Section 3.2 below), we assume that each kWh of electricity comes from the U.S. electric industry's mix of primary fuel (coal, gas, nuclear, etc.). This corresponds to 0.19 MtC/BkWh (0.7 Mt CO_2) in 1990. We also assume that the marginal fuel for heating buildings is natural gas and that the marginal fuel for cars is gasoline.

Table 3.1 shows recent and projected U.S. energy use and CO_2 emissions from combustion of fossil fuel (Shaw, 1991). Energy use rises 1.2% per year, an increase of 60% by 2030. Projections indicate that electricity consumption in the residential/commercial sector will increase by 80%, electricity use in the transportation sector will increase from negligible current needs to 25% of the residential/commercial demand, and industrial electricity use will increase by 300%. We estimate that natural gas demand for the residential/commercial sector will decrease by one third in 2030 and transportation use of petroleum products will decrease by 6% in 2030 (Shaw, 1991).

3.1.2 Efficiency Can Continue to Increase

During the 13 years of high-priced energy described above, the efficiency of new products actually doubled. New cars satisfied the Corporate Average Fuel Efficiency (CAFE) standards and jumped from 14 to 28 mpg (the fleet average is still only about 20 mpg because of the lag from the 10-year life of cars). Similarly, appliance efficiency has doubled, and the energy requirements of new homes

Table 3.1   U.S. energy use and carbon dioxide emissions.

	1980	1990	2000	2015	2030
Energy Use (Quads)	78.3	82.3	92.3	109.7	132.2
Total CO_2 Emissions (Gt)	5.1	5.5	5.9	6.6	7.3
Sectoral CO_2 Emissions	**Percent of Total**				
Residential/commercial	28.8	28.4	24.3	19.7	14.1
Transportation	30.7	28.0	25.6	25.2	23.1
Industrial					
Energy	22.8	25.2	27.5	30.6	32.9
Non-Energy	10.8	11.6	11.5	12.0	13.3
Refining Losses	7.0	6.9	11.1	12.4	16.7

Note: Includes those emissions due to electricity in consuming sectors.

and commercial buildings have been cut in half. However, due to the long lives of buildings, it will take 50 years before these latter gains are fully seen.

How many more times can we double efficiency? Efficiency of existing equipment (e.g., miles per gallon for cars, lumens/Watt for lamps) is still generally far from any theoretical physical limits. Thus, in 20 years it will probably be possible to buy a 90 mpg car, and windows will have become net sources of energy (i.e. will gain more solar heat during a winter day than they lose at night). If we include moderate lifestyle changes in energy services, we can further increase energy efficiency. For example, in 1975 the average 14 mpg car carried 1.1 commuters to work and thus got 15.5 passenger miles/gallon. Today, the 1992 Honda Civic gets 50 mpg and the prototype Volvo LCP 2000 gets 60 mpg in the city and 80 mpg on the highway. Assuming 4 passengers and 50 mpg, 200 passenger miles/gallon will be achieved, a factor of 13 increase.

3.1.3 Other Factors in Energy Demand

Energy demand is ultimately a function of three factors: (1) the efficiency of technology in providing energy services; (2) individual demand for technology and energy services (standard of living); and (3) population. Continued support for research and development by both the private and public sectors can improve the efficiency of technology options over time. However, getting the public to accept "new" technologies or different service characteristics may be difficult, especially if energy prices in the U.S. (and developing nations) remain low.

For example, out of 625 car models offered for sale in the U.S., there are already over 20 models of automobiles that achieve 40 mpg. However, with today's low gasoline prices, few Americans are motivated to put up with the downsizing inconveniences associated with current 40 mpg cars. The problem in the future may not be as much one of improving the efficiency of the technology as one of demand for technology and energy services. Proper pricing signals are critical to encourage appropriate individual or corporate behavior.

3.1.4 Doubling the Efficiency of New Products Every 20 Years

The above discussion illustrates that at any given time it seems possible to achieve a gain of 3 to 10 times in energy-service/kWh. We believe the U.S. can continue to double the energy efficiency of new products every 20 years for the foreseeable future, with an aggressive research, development, and demonstration (RD & D) program.

If this goal is accomplished, the implications for fulfilling the world's energy needs 40 years from now are astounding. Assuming that today's energy use is about 300 quads/year (i.e., 10 TW-years/year, referred to as "10 TW") and the world population is 5 billion, then current worldwide average consumption is about 2 kWh/person per year. The challenge is to maintain current energy consumption levels in the face of growing GNP and population. To understand this, consider Germany and Japan, where the standard of living is extremely high relative to most of the world. These countries now consume 5 kWh/person per year and if they meet our goal and improve their E/GNP by a factor of 4 in the next 40 years, their consumption will decrease to 1.25 kWh/person. Assuming that this level of energy consumption supporting a relatively high standard of living could be extended worldwide, a world population of 8 billion in 2031 (the current United Nations projection) would need only 10 TW, the same amount of energy consumed worldwide in 1990 (Goldemberg et al., 1988).

Thus, if these efficiency gains could be accomplished we would cap global energy growth. With some replacement of fossil fuels by renewables as discussed in Chapter 4, we might reduce CO_2 production enough to avert the risks of global warming.

3.2 EIGHT MAJOR RECENT STUDIES OF DEMAND REDUCTION POTENTIAL

Eight major recent studies have concluded that it is possible to save 30–50% in E/GNP, or CO_2/GNP. The National Academy of Sciences study found that such savings could be accomplished at a net economic benefit of around $50 billion/year (10% of the 1989 U.S. energy bill). These reports warn that we are likely to capture less than half of this potential unless we adopt strong policies motivated by heightened awareness of global warming potential. However, the debate on the economic benefits vs. costs of reducing CO_2 emissions has only begun. Other studies show potential economic costs to society (NAS, 1991).

The eight major studies produced between 1990 and 1992 are:

1. National Academy of Sciences, Policy Implications of Greenhouse Warming: Mitigation, Adaptation, and the Science Base, Washington D.C.
2. U.S. Congress, Office of Technology Assessment, Changing By Degrees: Steps to Reduce Greenhouse Gases, Washington, D.C.: U.S. GPO, 1991.
3. U.S. Environmental Protection Agency, Policy Options for Stabilizing Global Climate: Report to Congress, Washington, D.C.: U.S.
4. U.S. Department of Energy, National Energy Strategy, Washington, D.C.: Office of Science and Technical Information, 1991.
5. A. Faruqui, M. Mauldin, S. Schick, K. Seiden, G. Wikler, Efficient Electricity Use: Estimates of Maximum Energy Savings, prepared by Barakat & Chamberlin, Inc. for the Electric Power Research Institute (CU-6747) Palo Alto: Electric Power Research Institute, 1990.
6. International Energy Agency, Study of Energy Efficiency and Its Contribution to Environmental Goals, 1991.
7. ASE, ACEEE, NRDC, and UCS, America's Energy Choices: Investing in a Strong Economy and a Clean Environment. Alliance to Save Energy, American Council for an energy-Efficient Economy, Natural Resources Defense Council, and Union of Concerned Scientists, 1991.
8. Climate Change Action Plan.

3.3 U.S. SECTORAL ANALYSIS

In this section, we provide an analysis of the potential for reduction of energy demand in the buildings, industrial, and transportation sectors. We also discuss barriers that currently impede this reduction and present policies and research, development, and demonstration opportunities to hasten energy demand reduction in these sectors.

3.3.1 Buildings

The buildings sector includes residential, commercial, and industrial buildings. Residential buildings include single-family, multi-family, and mobile homes; commercial and industrial buildings include office buildings, retail, warehouses, and others. Because buildings typically have a long lifetime, on the order of 50 to 100 years, it is important that demand reduction strategies address ways to improve existing buildings, as well as design practices to maximize the efficient use of energy in new buildings.

3.3.1.1 Patterns and Trends for Energy Savings

In 1990, the U.S. buildings sector consumed approximately 31 quadrillion Btu, over 36% of 1990 total primary energy use (Energy Information Administration, 1991). Residential buildings consumed 55% of total building energy use, while commercial and industrial buildings consumed 45%. Energy consumption is projected by the U.S. Department of Energy to increase by approximately 7 quadrillion Btu by 2030 (0.5% per year) under baseline assumptions of GNP growth and efficiency improvements with existing and expected conservation efforts (Energy Information Administration, 1990).

In order to project future energy growth and the potential for demand reduction, the requirements for individual energy services or end uses within the buildings sector must be considered. These include space conditioning (heating and cooling), water heating, lighting, appliances, and equipment, etc. They reflect regional population growth, end use saturation rates and number of occupants. Figures 3.3 and 3.4 show current residential and commercial primary energy consumption by end use.

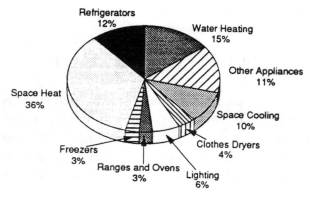

Figure 3.3   Residential consumption of primary energy by end use, 1987. (From Department of Energy, Conservation and Renewable Energy staff estimates, based on Energy Information Administration, *Household Energy Consumption and Expenditures,* 1987.)

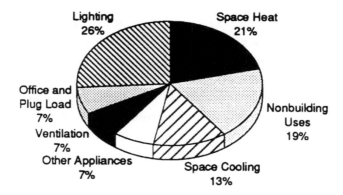

Figure 3.4   Commercial-sector consumption of primary energy by end use, 1990. (From Department of Energy, Office of Conservation and Renewable Energy, staff estimates, based on DoE/EIA Commercial PC-AEO Forecasting Model.)

For each energy service, a wide variety of conservation measures has been and will need to be developed to reduce energy use. Sectoral energy savings can be accomplished by a combination of relatively small, but collectively significant, improvements in building design, appliance efficiency, energy control systems, etc. Many highly cost-effective products and services could be adopted now that would reduce greenhouse gas emissions in the buildings sector. At the same time there are many other technologies on the horizon with a high probability for successful development that will, with sufficient public awareness and policy support, make it possible to improve energy efficiency in this sector by a factor of four over the next 40 years.

Concern over indoor air quality affects energy use in buildings. One response to this concern is to increase the outdoor air ventilation rates in commercial buildings, thereby increasing energy use to heat and cool buildings. A more efficient solution that also provides greater health benefits is to reduce the emission of dangerous or irritating gases from building materials, especially glues, carpets, plywood, etc. This need for safer building products represents a clear challenge to engineers and product designers.

3.3.1.2 Conservation Potential

Most of the existing stock of buildings and appliances was built before the current emphasis on energy efficiency. Upgrading these buildings and equipment to current cost-effective levels of efficiency represents a sizable opportunity for major short-term reductions in energy demand and CO_2. However it is very important that we continue to develop new and better technologies to reduce retrofit costs and to introduce market mechanisms that will induce consumers to make conservation improvements sooner rather than later.

Considerable opportunities exist for applying total system solutions to providing energy services in new buildings. Since some conservation measures are expensive and more difficult (or even impossible) to install once a building is constructed, it is worth installing efficiency measures at the time of construction in order to avoid the "lost opportunity" associated with the inability to install them later, when energy prices rise.

Many different types of efficiency technologies are available, or could be developed with relatively inexpensive research effort. However, just developing the technologies is not enough; they also need to be applied. Because of market barriers in the buildings sector (as described later in this section), various market mechanisms must also be used.

Figure 3.5 and Table 3.2 present a conservation supply curve of potential electricity and CO_2 savings and costs for conservation measures in the U.S. buildings sector (NAS, 1992). Conservation supply curves relate energy savings achieved by implementing a given efficiency measure to that measure's "cost of conserved energy" (CCE). CCE is the annualized investment divided by the annual energy saved (Meier et al., 1983). The curve in Figure 3.5 shows 12 steps graphed in order of increasing CCE. These add up to a savings of 734 kWh, which is 45% of the 1989 U.S. buildings' electricity consumption.

Annually, $140 billion is spent on electricity in U.S. buildings. The annual savings shown in Figure 3.5 (shaded) vary from $10 billion to nearly $40 billion, depending on the price of the avoided electricity. In this supply curve, the height of each step (y-axis) is the annualized cost of conserving each unit of electricity (expressed in ¢/kWh), and the width of each step (x-axis) is the savings (in kWh), so the area underneath each step is the annualized cost (in $ billions). The shaded area above each step (between the step and the horizontal lines labeled by price of avoided electricity) is the annual savings. Three different prices or costs are shown:

1. The top line is the 1990 average price of electricity to U.S. buildings, which represents the consumer's point of view. A rational consumer would invest through step 12.

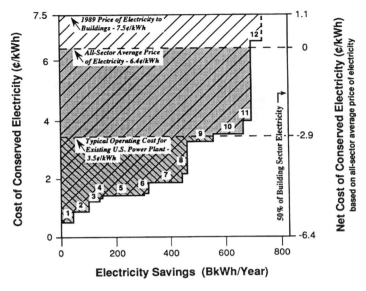

Potential Net Savings:

$37 B/yr. — Case 1: Below 7.5¢/kWh 1989 Price of Electricity to Buildings

$29 B/yr. — Case 1: Below 6.4¢/kWh All-Sector Average Price of Electricity

$10 B/yr. — Case 3: Below 3.5¢/kWh Typical Operating Cost for Existing
US Power Plant

Figure 3.5   Cost of conserved electricity (CCE) for buildings. The full x-axis corresponds to 813.5 BkWh, which is half of the total 1989 U.S. buildings electricity use of 1627 BkWh and which cost $140 billion. The net CCE scale is displaced by 6.4¢/kWh — the all-sector average price of the avoided electricity. All recommended measures that have a CCE of less than 6.4¢/kWh have a negative cost, i.e., save money. Areas between the CCE and a price line represent annual dollar savings. Case 1 (lightly hatched area) shows this potential annual net savings of $37 billion, based on the average price of the avoided electricity of 7.5¢/kWh. Case 2 (shaded area) represents the potential annual savings of $29 billion, based on the all-sector avareage price of 6.4¢/kWh (defined as Net CCE of 0 on the right-hand scale). To be extremely conservative, the net CCE can be referenced to the avoided cost of merely operating an existing plant — about 3.5¢/kWh at the meter. Case 3 (heavily hatched area) represents this most conservative estimate of savings of $10 billion/year. (From NAS, 1992. With permission.)

2. The middle line is the 1990 average all-sector price of electricity. Since conserved kWh cannot necessarily be assigned to the buildings sector, this shows a more societal point of view. A rational society would then invest up through step 11.

3. The bottom line is the marginal cost of operating an existing coal plant. This cost is relevant to a region that has excess

Table 3.2   The cost of saving electricity and carbon dioxide through conservation in buildings.

	CCE ¢/kWh d=0.06	Net CCE ¢/kWh (A-6.4[a])	Net CCCO$_2$ $/tonne (10xB/0.7[b])	Potential U.S. Elect. Savings *Cumulative TWh/yr*	Potential U.S. CO$_2$ Savings *Cumulative Mt CO$_2$/yr*
White Surfaces + Urban Trees[c]	0.5	-5.9	-84	45	32
Residential Lighting[d]	0.9	-5.5	-79	101	71
Residential Water Heating	1.3	-5.1	-74	139	97
Commercial Water Heating	1.4	-5.0	-72	149	104
Commercial Lighting	1.5	-5.0	-71	315	221
Commercial Cooking	1.5	-4.9	-70	322	225
Commercial Cooling	1.9	-4.5	-64	437	306
Commercial Refrigeration	2.2	-4.2	-60	459	321
Residential Appliances	3.3	-3.1	-44	562	393
Residential Space Heating	3.7	-2.8	-39	667	467
Com. & Ind. Space Heating	4.0	-2.4	-35	689	482
Commercial Ventilation	6.8	.4	6	734	514
US Buildings Sector 1989 Use				**1627**	**1140**

Note: The measures tabulated are the 11 steps from the EPRI curve, with an additional first step for white surfaces/urban trees (Faruqui, 1990). Net savings (Net CCE) are based on all-sector average price of electricity of 6.4¢/kWh. Columns A (Cost of Conserved Electricity-CCE) and D are plotted in Figure 3.5. Column C presents the savings in units of CO_2 (Net Cost of Conserved Carbon Dioxide-CC CO_2) and Column E presents the total potential CO2 savings (using the conversion 1 kWh = 0.7 kg CO2 so 1 TWh = 0.7 Mt CO_2).

[a] All-sector average price of electricity.

[b] Carbon emissions from J.A. Edmonds, A Preliminary Analysis of U.S. CO2 Emissions Reduction Potential from Energy Conservation and the Substitution of Natural Gas for Coal in the Period to 2010, DOE/NBB-0085, February 1989.

[c] Akbari, H., Huang, J., Martien, P., Rainer, L., Rosenfeld, A., and Taha, H. "The Impact of Summer Heat Islands on Cooling Energy Consumption and Global CO_2 Concentration," Proceeding of ACEEE 1988 Summer Study on Energy Efficiency in Buildings, Vol 5, pp. 11023, Asilomar, CA, August 1988; and Akbari, H., Garbesi, K., and Martien, P. Controlling Summer Heat Islands, Proceedings of the Workshop on Saving Energy and Reducing Atmospheric Pollution by Controlling Summer Heat Islands, Berkeley, CA, Lawrence Berkeley Laboratory, 1989.

[d] EPRI measures 2-12 have been converted from 5% discount rate to 6% discount rate by multiplying EPRI CCE values by 1.05. EPRI values for "Potential U.S. Electricity Savings (Cumulative)" have been expanded by a factor of 1.4 to account for EPRI's failure to include improvements for which they had no cost data (residential residual appliances and commercial miscellaneous equipment) and to adjust the EPRI savings, which were compared to utility projections that included 9% naturally occurring efficiency improvements, to frozen efficiency (See also Rosenfeld et al., 1991).

electricity, and that is not interested in avoiding the cost of new capacity. The local utility would then encourage its customers to invest only up through step 9.

Thus we see that even from a utility's perspective, there are significant electricity conservation opportunities available at a net negative cost.

3.3.1.3 Energy Technologies

Some recent advances in energy efficient technologies for six end uses are described below.

Lighting.

Commercial buildings use 30% of all U.S. electricity. Lighting is responsible for about 35% of electricity consumption in commercial buildings, about 10% in residential buildings, and 10% in industrial buildings.

Over the past decade, major advances have been made in energy efficient lighting technologies through public and private sector R & D programs. Applying existing lighting technology — more efficient lamps, ballasts, fixtures, and controls to turn off lights when a room is unoccupied or dim them when daylight is available, as well as daylighting design concepts — could substantially reduce commercial building energy use (Atkinson et al., 1992). Reducing lighting electricity use can also reduce cooling load uses resulting in savings from lower energy use and equipment downsizing. (In some climates these savings may be offset by increased heating.)

For example, compact fluorescent lamps, which in many cases can be used to replace incandescent lamps, cut lighting energy use up to 75% and reduce maintenance costs because they last 10 times longer. See Section 3.5 for a description of successful research and development contributions to the advancement of this technology.

A key impediment to achieving these savings is the lack of education for lighting designers, building operators, and others involved in lighting decisions. Effective lighting design requires a knowledge of the available options and how they can be used to achieve both proper lighting levels and energy savings. To address this problem, some colleges are developing illuminating engineering or architecture programs. Government agencies and professional organizations are conducting training programs for building operators. Utilities are establishing lighting demonstration laboratories where designers, owners, and occupants can learn about new technologies.

Window Systems.
   Window technology has advanced greatly in recent years. Many existing buildings were originally built with single pane windows with a thermal resistance (R) of 1. After World War II, "efficient" double pane (R-1.5 to R-2) windows were developed. These have now been enhanced by applying "low-emissivity" films to the glass surfaces or stretching low-emissivity treated films between glass panels, and filling the window cavity with a heavy insulating gas. These technologies have resulted in the development of "super windows" (R-8 at center of glass, R-5 to R-6 including the frame, depending on window size) that are being used in colder sections of the country. In Minneapolis in January, even a north-facing super window admits and retains more solar heat than it loses from inside, producing a net energy gain (Bevington and Rosenfeld, 1990).
   The low-emissivity film transmits the visible portion of sunlight but reflects the "near infrared" heat portion. These "selective windows" allow offices to take better advantage of daylighting while reducing cooling load (Rosenfeld and Price, 1991). Section 3.5.2 contains a description of successful research and development contributions to the advancement of this technology.

Control Systems.
   Electronic controls can reduce building energy use in many ways. Energy management computer systems in commercial buildings can efficiently control HVAC (heating, ventilation, air conditioning) equipment, lighting, and other systems. Photocell controls can be used to dim electronic ballasts to take advantage of daylight and maintain a constant light level. Occupancy sensors can turn off lights when a room is vacant, and CO_2 sensors can reduce ventilation rates when occupants are not present, especially in high occupancy areas such as assembly halls and conference rooms. With decreasing costs such systems may also be introduced in the residential sector.

Office Equipment.
   Increasingly, office equipment — computers, printers, faxes, copiers — is a major contributor to commercial building energy

growth. This equipment not only requires electricity, but also gives off excess heat that increases building cooling requirements. There is a great deal of variation in the energy use of this equipment, but consumers are typically not aware of this variation. The EPA "Energy Star" rating system helps make informed energy use decisions concerning purchase or efficient operation. Fortunately, the demand for laptop, battery-powered computers has increased their efficiency tenfold since there is motivation to extend operating time between recharges.

Refrigerators and Freezers.

The efficiency of refrigerators has varied greatly over the last 40 years. In 1950, refrigerators used about 650 kWh per year. This increased to about 1,800 kWh per year by 1975 because motor efficiency and wall thickness were reduced and defrost cycles and various convenience items were added. Since refrigerators use 15% of electricity in homes, such efficiency loss was significant. Since 1977, this trend has been reversed due to standards that first took effect in California, and then were adopted nationally in 1990. The California standard was 1000 kWh per year for a 20 cubic foot refrigerator. In 1993 a new national standard took effect that reduced the average energy use of new refrigerators to about 690 kWh per year.

Recent concern regarding ozone depletion and global warming has focused attention on the chlorofluorocarbons (CFCs) used in refrigerator insulation (CFC-11) and refrigerant fluids (CFC-12). These CFCs will be phased out by the U.S. Clean Air Act by 2000 or earlier. Fortunately, new technologies will be available to address this serious problem. For instance, HCFCs, which have lower environmental degradation properties, can replace the working fluid in refrigerators. Other refrigerants completely free of CFCs are being developed. CFCs can be further reduced by using better non-CFC insulation, also allowing the use of smaller compressors.

Heating and Cooling Systems.

Enormous efficiency gains have been made in the design and manufacture of almost all HVAC technologies. For instance, the highest available annual fuel utilization efficiency (AFUE) (Wilson, 1991) of new gas furnaces has risen from about 65% to above 90%. This was achieved by new combustion designs that incorporate pulsed fuel feed systems that improve heat transfer and recover heat from exhaust gases by condensing water. New air conditioning system efficiency has improved dramatically: the highest available seasonal energy efficiency ratio (SEER) (Wilson, 1991) has improved from 6 to 15 over the last 20 years. Increasingly, new technologies that may have major impacts are moving into the marketplace. For instance, ground source heat pumps will soon compete with new thermally-powered heat pumps now being field tested. Space cooling is growing because more construction is occurring in hot climates and added office equipment produces additional internal gains in commercial buildings. More efficient cooling systems such as evaporative coolers and natural gas-fired chillers for office buildings will make important contributions to reducing CO_2 emissions.

3.3.1.4 Barriers

In many ways energy efficient products are no different from other consumer products because market penetration is not instantaneous and requires good marketing techniques. However, there are a number of institutional and market barriers that inhibit the optimal use of energy efficiency measures in buildings. These barriers are the major reason that our stock of buildings and appliances is significantly less energy-efficient than would be economically optimal. Four of the major barriers are:

Improper Price Signals.

The cost of energy does not include the cost of environmental impacts from energy production and use, or the cost of national security considerations. To the extent that energy prices do not reflect such costs, consumers and planners undervalue the benefits of energy conservation.

Lack of Capital.

Builders and homeowners typically try to minimize the first cost of buildings and appliances. This is due to the lack of adequate capital at the time of purchase and the desire to spend available capital on amenities (upgraded carpets in new homes, through-the-door ice makers in refrigerators, etc.). The lack of capital is especially acute for low-income households, which tend to have high energy bills because of inability to purchase efficient homes and appliances. State, local, and the federal governments find it hard to pay for efficiency improvements when faced with competing demands for expenditures on schools, health care, and infrastructure.

Lack of Consumer Information, High Discount Rates.
Consumers lack reliable information on the energy and cost savings from efficiency options. While major appliances such as refrigerators are labeled, many other appliances and equipment are not. Since customers lack information, they typically require very fast paybacks on investments in efficiency. In other words, they discount future benefits, including energy savings from more expensive products, at a very high rate.

Landlord/Tenant and Builder/Owner Split.
Since landlords and builders do not pay the future utility bills, they have inadequate incentive to invest in efficient buildings or appliances. A requirement for energy rating of homes at time of sale, as well as on rental property, would help to overcome this problem.

3.3.1.5 Delivering Efficiency to Consumers
The tools for delivering efficiency — information programs, standards, and utility programs to beat the standards — are briefly discussed in this section.

Consumer Information Programs.
Consumers are seldom knowledgeable about the energy efficiency of products they purchase. In most cases, the lifetime energy cost of a major appliance is more than the original cost of the product. Consumers can be greatly assisted in making informed decisions through labeling and rating programs systems that provide clear information about equipment energy costs. As mentioned above, the EPA has sponsored an "Energy Star" labeling program for office equipment.

In order to capture more potential energy savings beyond those achieved through standards, it is necessary to overcome consumer emphasis on minimizing first cost by educating consumers on life-cycle costs (considering energy costs over a product's lifetime). Labeling and information programs for windows, lamps, and luminaries are included in the Energy Policy Act of 1992 (EPAct).

Residential energy efficiency rating and energy efficient mortgages are part of the legislation. Funding for energy efficient lighting and building energy centers, as well as federal energy efficiency information programs, are included.

The largest financial decision most consumers make is buying a home. After mortgage costs, energy is the next largest cost of home ownership. Home energy rating programs allow consumers to compare the relative energy cost implications of their home-buying decisions. These programs can also be linked to energy efficient home mortgage programs.

Training and Education

In order to actually achieve the savings discussed above, many participants in the building community must be educated about efficient design and operation. For effective implementation of efficiency standards, designers, builders and building officials must be trained regarding compliance and enforcement. Experience shows that setting standards without training and education greatly diminishes the value of efficiency standards.

Building Energy Standards

Most buildings have a useful life of 50 years or more, so the original building design has an enormous impact on the amount of energy consumed over its lifetime. Building standards (or codes) are an important means to ensure that buildings are designed and constructed with an eye towards energy efficiency. Over the past few years the engineering community has upgraded residential and commercial building standards through the American Society of Heating, Refrigerating, and Air-Conditioning Engineers (ASHRAE). The Council of American Building Officials (CABO) translates the ASHRAE standard into the Model Energy Code for use by local building departments for residences.

Appliance and Equipment Standards

In many cases, it may be far easier and more effective to overcome consumer emphasis on minimizing first cost by removing or discouraging inefficient products from the market rather than by trying to educate consumers on how to make individual energy purchase decisions. A recent study compared various policy options and found that setting standards results in more savings than other methods, including tax credits, rebates, and consumer education (U.S. Congress, 1991).

Currently, the federal government has set energy efficiency standards for 13 major home appliances and fluorescent ballasts. The EPAct sets standards on additional products such as lamps, motors, commercial heating, cooling, and water heating equipment, plumbing products, and distribution transformers. While initial savings will be modest from the first generation of standards, several of these standards will undergo periodic updates, resulting in greater energy savings.

Other Federal Programs

Other provisions of EPAct relating to energy efficiency include state adoption of commercial and residential building energy efficiency standards, federal building energy efficiency programs, district heating and cooling studies, grants to state regulatory agencies, assistance to state energy conservation and low-income programs, encouragement of efficiency investments by electric and gas utilities, advanced appliance development, study of early retirement of inefficient appliances, and industrial energy efficiency programs. These are good first steps toward comprehensive federal incentives toward U.S. energy efficiency in all sectors of the economy.

Electric and Gas Utility Programs

Many electric utilities and increasingly, some gas utilities are operating demand-side management programs to improve customer end-use efficiency in order to delay or eliminate the need for new supply investments. These programs are becoming major providers of energy conservation services. Utility regulations in some states (California and the Northeast) have been drastically reformed recently to provide utilities with enhanced profits from selling efficiency instead of energy. Utility conservation programs are likely to grow significantly as states encourage utilities to implement "integrated resource planning" in which supply and demand-side energy sources are evaluated on an equivalent basis.

3.3.1.6 Research, Development and Demonstration (RD & D) Opportunities

In the public hearings held in support of the development of the National Energy Strategy, the following goals were identified for residential and commercial buildings:

1. Achieve cost-effective reductions in energy requirements for the nation's new and existing buildings through improved designs, materials, and construction;
2. Achieve cost-effective energy efficiencies for building equipment and appliances; and
3. Maintain indoor environments conducive to the health, safety, productivity, and comfort of occupants (U.S. DoE, 1990).

Despite its importance to the U.S. economy (U.S. DoE, 1990), the building industry spends relatively little on research. (Building construction accounts for 10% of U.S. GNP, building finance for 25%, and building energy consumption is 4% of U.S. GNP.) The U.S. building industry invests less than 0.33% of its revenues on research, the lowest of any industrial sector. For comparison, the U.S. pharmaceutical industry spends about 7% of gross sales on research, the aerospace industry 4%, and motor vehicles 3.2%. The Japanese building industry reinvests 3% of its revenues in R & D (U.S. Congress, 1987).

The decentralized and cyclical nature of the building sector discourages firms from supporting long-term research and development. Innovation in U.S. industry is slow (Gilleard, 1989), and productivity is stagnant and may have even declined in recent years, in part due to insufficient R & D (National Research Council, 1986). The U.S. balance of trade in building products has deteriorated from approximate balance in 1980 to a negative $1 billion in 1987 (Kelly, 1989).

The small Federal R & D program ($35 million per year), leveraged through cost-sharing with industry, has led to major advances in building technologies and practices, producing significant savings in energy and consumer expenditure, with extremely high benefit/cost ratios.

The technologies discussed above can be developed much further. For example, the average lighting efficiency in the U.S. is about 25 lumens/watt, and scientists expect to have new technologies that will have efficiencies of 180 to 200 lumens/watt. Gas furnaces are now on the market that have efficiencies of almost 100%, but new gas heat pump technologies will shortly be available that push system efficiencies far above 100%.

The following summarizes both the near and long-term research and development needs in the buildings sector. For further discussion of technologies that have already been developed through successful RD & D programs, see Section 3.5 and Table 3.14 of this chapter.

Near term RD & D needs include the following:

- *Energy Conversion Technologies.* Continue to improve the efficiency of advanced heat pumps, and gas cooling and refrigeration systems; develop alternative, non-CFC refrigerants; increase engine durability and performance for gas cooling systems; and improve thermal storage systems.
- *Building Components and Systems.* Improve controls and electronics; develop low-energy appliances for cooking, clothes drying, and dish washing; develop non-CFC blowing agents for foam insulation, advanced windows, improved window frames, smart windows (electro-, photo-, and thermochromic), and walls (transparent or translucent insulation).
- *Design and Practice.* Develop "expert systems" for designing, commissioning, and operating buildings; perform social science research to understand consumer behavior.

Long-term RD & D needs include the following:

- *Energy Conversion Technologies.* Develop absorption heat pump, cooling, and refrigeration cycles; develop absorption cycles driven by solar energy or a combination of natural gas, biomass-derived methane, or hydrogen; develop photovoltaic systems, coatings for windows, advanced co-generation systems, heat engines, and solid oxide fuel cells. (More detail on supply technologies is found is Chapter 4).
- *Building Components and Systems.* Develop "super insulation" for walls, windows, and roofs, advanced refrigeration systems and hot water storage systems; advanced lighting technologies such as electrodeless fluorescent and high intensity discharge lamps, gas-filled panels for refrigerator walls, electronics for building controls and office equipment, and efficient construction materials.

- *Design and Practice.* Conduct research on building design and human productivity, diagnostic technologies for buildings, and community design planning techniques.

To emphasize the significant energy savings potential in the buildings sector, we quote a report prepared as supporting analysis for the National Energy Strategy: "with technology advances supported through effective research, development, and commercialization efforts, the new house of 2025 should not have to rely on an external power supply." (International Energy Agency, 1989).

3.3.2   Industrial Sector
3.3.2.1 Current Consumption and Historical Trends
In 1990 the industrial sector energy used 30 quads and was responsible for one third of U.S. CO_2 emissions. Half of these emissions is from the direct combustion of fuels and half comes from purchased electricity. Industry, the most heterogeneous of the demand sectors, is comprised of manufacturing, agriculture, mining and construction. Domestic manufacturing (defined by Standard Industrial Classification codes 20 to 39) consumes 80% of industrial energy use and therefore will be the focus of this discussion.

Energy is consumed for a variety of purposes in manufacturing. The largest end uses are steam, feedstocks, process heat, and machine drive. Each of these has different characteristics in terms of distribution among industries and fuel mixes. Some end uses (e.g., heat for glass melting or electrolysis for aluminum) are very process-specific. Other services, such as steam, are more generic, and apply to most types of manufacturing. Machine drive is supplied primarily by electricity. Feedstocks (e.g., natural gas), used in the chemical industry, do not contribute to carbon emissions since the carbon is bound into a product rather than released through combustion.

The energy intensity of manufacturing processes varies by over a factor of 15, from 0.9 to 16.6 thousand BTUs per 1980 dollar value of shipments. The four most energy-intensive industries (paper, chemicals, petroleum refining, and primary metals) together account for 75% of manufacturing energy use, and each produce 14 to 19% of manufacturing CO_2 emissions. Electricity consumption is not as concentrated in specific industries, and the four most energy intensive industries account for just less than 60% of all direct electricity consumption.

Since 1972 the energy intensity of manufacturing has declined by 40%. Manufacturing output increased by 45%, while energy use declined by 12% over the period. The drop in fossil fuel intensity was about 50%, while electricity intensity remained fairly constant. Several factors have contributed to these trends, including structural change, process change, and efficiency improvements.

Structural change in the U.S. economy, which has resulted in a shift in growth away from energy intensive industries, is estimated to account for one third to one half of this intensity drop. In part, this was due to a shift in the product mix of demand. For example, plastics have replaced metals in many applications, including automobiles; the chemicals industry is producing more high value/low energy intensive goods, such as drugs, rather than energy intensive commodity chemicals; and the electronics industry has boomed. In part, structural change resulted from greater imports of energy intensive goods, thus reducing U.S. energy use and CO_2 emissions but not necessarily global emissions. The continued trend toward global markets for fuels and products blurs the accounting of CO_2 emissions to different nations. The energy embedded in U.S. non-energy imports rose from approximately 3 quads in 1972 to 7 quads in 1985. More recent evidence from the period 1985 to 1988 suggests that these structural changes may be shifting back toward more domestically produced, energy intensive products and industries (U.S. Congress, 1991).

The processes used to manufacture goods are continually changing with technological innovation. Over the last two decades this has led to energy efficiency gains. These changes often are initiated in order to improve overall production efficiency, not necessarily energy efficiency per se. There has been greater use of input recycling, such as using waste biomass or pulping liquor in the paper industry. The use of co-generation (the simultaneous production of heat and electricity) has increased dramatically, leading to lower energy consumption. On the other hand, many process changes involve greater use of electricity. These include thermo-mechanical pulping in the paper industry, electric arc furnaces in steel production, microwave drying, and robotics. This electrification tended to offset structural shift and energy efficiency improvements and has led to a slight increase in electricity intensity rather than the decline in intensity experienced by fossil fuel use.

Equipment energy efficiency improvements, as opposed to efficiency improvements resulting from more efficient processes, have also played a major role in decreasing energy intensity. A variety of new technologies such as high efficiency motors, adjustable speed drives, industrial heat pumps, and advanced sensors and controls have become available. Simple energy management measures for processes and buildings have also produced energy savings.

3.3.2.2 Technical Potential For Energy Savings

Significant opportunities exist in the manufacturing sector to reduce future annual emission levels of CO_2 through a combination of options including enhanced end use efficiency, fuel switching co-generation of heat and power, and process conversion and redesign.

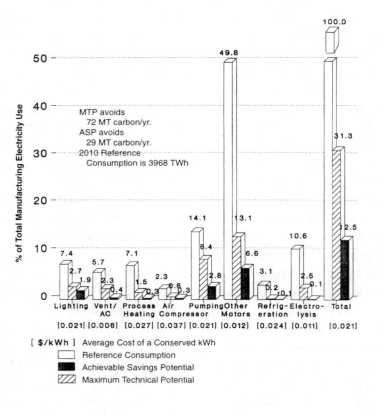

Figure 3.6   Electricity savings potential in manufacturing — 2010. (From R. Ciliano, F. Stern, and C. Lang, Electric and Gas Utility Modeling System (EGUMS), "DSM as Best Available Control Technology (BACT)" Policy Analysis, Prepared for the U.S. Environmental Protection Agency by RCG/Hagler, Bailly; May 1991.)

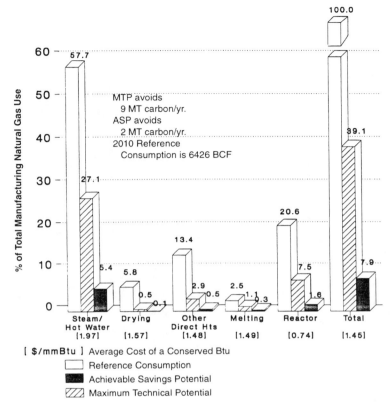

Figure 3.7   Natural gas savings potential in manufacturing — 2010. (From R. Ciliano, F. Stern, and C. Lang, Electric and Gas Utility Modeling System (EGUMS), "DSM as Best Available Control Technology (BACT)" Policy Analysis, Prepared for the U.S. Environmental Protection Agency by RCG/Hagler, Bailly; May 1991.)

Enhanced End Use Efficiency

Enhanced end use efficiency can be achieved through a variety of measures that encompass technology substitution and changes to facility and equipment operating and maintenance practices. Estimates recently developed by the U.S. Environmental Protection Agency (EPA) suggest a large potential to conserve both electricity and natural gas in manufacturing (Ciliano et al., 1990).

Figures 3.6 and 3.7 show the sector-wide maximum technical savings potential (MTSP) by major manufacturing end-uses for electricity and natural gas, and the fraction believed realizable through a combination of market forces, government policies and utility programs. It should be noted that the government policies examined in this analysis did not include mandatory equipment efficiency standards.

Table 3.3 Industry and end-use specific electricity savings potential. (From R. Ciliano, F. Stern, and C. Lang, Electric and Gas Utility Modeling System (EGUMS), "DSM as Best Available Control Technology (BACT)" Policy Analysis, Prepared for the U.S. Environmental Protection Agency by RCG/Hagler, Bailly; May 1991.)

Technical Potential Savings
(as a Percentage of Reference Consumption in 2010)

End–Use	Food 20	Paper & Pulp 26	Chemicals 28	Petroleum Products 29	Primary Metals 33	Non-Electric Machinery 35	Transportation Equipment 37	All Manufacturing 20–39
Lighting	37.2	37.2	37.2	37.2	37.2	37.2	37.2	37.2
Vent/AC	39.6	39.6	39.6	39.6	39.6	39.6	39.6	39.6
Process Heating	9.0	9.0	27.1	24.0	40.5	9.0	9.0	22.4
Air Compression	27.7	--	27.7	--	--	27.7	27.7	27.7
Pumping	59.9	59.9	59.9	59.9	59.9	59.9	59.9	59.9
Refrigeration	4.8	--	4.8	4.8	--	--	--	4.8
Motors, Other	26.2	26.2	26.2	26.2	26.2	26.2	26.2	26.2
Electrolysis	--	--	24.0	--	24.0	--	--	24.0
Total	29.9	36.5	41.2	48.6	26.4	28.6	27.8	32.3

Achievable Potential Savings
(as a Percentage of Reference Consumption in 2010)

End–Use	Food 20	Paper & Pulp 26	Chemicals 28	Petroleum Products 29	Primary Metals 33	Non-Electric Machinery 35	Transportation Equipment 37	All Manufacturing 20–39
Lighting	25.7	25.7	25.7	25.7	25.7	25.7	25.7	25.7
Vent/AC	6.6	6.6	6.6	6.6	6.6	6.6	6.6	6.6
Process Heating	1.3	1.3	4.5	3.9	7.3	1.3	1.3	3.7
Air Compression	16.6	--	16.8	--	--	16.8	16.8	16.8
Pumping	19.6	19.6	19.6	19.6	19.6	19.6	19.6	19.6
Refrigeration	1.1	--	1.1	1.1	--	--	--	1.1
Motors, Other	13.3	13.3	13.3	13.3	13.3	13.3	13.3	13.3
Electrolysis	--	--	10.1	--	10.1	--	--	10.1
Total	12.6	15.6	15.2	17.1	11.4	14.3	14.2	13.7

Table 3.4   Industry and end-use specific natural gas savings potential. (From R. Ciliano, F. Stern, and C. Lang, Electric and Gas Utility Modeling System (EGUMS), "DSM as Best Available Control Technology (BACT)" Policy Analysis, Prepared for the U.S. Environmental Protection Agency by RCG/Hagler, Bailly; May 1991.)

Technical Potential Savings
(as a Percentage of Reference Consumption in 2010)

End-Use	Food 20	Paper & Pulp 26	Chemicals 28	Petroleum Products 29	Stone Clay Glass 32	Primary Metals 33	All Manu-facturing 20-39
Direct Heating	13.1	8.9	13.3	13.3	23.1	28.8	21.4
Drying	18.3	6.0	11.6	--	--	--	9.1
Melting	5.0	5.0	--	--	35.6	52.4	43.8
Reactor	--	5.0	37.5	37.5	47.8	37.5	36.3
Boiler	53.4	49.5	45.1	45.1	45.1	45.1	46.9
Total	44.1	34.1	39.2	42.7	39.9	37.6	39.0

Achievable Potential Savings
(as a Percentage of Reference Consumption in 2010)

End-Use	Food 20	Paper & Pulp 26	Chemicals 28	Petroleum Products 29	Stone Clay Glass 32	Primary Metals 33	All Manu-facturing 20-39
Direct Heating	2.3	1.7	2.2	2.2	4.2	5.1	3.8
Drying	2.9	0.9	1.8	--	--	--	1.4
Melting	1.2	1.2	--	--	7.5	12.1	9.9
Reactor	--	1.2	8.1	8.1	10.4	8.1	7.9
Boiler	10.2	9.7	9.1	9.1	9.1	9.1	9.4
Total	8.3	6.8	8.1	8.7	8.4	7.5	7.9

These figures show savings in terms of the cost of conserved energy (CCE); that is, as the average cost of a conserved kWh or of a conserved Btu. CCE is the annualized investment divided by the annual energy saved (Meier et al., 1983). The CCEs can then be compared to the 1990 average industrial price of electricity of 0.047 $/kWh and the 1990 average industrial price of natural gas of $3.00/mmBtu. Tables 3.3 and 3.4 are matrices assigning the savings of Figures 3.6 and 3.7 to individual major industries.

Aggressive nationwide implementation of efficiency programs in manufacturing could result in a reduction of slightly over one third of projected sector-wide consumption levels by 2010. However, the lower level of "attainable" savings shown reflects the inhibiting effect of important market barriers that need to be ameliorated before savings levels approaching MTSP can be achieved.

Fuel Switching

The major fuel-switching opportunities in manufacturing are from replacing coal-fired boiler consumption with oil and gas. Depending on the quantity of coal that can be feasibly switched and the amount of oil and gas that replaces that coal, this could result in annual reductions of 10–20 MT of CO_2 out of total U.S. emissions of 5 GT. The cost-effectiveness of such conversions is heavily influenced by current and prospective oil and gas prices relative to coal and by the numerous site-specific modifications that would be necessary to effect dual-fuel firing and to install storage capability for oil.

Switching from fossil fuels to biomass wastes or process related by-products appears to have limited potential beyond industries that extensively use such sources already (e.g., food processing, paper and pulp). Further expansion of this potential would require design of comprehensive regional networks to logistically gather and transport forest and field crop residues or industrial waste products. Such movements of high-volume, low-value materials coupled with the poor combustion efficiency of biomass fuels renders this option relatively implausible given the current infrastructure and technology.

Co-generation

Figure 3.8 summarizes the distribution of projected installed co-generation capacity in 2010 by industry and the corresponding steam production and reduced CO_2 emissions this level of implementation

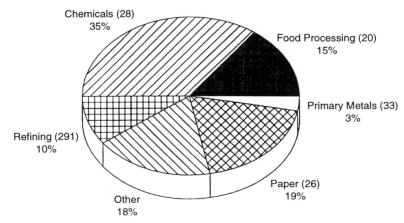

Chemicals (28)
35%

Food Processing (20)
15%

Primary Metals (33)
3%

Refining (291)
10%

Paper (26)
19%

Other
18%

Figure 3.8   Installed cogeneration potential in manufacturing — 2010. [Distribution of installed capacity (by SIC) • 64 GW reference installed capacity • 2.2 TBTu steam production • 75 Mt/year CO_2 avoided]. (From R. Ciliano, W. Steigelmann, and S. Williams, "Customer Purchase Criteria for Small-Scale Cogeneration — A Historical Perspective," prepared for Electric Power Research Institute, Customer Systems Division by RCG/Hagler, Bailly; August 1991.)

would yield. This rough estimate of 75 MT per year of avoided emissions is based upon the assumption that gas-fired co-generation (historically the fuel of choice by cogenerators) replaces gas-fired central-station generation, and wherein co-generation system sizing is predicated upon thermal load-following operation.

Process Redesign

A number of studies have identified the potential to displace significant quantities of fossil fuel with electricity through process-wide modifications in a way that could result in lowered CO_2 emissions (U.S. DoE, 1990; Ciliano et al., 1991; Felix et al., 1990; Ciliano, 1987; Miller et al., 1989). Table 3.5 displays results from an earlier analysis conducted as part of an electricity options study conducted in Michigan (Ross, 1987). This study analyzed in part the net "fossil energy displacement rate" per unit of added electric load. This displacement rate is adjusted upward by 8–10% for transmission and distribution losses to reflect a primary consumption equivalent at the generator. Even then, in about half the industrial sectors analyzed, the maximum process-wide substitution of electro-technologies would have resulted in a net reduction of primary energy use. For example, in primary metals one can displace up to 80,000 Btu/kWh of natural gas with only one kWh of electricity at 11,500 Btu/kWh. The corresponding impact in lowered CO_2 emissions depends on the mix of fossil fuels used at the affected facilities vis-a-vis the fuel mix of the electric generating plants supplying the increased power to those plants.

DoE studies document that of the 30 quads of energy consumed by industry, 9 are associated with "waste heat" and other inefficiencies (U.S. DoE, Industrial Technologies, 1991). The large volume of industrial wastes generated as solid waste, contaminated water, and air pollutants creates considerable energy losses as (1) energy embedded in the waste material; (2) energy waste in the form of poorly used raw materials; and (3) further indirect energy requirements. Significant opportunities exist for waste minimization through process optimization, and waste utilization through recycling, especially in the production of paper, glass, organic chemicals, and petroleum refining. Such changes can help industry reduce the costs of meeting EPA environmental standards.

The feasibility of process reconfiguration and the time line over which it may occur will be highly conditioned by the specifics of each conversion. In some instances recycled materials contain impurities that will require significant pre-processing and treatment. In other cases, because of recycling rates that are higher than product demand, new markets will have to be created for the products manufactured from recycled materials.

3.3.2.3 Barriers

Although industry has made significant energy reductions in the past and there is a large technical potential for further reductions, the full potential may not be realized due to barriers to new energy technology adoption. Market diffusion dynamics slow implementation. Because it is generally not cost effective to replace operating equipment before the end of its useful life, efficiency improvement investment decisions closely track normal equipment replacement cycles. If technologies are not available or are not deemed cost effective at those replacement times, then that opportunity for technology/process efficiency improvement will be lost.

In adopting new energy efficiency improvements, a total engineered systems approach is often necessary, rather than focusing on individual pieces of equipment. For example, more savings can often be achieved by redesign or improved maintenance of all driven equipment in a process line than by simple installation of a high efficiency motor.

For new technologies there is also a penetration lag due to the time needed for information to be disseminated from researchers to production managers. Furthermore, the technology must be well proven before it will be widely adopted. Implementation that might disrupt production processes often imposes unacceptably high levels of risk to would-be adopters. For this reason demonstration projects can be very important.

Industrial companies also may not implement all cost effective new energy efficient technology options available to them. Acceptable payback times may be only one to three years despite the fact that the equipment will yield energy savings over a much longer period. Cost cutting measures, especially energy investments, usually receive a lower priority for scarce investment funds than improvements that boost output or expand markets. For many industries this is particularly true because energy expenditures are not a large percentage of total production costs.

The maximum technical potential for energy savings probably exceeds that which is cost effective. "Transaction costs" and other associated costs are often quite high and these are sometimes ignored by those creating engineering estimates. On the other hand, energy prices that are used in determining payback times do not reflect the externality costs associated with using these fuels.

3.3.2.4 Research, Development and Demonstration (RD & D) Opportunities

In order to achieve the identified potential for energy efficiency improvements in the industrial manufacturing sector, and in response to the barriers to implementation of these improvements, aggressive technology and process R & D is required. Options open to industry to improve efficiency and reduce greenhouse gas emissions appear to generally fall into the four areas generally discussed above, which include the following:

1. Innovative technology implementation (including co-generation)

2. Improved operating procedures and inherently more efficient manufacturing processes
3. Fuel switching (including use of bio-mass fuels and feedstocks)
4. Recycling of materials for use as feedstocks.

In order for any of these options to be implemented, appropriate and cost-effective technologies, including more efficient process options, are required. Research and development opportunities leading to these technologies can be generally defined as either (1) mission-oriented, or (2) enabling. Mission-oriented research and development is defined as technology or process development and implementation that focuses on a specific opportunity (e.g., major process changes in aluminum production to eliminate either the Bayer process or the Hall-Heroult process, or both). Enabling technology R & D includes targeted, interdisciplinary research that will lead to improved understanding of the fundamental phenomena underlying industrial processes and unit operations. Enabling technology research, which is generally longer-term and higher-risk, serves as a foundation on which innovative, revolutionary, and advanced concepts can evolve. In addition, the results of enabling technology research can often be immediately useful for optimization of emerging technologies.

Industrial heat pump development and implementation provides an interesting example of the need for both mission-oriented and enabling R & D. The extensive, mission-oriented application of second-law type analysis techniques (such as "pinch" technology) would result in the identification of efficient, cost-effective placement opportunities and significant near-term enhancement of the utility of low grade waste heat. Table 3.6 illustrates energy savings resulting from this approach already identified by studies conducted by DoE and industry. In concert with this near-term, mission-oriented approach, the longer-term enabling development of innovative, CFC-free chemical working media alternatives would lead to advanced, higher-lift heat pump/refrigeration technologies capable of generally improving energy efficiency and reducing greenhouse gases throughout industry.

Another example of the potential for both mission-oriented and enabling technology R & D to work in concert can be found in the pulp and paper industry. In this industry conventional chemical pulping is dominated by the very energy-intensive kraft process (Oak Ridge National laboratory, 1989). Opportunities for higher process

Table 3.6 Summary of state-of-the-art heat pumps in industrial processes.

Process	Type of Heat Pump	Lift (°F)	Delivery Temp (°F)	% Potential Energy Saving*	Payback Years
Sugar Refining	Semi-Open	35	280	40	1.6
Liquor Distilling	Semi-Open	40	240	25	1.2
Chlor-Alkali	**No Feasible Pump**	300	300		
Aromatics: BTX Unit	Semi-Open	45	320	20	1.0
Ethylene: Hot End	Closed	45	200	13	2.6
Food: Evaporation and Drying	Semi-Open	14	230	42	0.7
Petroleum: Crude Unit	Closed	60	370	4	2.7
Urea	Closed	60	280	15	3.2
Pulp and Paper Digester	Semi-Open	40	290	11	1.3
Pulp and Paper Evaporator	Semi-Open	25	150	3	0.7

*Potential energy savings as percentage of residual load after optimal heat exchange.

efficiency exist through improved process control (better automation and improved sensors) and improved process physics and chemistry (higher production speeds). Also, the development of materials resulting in higher pressure rollers would further contribute to process efficiency improvements. Thus, in the pulp and paper industry, which uses about 3/4 ton of oil per ton of paper produced, opportunities exist for energy demand reductions through application of more efficient technologies in existing processes and the development of new processes that are inherently more efficient. Ideally, as a result of the application of both mission-oriented and enabling technologies, the industry could potentially supply all its own energy from lignin wastes and not require any purchased fuel.

The above are only two of many examples where substantial near- and longer-term energy efficiency improvements could be realized through the application of the results of a cooperative RD & D program (involving industry, government, and universities) with a focus

Table 3.7 R & D agenda for the industrial sector outlines in the 1991 DoE Multi-Year Program Plan.

R & D Category	Mission-Oriented	Enabling Research	Potential (Quads)
Advanced Industrial Concepts Combustion, materials, catalysis, thermal sciences	XX	X	
Alternative Feedstocks	XX	XX	4
Industrial Wastes Reduction, utilization and conversion	XX	XX	3–6
Enabling Materials High-temperature ceramics, etc.		XX	
Separations Distillation and drying technologies (e.g., membranes)	XX	XX	4–5
Cogeneration	XX		3
Process Heating and Cooling Combustion, heat pumps, recuperators	XX	X	4–5
Materials Processing Metals, glass, etc.	XX	XX	1
1990 Industrial Consumption			**30**

Note: X and XX indicate relative degree of emphasis

upon both mission-oriented and enabling technologies. The fact that DoE has recognized the need for both components is illustrated by the R & D agenda outlined in the Industrial Sector Multi-Year Plan (U.S. DoE, Industrial Technologies, 1991), which has had significant input from industry. Table 3.7 is a synopsis of elements of this plan, with emphasis upon both mission-oriented and enabling research that has the most significant potential for improvements to energy efficiency in the U.S. manufacturing sector. Where data are available, the potential for energy savings (as estimated by DoE) is included.

3.3.3 Transportation

3.3.3.1 Transportation Energy Demand and Engineering Concerns

U.S. transportation energy consumption in 1989 was 27.3% of national energy consumption and 63.2% of petroleum consumption. Transportation energy consumption in 1989 is shown in Table 3.8 which summarizes transportation vehicle stocks, transportation movement, and energy consumption by end use (Davis and Hu, 1991). Automotive energy use is the most significant element, although energy consumption by aircraft has been growing rapidly. U.S. automotive fuel economy has improved dramatically since the late 1970s. New car fuel economy statistics are shown in Table 3.9 for Germany, Italy, Japan and the U.S. for 1978–1988 (Davis and Hu, 1991). Figure 3.9, taken from the DoE National Energy Strategy, shows the wide range of U.S. transportation energy use forecasts (see U.S. DoE, 1991/1992).

Personal passenger vehicles (both automobiles and light trucks) account for a majority of the CO_2-related emissions from petroleum-based fuels in the transportation sector. Forecasts of sectoral energy use vary, but total energy use is expected to grow in the U.S. and especially in developing nations. In those countries the number of vehicles per capita is expected to grow dramatically as a consequence of improved living standards.

3.3.3.2 Technological Potential for More Efficient Transportation

Over 80% of vehicle energy is not used as shaft power, but is lost in overcoming internal friction in auxiliary items and in thermodynamic losses incurred as the engine converts stored chemical energy into mechanical work. The engine losses manifest as heat rejected to the coolant and exhaust. These losses (for a mix of urban, rural and highway travel) are shown in Table 3.10.

Table 3.8   Statistical summary. (From Oak Ridge National Laboratory, *Transportation Energy Data Bank: Edition 11*, (ORNL-6649), 1991.)

Vehicle stock (thousands)

	New sales	In use
Automobiles, 1988	**10,625**	**121,519**
Two seater	1.7%	2.3%
Minicompact	0.8%	3.6%
Subcompact	19.0%	23.3%
Compact	39.2%	22.7%
Midsize	26.2%	28.3%
Large	13.1%	19.8%
Trucks, 1988	**5,149**	**50,222**
Light	93.2%	91.1%
Medium	0.9%	2.3%
Light-heavy	1.0%	1.7%
Heavy-heavy	4.9%	4.1%
Motorcycles, 1988	**710**	**4,584**
Buses, 1988	**580**	
Aircraft[b], 1988	**210**	
Commercial water vessels, 1988	**39**	
Recreational boats, 1988	**9,511**	
Railroad cars, 1988		**725**

Transportation movement (billions)

Domestic passenger travel, 1988		
Vehicle-miles		**1,713**
Automobiles		83.5%
Motorcycles		0.6%
Personal trucks		15.1%
Buses[a]		0.3%
Air[b]		0.5%
Rail		0.1%
Passenger-miles		**3,493**
Automobiles		71.4%
Motorcycles		0.3%
Personal trucks		14.1%
Buses[c]		3.7%
Air[d]		9.9%
Rail		0.7%
Domestic intercity freight movement, 1988		
Ton-miles		**3,204**
Trucks		22.0%
Water		27.8%
Pipeline[e]		19.1%
Rail[f]		31.1%

Energy use (trillion Btu)

Total United States energy use, 1989	**81,280**
Total U.S. petroleum use, 1989	**34,030**
Percentage of U.S. energy use	41.9%
Percentage used in transportation	63.2%
Percentage of transportation energy	97.1%
Total transportation energy use, 1989	**22,150**
Percentage of U.S. energy use, 1989	27.3%
Highway, 1988	**72.9%**
Automobiles	39.5%
Motorcycles	0.1%
Trucks	32.6%
Buses	0.7%
Nonhighway, 1988	**20.8%**
Air	8.7%
Water	5.9%
Pipeline	3.9%
Rail	2.3%
Off-highway, 1985	**2.9%**
Military operations, 1988	**3.5%**

[a]Transit bus only; [b]Certified route air carriers and general aviation; [c]Transit and intercity bus only; [d]General aviation aircraft only; [e]Coal slurry and crude oil and products pipeline only; [f]Class I rail only.

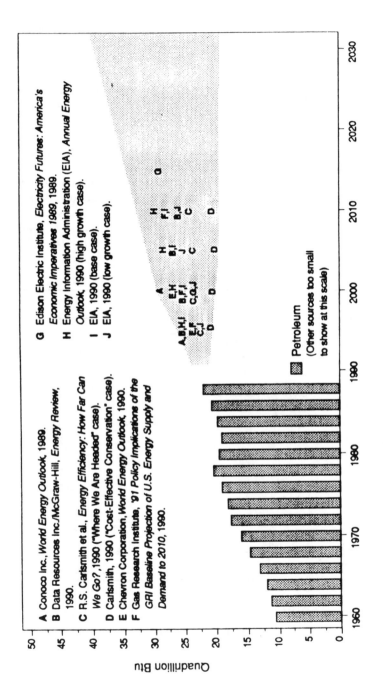

Figure 3.9 Projections of U.S. energy consumption — transportation sector. Note: differences in projections are caused, in part, by varying assumptions concerning energy prices, economic growth, consumer and producer behavior, and rates of technological change, including replacement of capital stock. The shaded area represents an envelope bracketing these differences. (From U.S. Department of Energy, Office of Science and Technical Information, *National Energy Strategy*, 1991.)

Table 3.9   New Car Fuel Economy for Selected Countries, 1978–1988. According to the best available data, new cars in Italy have the highest fuel economy of the listed countries. Caution should be used, however, when comparing fuel economy data between countries because each country may use different methods of calculating new car fuel economy. The data, therefore, may not be directly compatible. [a]Combined fuel economy for gasoline and diesel automobiles; [b]Data not available; [c]Average annual percentage changes are for years 1978–85 and 1982–85; [d]Average annual percentage change is for years 1980–88. (From U.S. data – Williams, Linda S. and Patricia S. Hu, *Light Duty Vehicle MPG and Market Shares Report: Model Year 1989*, Oak Ridge National Laboratory, ORNL-6626, Oak Ridge, TN, April 1990.) Other countries – personal communication with Lee Schipper, Lawrence Berkeley Laboratory, Berkeley, CA, 1990. Data were compiled from country sources, such as oil companies, energy economics institutes, and government ministries. Oak Ridge National Laboratry, *Transportation Energy Data Book: Edition 11*, (ORNL-6649), 1991.)

					(miles per gallon)	
	Germany		Italy		Japan	United States
Year	Gasoline	Diesel	Gasoline	Diesel	Combined[a]	Combined[a]
1978	20.5	25.0	[b]	[b]	25.4	19.7
1979	20.6	24.8	[b]	[b]	25.9	20.5
1980	21.2	25.0	28.2	29.4	25.9	23.3
1981	21.4	28.0	28.7	31.0	27.5	25.3
1982	22.0	27.4	29.4	32.2	29.0	26.3
1983	22.8	28.7	31.7	34.6	28.6	26.1
1984	23.5	30.2	32.7	36.2	28.6	26.3
1985	23.5	30.6	32.7	36.7	27.7	27.0
1986	[b]	[b]	33.7	36.7	26.8	27.9
1987	[b]	[b]	34.1	36.7	26.3	28.1
1988	[b]	[b]	34.1	36.7	25.9	28.5
	Average annual percentage change					
1978–88	2.0%[c]	2.9%[c]	2.4%[d]	2.8%[d]	0.2%	3.8%
1982–88	2.2%[c]	3.8%[c]	2.5%	2.2%	-1.9%	1.3%

Fuel economy is affected by engine type, vehicle weight, aerodynamics, tires, steering, suspension, transmission, and how people drive their vehicles. CO_2 emissions from vehicles vary directly with fuel consumption. All the losses in Table 3.10 can be reduced with the help of research and development. For example, new materials for transportation applications that operate at higher temperatures can

Table 3.10   Energy losses in typical car. (From International Energy Agency, *Study of Energy Efficiency and Its Contribution to Environmental Goals,* 1992. This is based on 1982 data for a non-air-conditioned vehicle.)

Thermodynamic and Mechnical Losses	Remaining Percentage	Percentage
Engine Losses		**100**
Radiation	12	88
Exhaust	20	68
Coolant	40	28
Auxiliaries	8	26
Transmission	2	18
Losses in Overcoming Forces Associated with the Motion		
Aerodynamic Drag	4	14
Rolling Resistance	6	8
Braking	8	0

reduce engine losses and friction. Advanced materials, most likely ceramics for motor vehicles, hold the potential for higher temperature engines that can improve efficiency by reducing energy losses due to coolant requirements. The development of plastics and composites can lead to significant weight savings and at the same time achieve better durability. For safety purposes, it is important to develop better energy-absorbing materials for the structural components. These materials are needed to compensate for any downsizing of automobiles in order to reduce weight. Before these new materials can be widely used additional research must address the recycling or reuse potentials of the new materials.

Table 3.11 shows examples of existing high fuel economy "prototype" vehicles and suggests a wide variety of innovative features that may be feasible in the future (International Energy Agency, 1989). However, the IEA report notes that "the impact of this research on the fuel economy of real vehicles is difficult to assess and so is its timeframe." Despite the high potential savings shown in this table, IEA estimated that fuel efficiency improvements of only 10 to 40% can be anticipated for new vehicles by 2010. Many of these vehicles are experimental, one-of-a-kind concept cars that sometimes do not meet all emission and safety requirements and are not ready for production. General Motors, for example, developed its TPC experimental two-passenger car in the early 1980s powered by a gasoline engine. It was

Table 3.11   High fuel economy prototype vehicles. *L/100 km = 234/mpg. Note: DI = direct injection; hwy = highway; CVT = Continuously Variable Transmission. (From IEA, *Study of Energy Efficiency and Its Contribution to Environmental Goals,* 1991. Table based on Johansson and Williams, 1987; Bleviss, 1988; and Delsey, 1990.)

Company	Model	# of passengers	Aero-dynamic Drag Coefficient	Curb Weight (kg)	Maximum Power (hp)	Fuel Economy (mpg)*	Innovative Features	Development Status
General Motors	TPC (gasoline)	2	.31	472	38	60 city 73 hwy	Aluminium body and engine	Prototype complete, no production plans
British Leyland	ECV-3 (gasoline)	4-5	.24-.25	662	72	40 city 52 hwy	High use of aluminium and plastics	Prototype complete
Volkswagon	Auto 2000 (diesel)	4-5	.25	778	53	62 city 71 hwy	DI with plastic and aluminium parts, fly-wheel stop-start	Prototype complete
Volkswagon	VW-E80 (diesel)	4	.35	698	51	73 city 98 hwy	Modified DI 3-cyl. Polo, fly-wheel stop-start, super-charger	Ongoing research, possibility of production
Volvo	LCP 2000 (diesel)	2-4	.25-.28	705	52, 88	62 city 81 hwy	High magnesium use; 2 DI engines developed, 1 heat insulated	Prototype complete, adaptable to production
Renault	EVE+ (diesel)	4-5	.225	853	50	62 city 81 hwy	Supercharged DI with stop-start	Prototype complete
Renault	VESTA2 (gasoline)	2-4	.186	475	27	78 city 106 hwy	High use of light material	Programme complete
Peugot	VERA+ (diesel)	4-5	.22	700	50	56 city 87 hwy	DI engine, high use of light materials	Ongoing development
Peugot	ECO 2000 (gasoline)	4	.21	458	28	67 city 73 hwy	2-cylinder engine, high use of light material	Ongoing development
Ford	— (diesel)	4-5	.40	850	40	57 city 90 hwy	DI engine	Research
Toyota	AXV (diesel)	4-5	.26	649	56	90 city 106 hwy	Weight is 15% plastic, 6% aluminium, has CVT and DI engine	Ongoing development

designed and built to compare the practical limits of fuel economy with other worldwide designs. This vehicle achieved a 68 mpg/city, 95 mpg/highway (78 mpg composite) fuel efficiency, but it did not meet federal safety and environmental standards. Table 3.12 shows 1991 models in the U.S. that achieve 40 mpg. Overall there are over 625 car models offered for sale in the U.S. with the average car chosen by consumers achieving about 28 mpg and weighing about 3200 pounds.

Others have also examined potential fuel economy improvements without sacrifice of vehicle performance or size (Difiglio et al. 1990). They estimate that in the U.S. 30.9 mpg for on the road new car fuel economy could be reached by 2000 (this number differs from EPA CAFE numbers due to differences in driving cycle and driving conditions, including congestion). On-road fuel economy is shown in Figure 3.10, which shows current gasoline prices in five nations and estimates of the cost of different fuel efficiency levels.

Shepard suggests that there is available technology, such as increased use of front-wheel drive, four valves per engine cylinder,

Table 3.12   1991 automobile models achieving at least 40 miles per gallon. (From IEA, 1992.)

Nameplate	Number of Models	Unadjusted 55/45 MPG	Test Weight (lbs)
Geo Metro	7	45–65	1875–2125
Honda Civic	3	41–59	2250–2750
Suzuki Swift	3	44–55	2000–2250
VW Jetta (Diesel)	2	45–47	2750
Daihatsu Charade	2	43–46	2125–2500
Ford Festiva	1	45	2125–2250
Ford Escort	1	41	2125–2375
Subaru Justy	3	40–41	2250–2500
Toyota Tercel	1	40	2250–2500
Pontiac LeMans	1	40	2500–2750
Average U.S. Model	625	28	3200

better aerodynamics, new intake-valve control systems, continuously variable transmissions, multipoint fuel injection, etc., that can be implemented to raise the fuel efficiency of the new U.S. car fleet to 44 mpg by the year 2000 (Shepard, 1991). Ledbetter and Ross (1990) estimate that increasing efficiency from the current 27.5 to 44 mpg would cost the consumer $0.55/saved gallon. To put this number in perspective, one can go back to the changes that occurred between 1978 and 1985. During that period, standards were raised from 18.7 to 25.8 mpg at a cost of $400 per car. This translates to $0.30/gallon saved over the lifetime of the car with typical miles traveled. Clearly, these costs and savings are a good buy for the consumer. A more remarkable number is the net savings in gasoline due to improved efficiency. The U.S. is spending annually some $40 billion less at the gas pump now than it would have if efficiency had remained at 1975 levels.

A major objection to efficiency improvements has been the need to reduce the weight of automobiles, which often results in smaller and presumably less safe cars. However, Shepard (1991) points out that in the period 1975 to 1986, while fuel efficiency doubled and the average weight of new cars decreased by 3,000 pounds, traffic fatalities decreased by 40%.

Both market conditions and fuel prices have caused consumer preferences for vehicles to change over time. Corporate Average Fuel Economy (CAFE) standards that are not encouraged by energy price

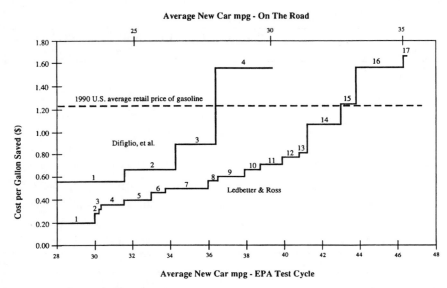

Figure 3.10   Conservation supply curves for automobile fuel efficiency. Each step on the two curves represents a specific efficiency measure. The verticle dimension shows the investment required per unit of energy saved, assuming each measure has a lifetime of 10 years and is amortized using a 7% real discount rate. The horizontal dimension is the improvement in average mpg, assuming each measure is implemented to the maximum extent feasible in the new car fleet in the year 2000. Average interior volume and acceleration capability are held at their 1987 levels (no downsizing). (From M. Ledbetter and M. Ross, *Supply Curves of Conserved Energy for Automobiles,* American Council for an Energy Efficient Economy, Washington, D.C., 1990 (on-the-road values taken from this study); C. Difigio, K.G. Duleep, and D.L. Green, "Cost Effectiveness of Future Fuel Economy Improvements," *Energy Journal,* 11:1 1990.

signals may be resisted by the public by not purchasing new vehicles or shifting to light-duty trucks or vans. The NAS report describes the disagreement on the relative impact of fuel prices and regulations on the supply of fuel efficient car.

3.3.3.3 Technical Potential for Energy Savings

The Department of Energy's Energy Information Administration (EIA) considered several energy conservation "excursions " (i.e., scenarios) as backup for the 1991/1992 National Energy Strategy (Energy Information Administration, 1990). The High Conservation excursion makes assumptions concerning alternatively-fueled vehicles and travel trends that result in a significant reduction in total energy use, reliance on oil, and emissions, compared with Reference and Constant Efficiency cases. It also examined a Very High Conservation excursion in which alternative fuels become the major source of transportation energy by 2030 and improved system efficiencies significantly reduce vehicle miles traveled. The current gasoline infrastructure is basically replaced by an alcohol distribution network. The results are shown in Figure 3.11.

As EIA noted, "the results depend heavily on the success of a variety of developmental projects to make high technology vehicles succeed in the marketplace." These assumptions are all based on the development and successful market penetration of new technologies such as telecommunications, "smart roads," novel mass transit approaches, and high speed trains. As EIA noted, the assumptions

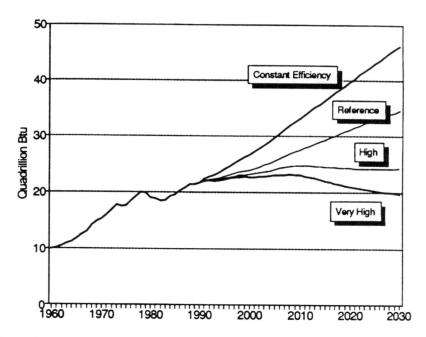

Figure 3.11   Energy conservation in the transportation sector excursions. (From Historical Data—*Annual Energy Review 1989,* Energy Information Administration. Forecasts—Transportation Sector, constant efficiency case, reference case, and High and Very High Conservation excursions. Energy Information Administration, *Energy Consumption and Conservation Potential: Supporting Analysis for the National Energy Strategy,* (SR/NES/90-02), 1990.)

used to develop these two excursions may seem extreme compared to the Reference case but "they seem far less extreme, given the perspective of the gains in overall transportation sector efficiency made between the mid-1970s and mid-1980s."

Telecommunications/Computers

Accelerated implementation of telecommunications/computer innovations in society could mitigate travel growth in the future as well as replace many current uses for paper. Telecommunications/computer cost reductions and increased capability continue to expand the range of services available to the public and to industry. Powerful computing capacity combined with novel sensor technologies is changing manufacturing technologies, home entertainment, games and educational systems. Vast potential changes in education, industry, homes and travel are becoming technologically and economically feasible. Electronically "smart" homes can manage heating, cooling and lighting systems. Computers and telecommunications are already revolutionizing the communications industry; cable TV, satellites, facsimile, fiber optics, local area computer networks, super-computers, video-teleconferencing, etc., offer potential options to reduce worldwide growth in physical travel. Technologies such as smart highways and traffic lights, vehicle onboard navigation and traffic warning systems, improved ridesharing and mass transit systems, home shopping, automatic vehicle time-of-use bridge and roadway pricing, access to data and information bases and home entertainment can increase vehicle operating efficiency as well as reduce passenger miles traveled.

Current regulatory and legal barriers must be removed to encourage more home and business applications. However, continuing technology advances and cost reductions are already acting as powerful stimuli to innovation and commercial exploitation by the computer, telecommunications and service industries.

Alternative Fueled Vehicles

Interest in alternative fueled vehicles originated after the oil shortages in the 1970s as a means to diversify transportation fuel supply. More recently, alternative fueled vehicles are becoming an important strategy to reducing air pollution from gasoline and diesel vehicles. A wide range of fuels is being developed: alcohol (methanol and ethanol), compressed natural gas, electricity and propane. In the longer term hydrogen can be an extremely clean fuel, but production is hindered by the lack of cheap sources of electricity and technical barriers related to hydrogen storage and use. Given the environmental benefits of hydrogen, more research and demonstration projects should be emphasized.

The NAS mitigation panel report examined alternative transportation fuels and indicated that they have made inroads in other countries, including Canada, New Zealand and Brazil. In the U.S., the state of California has started a program to encourage alternative transportation fuels including methanol, compressed natural gas, cleaner gasoline fuels, and electricity to help improve air quality. These fuels must meet engineering criteria in terms of economic competitiveness, and environmental impacts. To be competitive, they must meet energy density requirements to allow reasonable vehicle range between refilling the tank or recharging the batteries. With the exception of cleaner automotive fuels, nearer term markets are hindered by their relatively short driving range. However, these vehicles are appropriate in fleet applications, especially in urban areas where air pollution emissions need to be reduced.

Optimal energy efficiency is only achieved when the engine and fuel are specifically designed to be compatible with each other. Thus, as new markets are developed for alternative fuels, automobiles and trucks will be needed that were designed to utilize them.

Acceptance of alternative fueled vehicles is burdened by the "chicken and egg" problem of vehicle and fueling station availability. California has aggressively sought to use flexible fuel vehicles that can use methanol, ethanol or gasoline, or dual fuel vehicles that use compressed natural gas or gasoline as a way to bring vehicles in to the market at the same time the fueling station infrastructure is being developed.

Other methods of propulsion utilize battery or fuel cell vehicles. In many respects, these cars are similar to some of the alternative fuel cars in that a major concern is vehicle range. The key problem is that the weight, volume, and other losses of the batteries must be reduced and overall battery capacity and durability increased. Fuel cells use either hydrogen or natural gas as the fuel. Thus, they have range limitations unless fuel cells are developed that use higher density fuels such as alcohols.

Electric vehicles (EV) can be attractive in urban areas because they have no direct emissions, and can contribute significant to reductions in greenhouse gases if the electricity is generated by non-fossil fuel sources. Additional development is needed to extend EV range (currently about 65 miles per charge), but future improvements in range with internal combustion engine (ICE) performance could broaden applications. One possible solution is hybrid vehicles that use a smaller ICE engine or fuel cell in the vehicle to generate electricity. EVs utilizing electricity derived from renewable or nuclear energy are particularly attractive for reducing transportation related greenhouse gas emissions.

The NAS report examined methods of reducing transportation greenhouse gas emissions through improvements in vehicle efficiency, development of alternative transportation fuels, and improvements in transportation system management. It concludes that: "In the short run, transportation energy consumption can change rapidly as consumers adjust their demands concerning when and how to travel. On a slightly longer time scale, higher vehicle and fuel prices, along with shifts in vehicle and transportation demand, will lead to changes in the types of vehicles in use. On a significantly longer time scale, investments can be made in alternative transportation fuels, construction of new mass transit facilities and high-occupancy-vehicle (HOV) lanes, land use planning and jobs/housing balance, research and development, and tooling for technological improvements." The NAS report also found significant energy efficiency improvements possible in trucks and aircraft.

Mass Transit

Mass transit options include new electrically powered light rail and high speed train systems. Light rail transportation options in urban areas and corridors could best be developed in high population density areas and could be extended into suburban areas with hubs connecting with other types of transportation (buses, cars, etc.) High speed rail including electrified rail and magnetically-levitated vehicles have been proposed between major urban areas of less than 500 miles distance. Such linkages could help reduce major urban congestion around airports and replace some future air travel. High capital costs and relatively low ridership in the U.S. have been a disincentive to the construction of new advanced mass transit systems.

3.3.3.4 Barriers

Energy costs have decreased in real dollars and as a portion of total vehicle operating costs in the U.S. since the mid-1970s. Thus, energy prices have not been a very significant determinant of consumer choice. Gasoline prices in the U.S. are back down to their historic lows. Energy price signals in the U.S. are not conducive to large investments in energy efficiency.

The NAS study noted that, "As fuel prices increase, many technologies become cost-effective, and manufacturers supplying these technologies gain a competitive advantage." Clearly, energy price signals in the U.S. continue to differ significantly from those in most other developed nations, while new car fuel economy differs much less. The NAS study suggested three major barriers:

1. *Uncertainty in technology availability.* Whether a technology will ever be available that can produce desired cost-effective efficiency gains is uncertain. Increased research, development and testing could help overcome this barrier. In addition, there are lags in implementing new technology associated with turnover times for various products or lags in the distribution of information, infrastructures, and products around the world.
2. *Consumer resistance.* Some consumers resist buying efficient vehicles at the perceived expense of other attributes like safety and performance.
3. *Consumer perception.* Consumer buying decisions are significantly impacted by the initial capital investment for the vehicle instead of a life-cycle cost. As noted previously the energy component in total operating costs is significantly lower today in the U.S. than it was in the mid-1970s.

In the near term, behavioral changes in driving habits, driving behavior (e.g., driving speeds), ride-sharing, use of new telecommunications alternatives to travel, and use of mass-transit could significantly impact energy consumption and greenhouse gas emissions. Many approaches are being attempted in an effort to affect transportation choices and behavior.

A variety of approaches have been utilized or suggested to overcome these barriers, including Corporate Average Fuel Economy Standards in the U.S., carbon or increased gasoline taxes and "fee/rebate " schemes for vehicles. The California legislature has considered a proposal called "DRIVE +" to increase the sales tax on inefficient, high-polluting cars, and to use the revenues to reduce the sales tax on more efficient, less polluting cars (Gordon and Levenson, 1990). However, this approach does not affect vehicle miles traveled. A gas guzzler tax exists for new automobiles that get less than 22.5 mpg, but the tax has had relatively little effect on consumer behavior because it is small in relationship to the purchase prices of these generally high performance vehicles. However, the tax was doubled in 1990 and may have a significant impact in the future (U.S. DoE, DOE/EH-0103, 1989).

The Federal and California Clean Air Acts contain provisions for "transportation control measures" — strategies to reduce vehicle trips, vehicle use, vehicle miles traveled, vehicle idling, or traffic congestion for the purpose of reducing motor vehicle emissions. In the Federal Act, EPA is charged with providing information to the states on a wide variety of transportation control measures. In the California Act, the requirements vary with pollution severity. They range from providing information to a requirement that transportation control measures achieve by 1999 an average of 1.5 or more persons per passenger vehicle during weekday commute hours in areas with severe air pollution. The experience with various transportation control measures to reduce traditional urban pollutants will go a long way toward understanding what kind of measures are cost effective and publicly acceptable to reduce vehicle miles traveled and increase average ridership.

New approaches are being attempted in the U.S. to support cooperative pre-commercial transportation industry research relative to technical barriers to improved efficiency and alternative fuel use and safety. Such joint efforts with industry, government, and universities can help speed the commercialization of new technologies.

3.3.3.5 Research, Development and Demonstration (RD & D) Opportunities

The U.S. Department of Transportation (DoT) recently discussed the importance of technological innovation in transportation.

"The United States has a long and proud history of breakthroughs in transportation technology and innovations in transportation management and operations. We must renew and strengthen our focus on technology and innovation if we are to meet the expectations and needs of the Nation and maintain U.S. technological leadership in the world." (U.S. DoT, 1990).

The DoT report concludes that there are many opportunities to improve transportation through research and innovation and that there is great future potential in the area of new forms of freight and passenger transportation. To translate this into real technology options for the future, a great deal of research will be needed.

The National Research Council's Transportation Research Board also addressed the issue of R & D opportunities in the transportation sector (National Research Council, 1990). It noted that "the technologies to improve new car and light truck fuel efficiency over the next 10 years are limited to those already in hand and those that are nearly ready for commercial application." Really significant innovations are probably not possible in the near- or mid-term. Technology-based research on vehicles is needed to make longer-term gains including those identified in both the NRC study and in the NAS Mitigation Panel Report. Important R & D areas are listed below.

Technology Base
- Efficient utilization of alternative fuels in spark ignition and diesel engines.
- Prediction of regulated emissions to meet projected environmental standards.
- High-temperature structural and engine materials such as advanced ceramics and metals.
- Economical lightweight composite structural materials to reduce vehicle weight.
- Structural materials applicable to engines, transmissions, and load-bearing parts.
- Computational aerodynamics and fluid mechanics addressing vehicle drag reduction and improvement in components such as torque converters.
- Tribology, aimed at reducing friction in engines and drive trains, particularly for engines operating at very high temperatures where conventional hydrocarbon lubricants may not be applicable.

Vehicle Components and Systems
- Battery components.
- Hybrid vehicles, including combinations of small heat engines

operating near full loads, and energy storage devices.

- Onboard storage mechanisms for alternative fuels, hydrogen, liquified natural gas and distribution and storage systems for hydrogen.
- Onboard fuel cells for generation of propulsive electric power.
- Onboard photovoltaics for accessory power.
- Engine systems to achieve better integration of engine, transmission, and ancillary components to improve part-load fuel economy.
- Optimization of engines and vehicles for alternative fuels, including work on materials compatibility with methanol and ethanol.
- Vehicle system studies to address such issues as construction of crash-worthy cars of light weight structural material, and control of emissions in superefficient engines.
- New innovative systems approaches to major transportation problems (e.g., electrified highways to couple (inductively or otherwise) electric vehicles to external sources of energy).
- Improvements in the efficiency of non-highway travel modes including aircraft, mass transit and boats.
- Encouragement of telecommunications options to reduce future travel requirements and on a wide range of behavioral topics that affect transportation choices.

3.4 ACHIEVABLE SAVINGS AND REMAINING BARRIERS

The seven studies used in this chapter have documented very large potential savings from efficiency measures. The NAS 1992 study reported potential carbon savings of 1.75 billion tons of CO_2 equivalent per year for the U.S. from various efficiency measures in the sectors discussed in this chapter (sum of residential and commercial buildings, industrial, and transportation energy savings potential). As shown in Table 3.13, this amount is 36% of the total of 4.8 billion tons per year identified in the report. All of the efficiency measures can be called "no regrets" measures because they result in either net benefit or very low cost. In other words, they either completely or nearly pay for themselves from energy savings alone, and hence the carbon reduction is either free or inexpensive. Thus, progress can be made

Table 3.13 Comparison of Selected Mitigation Options in the U.S. (From NAS, *Policy Implications of Greenhouse Warming*, National Academy Press, Washington, D.C., 1991. With permission.)

Mitigation Method	Net Implementation Cost Mid Range $/tCO$_2$ eq	Maximum Potential Emission BtCO$_2$ eq/yr	Reductions in 1988 CO$_2$ Emissions		Economic Svaings ($B/yr)	
			Per Measure	Cumulative	Per Measure	Cumulative
Building Energy Efficiency	−62	0.9	18	18	56	56
Vehicle Efficiency	−40	0.3	6	24	12	68
Industrial Efficiency	−25	0.5	11	35	13	81
Transportation System Mgmt	−22	0.05	1	36	1	82
Base Case (1988 GtCO2)		**4.8**		**100**		**$450 B**

toward reducing global warming without imposing additional costs on the economy, and in fact some measures will provide net savings.

The NAS study did not produce a conventional scenario (of 20 or 40 years) that assumes a role for GNP growth. Instead, NAS simply assumed that all "hardware" (appliance, cars, manufacturing plants) is replaced with today's optimally efficient model as it wears out, and that building shells are retrofit to efficient technologies over the next decade. Since cars and most household appliances as well as industrial processes and plants are replaced over a 10 to 20 year period, the potential improvements are completed between 2000 and 2010.

NAS did not correct for the fact that even by 2000, at 2.5% per year GNP growth, the demand for energy services will increase by 28%, and building area floor space will increase by about 28%. Instead, the study assumed that the 1.75 billion tons of potential savings in Table 3.13 applies to 1990, a year when we used 82 quads and emitted 4.8 billion tons of CO_2 equivalent. This allows us to calculate the potential savings in E/GNP and CO_2/GNP without having to assume growth in GNP, or escalation in energy prices, but it underestimates the large savings potential from energy efficiency.

For energy-efficient technologies, the payback period is the amount of time (in months or years) that it takes to recover the money invested in a technology through its savings in energy bills. There is a great disparity between acceptable payback periods for consumer investments in energy efficiency versus business investments in energy production. Typically, electric utilities will accept payback periods of 10 to 15 years for investments in energy production facilities. In contrast, consumers and commercial customers strongly prefer to invest capital in energy efficiency technologies with payback periods of three years or less. Of course, more rational energy prices would lower payback periods.

Some analysts argue that if these savings were truly cost-effective, then they would be adopted by consumers and businesses who would exploit cost-effective opportunities to reduce their energy bills. Figure 3.12 illustrates the difference between the economist's "energy modeling" and the engineer's "technological costing" views. The "100% implementation/low cost" line in this figure shows the optimistic view that a great deal of the savings provide net benefits or are very cheap. The "25% implementation/high cost" shows the view that since the savings are not realized in the market they must not be as cheap (or as plentiful) as estimated in the NAS study.

The difference between these two views is due to barriers and the resulting market failures. Some of these barriers have been addressed earlier in this chapter: lack of information, lack of capital, and externalities (environmental, social, and national security) that are not reflected in price. Other barriers are less obvious: the U.S. enjoys low energy prices, and as a result many other costs in homes and businesses are much larger and hence get more attention than energy bills. Low energy prices exacerbate the other barriers because if prices are low consumers do not take the time to obtain the information needed and do not shop around for the most efficient products. Hence, we observe unexploited opportunities to save money and energy and reduce carbon emissions. There are not engineering solutions to these problems, but engineers should understand the public policy issues and the magnitude of both the opportunities and challenges to achieving efficiency potential.

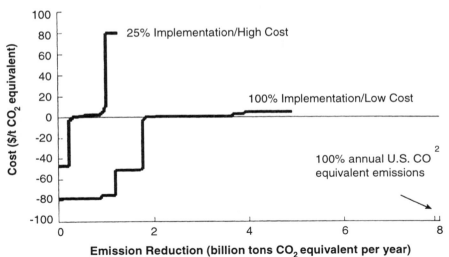

Figure 3.12   Comparison of mitigation options. Total potential reduction of CO_2 equivalent emissions is compared to the cost in dollars per ton of CO_2 reduction. Options are ranked from left to right in CO_2 emissions according to cost. Some options show the possibility of reductions of CO_2 emissions at a net savings. (From NAS, *Policy Implications of Greenhouse Warming: Report of the Mitigation Panel,* National Academy Press, Washington, D.C., 1991. With permission.)

3.5 EXAMPLES OF SUCCESSES

During the past decade, significant strides have been made in the development of energy-efficient technologies. Table 3.14 provides a summary of the characteristics and economics of three technologies: high frequency electronic ballasts, compact fluorescent lamps, and low-emissivity windows (Rosenfeld, 1991). These technologies, which were developed with DOE R & D funding, illustrate the remarkable benefits-to-R & D-cost ratios that can be realized with energy-efficient technologies. We will also discuss the value of energy efficiency standards for appliances.

3.5.1 Lighting

Prior to the 1973 OPEC oil embargo, fluorescent lamps predominated in U.S. commercial buildings, most frequently with "cool white" (bluer) phosphors. (The light output of a lamp is measured in lumens. A highly efficient modern fluorescent lamp with an electronic ballast emits about 85 lumens/watt, and a 60-watt incandescent lamp emits only about 15 lumens/watt.) A typical fluorescent fixture, with 2 lamps of 40 watts each, yielded 65 lumens/watt (i.e. about 6000 lumens), equivalent to that of 4 incandescent at 75 watts each. The smallest fluorescent on the market at the time was the 22-watt "circline" with 890 lumen outfit equivalent to a 60-watt incandescent. With end losses its efficacy was down from 65 to 40 lumens/watt. These fluorescents were fiercely resisted by homeowners and interior decorators who preferred the familiar, redder, but less efficacious incandescents.

In 1973, about 200 billion kilowatt-hours (BkWh) of U.S. electric sales of 2000 BkWh, were consumed by fluorescent lighting (for 80% of the lumens) and 200 more BkWh went to incandescent (for most of the remaining lumens). As electricity prices shot up, two developments became inevitable: (1) Improvement of fluorescent efficacy by use with high-frequency ballasts; and (2) development of compact fluorescents (with better color rendition) to screw into incandescent sockets. This compact fluorescent lamp development was jump-started when high-frequency ballasts were shown to economically cut end losses and ballast losses in half. (See Section 3.5.1.2.)

Table 3.14   Economics of three new energy efficient technologies and appliance standards. An update of Tables 1 & 4 of Geller et al., *Annual Review of Energy,* 1987, Vol. 12. [1]Calculations for CFLs based on one 16-watt CFL replacing thirteen 60-watt incandescents, burning about 3300 hours/year, assuming that a CFL costs $9 wholesale, or $5 more than the wholesale cost of thirteen incandescents. For retail we take $18–$8; [2]Assuming price of 7.5¢/kWh for commercial sector electricity and a retail natural gas price of $7/MBtu (70¢/therm.); [3]For hot weather applications where low-e windows substantially reduce cooling loads, air conditioners in new buildings can be down-sized, saving more than the initial cost of the low-e window; [4]Assuming marginal electricity comes from oil or gas at 11,600 Btu/kWh, thermally equivalent to 0.08 gallons of gasoline; [5]Net annual savings are in 1990 dollars, uncorrected for growth in building stock, changes in real energy costs, or discounted future values. See Geller et al., Table 1; [6]One 1000 MW baseload power plant supplying about 5BkWh/year; [7]1989 U.S. emissions of CO_2 were 5000 Mt; [8]This was augmented by private R&D; [9]Descended from high-frequency ballasts (only DoE assistance was in testing). From "The Role of Federal Research and Development in Advancing Energy Efficiency," Statement of Arthur H. Rosenfeld before James H. Scheuer, Chairman, Subcommittee on Environment, Committee on Science, Space, and Technology, U.S. House of Representatives, April 1991.)

	RESEARCH & DEVELOPMENT				STANDARDS
	HIGH FREQUENCY BALLASTS VS. CORE COIL BALLASTS	COMPACT FLUORESCENT LAMPS (1) VS. INCANDESCENTS	LOW-E (R-4) WINDOWS VS. DOUBLE GLAZED WINDOWS Per small window (10 ft2)	TOTAL	REFRIGERATORS AND FREEZERS '76 base case vs. '85 CA Stds.
1. UNIT COST PREMIUM					
a. Wholesale	$8	$5	$10		
b. Retail	($12)	($10)	($20)		($100)
2. CHARACTERISTICS					
a. % Energy Saved	33%	75%	50%		66%
b. Useful Life	10 years	3 years	20 years		20 years
c. Simple Payback Time (SPT)	2 years	1 year	2 years		1 year
3. UNIT LIFETIME SAVINGS					
a. Gross Energy	1330 kWh	440 kWh	10 MBtu		24,000 kWh
b. Gross $	$100 (2)	$33 (2)	$70 (2)		$1800
c. Net $ [3b-1a]	$92	$28	$60 (3)		$1700
d. Gross Equivalent Gallons (4)	100	40	80		1920
e. Miles in 25 mpg car	2500	1000	2000		48,000
4. SAVINGS 1985-1990					
a. 1990 Sales	3M	20M	20M		not
b. Sales 1985 through 1990	8M	50M	50M		ramping
c. Cum. Net Savings [4b x 3c]	$750 M	$1.4B	$3B	$5B/5yr	up
5. SAVINGS AT SATURATION					
a. U.S. Units	600M	750M	1400M		100M
b. U.S. Annual Sales	60M	250M	70M		6M
c. Annual Energy Savings [5b x 3a]	80 BkWh	110 BkWh	0.3 Mbod		144 BkWh
d. Annual Net $ Savings [5b x 3c](5)	$5.5B	$7B	$4B	$16.5B/yr	$10 B
e. Equivalent power plants (6)	16	22			29
f. Annual CO_2 savings(7)	55 Mt	80 Mt	18 Mt	153 Mt	100 Mt
6. PROJECT BENEFITS					
a. Advance in Commercialization	5 years	5 years	5 years		5 years
b. Net Project Savings [6a x 5d]	$27.5 B	$35B	$20B	$82.5B	$50 B
7. COST TO DOE FOR R&D(8)	$3M	$0(9)	$3M	$6M	$2M
8. BENEFITS/ R&D COST [6b/7]	9000:1	∞	6500:1	14,000:1	25,000:1

3.5.1.1 Electronic Ballasts

High frequency electronic ballasts are much more efficient than core-coil ballasts. Specifically, the power usage of ballasts that operate 2 40-watt lamps is as follows: an outmoded "standard" core-coil ballast draws 16 watts, an "efficient" core-coil ballast is down to 10 watts, and an electronic ballast uses only 4 watts. A further benefit of electronic ballasts is that they are easier to control electronically, permitting dimming to save electricity when daylight is available or lower light levels are desired. This raises the system efficacy of electronic ballasts, averaged over an entire floor of an office building, easily 30 to 40% above undimmed "standard" ballasts.

The electronic ballast was developed through DOE-sponsored research assisted by Lawrence Berkeley Laboratory (LBL) in the late 1970s. Assuming electronic ballasts are commercially available for about $8 each wholesale, would save 1330 kWh and $100 each over its 10-year lifetime (See Table 3.14). (This savings is equivalent to 100 gallons of gasoline, enough to drive 2500 miles in a 25 mpg car.) Between 1985 and 1990, 8 million electronic ballasts were sold in the U.S. Based on the net lifetime savings of $92 per ballast, cumulative net lifetime savings for these 8 million ballasts is $750 million. It is expected that 600 million electronic ballasts will have been sold when market saturation is reached, saving production of 80 BkWh, emissions of 55 Mt CO_2, and expenditures of $5.5 billion annually. The initial DOE project to develop electronic ballasts cost $3 million and is estimated to have advanced commercialization by 5 years, for a net project savings of $27.5 billion. This represents over an 9000:1 return on DoE's investment.

3.5.1.2 Compact Fluorescent Lamps (CFLs)

The economics of CFLs are shown in Table 3.15. An individual CFL rapidly pays for itself through reduced energy bills. For example, one 15-watt CFL replaces a series of about 10 60-watt incandescents since it burns 10 times longer than the incandescents. This CFL would save 440 kWh and $33 in electricity costs over its 40-month life in a commercial building.

A modern automated CFL production plant costs $7.5 million and can produce 6 million lamps annually, each of which will save 440 kWh over its service life. This represents a total savings of approximately 2.5 kWh per year, equivalent to the sales of a 500 MW intermediate or baseload power plant that costs up to $1 billion to construct and $200 million per year to operate.

Within a decade, if CFLs penetrate enough of the U.S. market, they will save over half of the 200 kWh used annually by incandescents. As shown in Table 3.11, when CFLs have saturated the market they will save production of 110 BkWh, emissions of 80 Mt CO_2, and expenditures of $7 billion annually. DOE spent no R & D money on CFLs; they descended directly from the development of electronic ballasts. But because of the electronic ballasts, commercialization of CFLs was also advanced by 5 years for a net savings of $35 billion. Thus, DOE R & D expenditures of $3 million for electronic ballasts actually resulted in total savings from advancing commercialization of both electronic ballasts and CFLs of $62.5 billion, an incredible 20,000:1 benefit-to-R & D-cost ratio.

3.5.2 Low-Emissivity Windows

In the late 1970s, research at LBL, combined with private industry efforts, resulted in the development of double-glazed windows with low-E coatings filled with low-conductivity gas such as argon. These low-E windows resemble the familiar double-glazed ("thermopane") windows but have thin coatings applied to at least one of the inner surfaces. These films are transparent to light but reflect 85% of room-temperature heat back into the room in winter or outside in summer, significantly reducing winter heating or summer cooling needs. This cost-effective film is slowly penetrating the market, and has doubled the thermal resistance (R) (a single-glazed window is R-1, double-glazed is R-1.5 to R-2, and a wall insulated with 10 millimeters (4 inches) of fiberglass is typically R-11) of high quality thermopane windows (from R-2 to R-4) (Rosenfeld and Price, 1991).

As shown in Table 3.14, a small (one square meter) low-E window costs $10 more than a conventional double-glazed window (wholesale), but in a winter application saves 10 gigajoules 10 million Btu) of natural gas worth about $70 over its 20-year lifetime. Many major window manufacturers such as Andersen, Pella, and Marvin now offer low-E windows. During the past 5 years, 50 million of these windows were sold in the U.S. When market saturation is reached and 70 million low-E windows are sold annually, the yearly net savings from these windows will be $4 billion. This savings will displace energy equivalent to the production of 11 baseload (1000 MW) power plants.

Although low-E windows were originally designed to reduce thermal conduction in winter, they can be equally effective in hot summer applications. As developing countries in hot climates become urbanized and industrialized, afternoon peak electrical demands occur because of air-conditioning and industrial requirements. If new buildings are constructed with low-E windows in these countries, air conditioning demands will be significantly reduced, and smaller chillers can be installed in buildings. This will save more than the initial cost of the low-E windows and result in a negative cost of conserved energy.

3.5.3 Appliance Standards

As discussed in Section 3.3.1.5, appliance standards are often the easiest way to remove or discourage inefficient products from the market. Along with R & D funding, legislatively-enacted standards also perform the function of advancing technology development.

An example of the benefits of such standards in improving the energy efficiency of appliances is provided in the last column of Table 3.14. This column illustrates the energy and economic savings attributable to the 1985 California refrigerator and freezer appliance standards when compared to 1976 base case appliances. California standards were enacted nationwide; annual energy saving are now over 140 BkWh (which represents over 1.5 quads of primary energy), valued at $10 billion and equivalent to the electricity produced by 29 baseload (1000 MW) power plants.

As Table 3.14 shows, between 1976 and 1985 the large available savings from increasing refrigerator/freezer efficiency nationwide to the level of California's 1985 standards would have saved 1.7 quads per year. Equipment manufacturers began to improve their products' efficiency in order to sell in California. The U.S. enacted national refrigerator/freezer standards effective in 1990, estimated to save 0.6 quads per year between 1990 and 2015 and new standards on these and other appliances effective in 1993 that will save another 0.3 quads per year (McMahon et al., 1990).

3.6 PROPOSED RESEARCH
DEVELOPMENT AND DEMONSTRATION

Recommendations for RD & D in the buildings, industry, and transportation sectors are provided in the Sections 3.3.1, 3.3.2, and 3.3.3 of this chapter. We also discuss RD & D successes in Section 3.5 of this chapter. Table 3.10 in Section 3.5 summarizes RD & D successes adding to $16.5 billion per year.

In addition to the information provided in this chapter, extensive lists of proposed RD & D can be found in the National Research Council's Confronting Climate Change: Strategies for Energy Research and Development (National Research Council, Confronting Climate Change, 1990).

One area that deserves additional RD & D is the mitigation of summer heat islands through the use of high solar reflectance surfaces in cities and planting of shade trees. On the conservation supply curve for electricity conservation in the buildings sector (Figure 3.4), the most rewarding first step is the reduction of air conditioning loads by returning to measures that were well known before air conditioning became very cheap. A home or single-story building with a light colored roof and shade trees or vines can have an air conditioning bill which is only about two thirds that of an unshaded, dark building. Such shading and "cool" roofs should be required in building codes in hot climates, and utilities should offer incentives to exceed the codes.

We know that in the summer cities are 2 to 5°C hotter than their surroundings, but it is less well recognized that this "summer heat island" arises partly because unshaded asphalt on roads, parking lots, and roofs gets very hot in the sun and warms the urban air. Hence white roofs and shade trees, implemented to save air conditioning for individual buildings, will also cool a city 1–2°C. Lightening the color of asphalt roadways as they are resurfaced (for example by topping them with light sand or oyster shells) will further reduce urban temperatures

The mean ambient temperature in all California cities fell in the summers before 1940, as irrigation and orchards spread. As the trees were replaced with asphalt, temperatures rose. Before 1940, Los Angeles cooled 1°C, but the temperature is now 3°C above the minimum, and rises 1°C every 20 years. (This 3°C explains about 20% of the smog "incidents.") Recommended research could show how to cool all U.S. cities below their surroundings. Annual savings should be about $1 billion/year, and energy savings about 0.5 quad.

The following are examples of research projects that are needed before we can efficiently introduce light surfaces and vegetation:

- Labeling paints. Testing and labeling of light-colored paint by its albedo, i.e. its reflectivity to sunshine, is needed. The energy in sunshine is distributed about half in the visible and half in the 'near infrared.' Thus a 'cool' paint should have excellent reflectivity for heat in the near infrared, but its visible color should be close to that of concrete to avoid glare. Given the market push that will accompany albedo labels, it may be possible to develop better paints than are now available.

- Development and labeling of high solar reflectance surfaces for roads and parking lots. It is important to develop high solar reflectance asphalt (for areas where switching to concrete) is not appropriate, such as where, city services are underneath the street. Light-colored, or solar reflective asphalt will probably also last longer than black asphalt, because it will have a topping of sand or oyster shells that will both protect it from ultraviolet degradation and keep it cooler and less viscous, so that it will have less tendency to flow out from underneath tires in hot weather.

- Quantifying the connection between air temperature and smog. Smog formation is temperature dependent and thus, rising urban temperatures lead to increasing numbers of "official smog episodes" that exceed air quality standard. For example, in Los Angeles, such episodes (which often require restricting economic activities such as transportation and industrial production) rarely occur when the temperature is below 74°F (24°C). However, when temperatures reach 90°F in the mid-afternoon, the probability of a smog episode increases dramatically (less than two thirds). This 16°F "window" is being eaten into by the urban heat island temperature increase of 6°F that has already occurred in Los Angeles. The details of this relation between smog and urban heat islands should be studied in detail and incorporated into the air quality model for urban airshed.

• Quantifying the indirect effect of vegetation and "cool" surfaces on energy use at the neighborhood scale. It is well established that shade trees and cool roofs can reduce air-conditioning use at the building scale (Akbara et al., 1990). More research on urban climate is needed, however, before we can identify and quantify such effects at the local scale.

3.7 INTERNATIONAL ISSUES

Because global warming is an international problem, the U.S. (emitting 25% of the world's CO_2) and other OECD countries (emitting another 25% of the world's CO_2) cannot adequately respond to the challenge alone. However, the policies of the industrialized countries play the largest role in current greenhouse gas emissions and can exercise significant influence on reducing future emissions. International negotiations involve many complex issues including important ethical and equity questions. Developing countries will be most likely to participate in efforts to combat global warming if the industrialized countries take the first steps, and if international agreements provide them with the necessary financial and technical resources to support their own programs.

To date, the industrialized countries have been responsible for the majority of greenhouse gas emissions from energy consumption. In 1985, about 48% of total carbon from CO_2 was emitted by the US/OECD, 36% by the USSR, Eastern Europe and China, and 16% by the developing countries (IPCC, 1990). See Figures 3.13 and 3.14 for illustrations of relative per capita electricity consumption's and carbon emissions for industrialized and developing nations.

Much attention has been focused recently on developing countries because of the impending growth in their emissions. While energy consumption is still 5 to 100 times lower per capita than in OECD countries, its rate of growth has been 7 times higher over the past two decades (see Levine, et al., 1991). If standards of living as well as population grow as projected, developing country greenhouse gas emissions could grow three- to fourfold by 2025, and their contribution to global warming would then rise to 30% of the total, with the US/OECD at 34%, and the Soviet Union, Eastern Europe, and China at 37% (IPCC, 1990).

Energy efficiency can be especially important to developing countries to reduce costs and environmental damage and to increase energy security. Significant energy savings, on the order of 20–25%, are available with present technologies at a payback period of two years or less (Levine et al., 1991).

3.7.1 Sectoral Savings Potentials

Growth in residential electricity consumption in the developing countries is expected to result from switching from biomass fuels to commercial fuels, especially electricity, due to urbanization and rural electrification. In this sector, efficiency improvement opportunities exist for appliances, lighting, and space heating. In the commercial sector, lighting, equipment, and building design and maintenance improvements offer savings. The industrial sector is often two to three times less energy-efficient than the same sector in developed countries. Savings can be achieved through more efficient processes, less energy-intensive and more durable materials, improved capacity utilization, rational fuel choice, and infrastructure improvements that facilitate delivery and operations. Transportation is a large and growing end use whose consumption can be reduced from current forecasts by effective transit policies and vehicle efficiency improvements. Rapid growth rates make infrastructure choices critical to reducing future transportation energy needs. The power sector can renovate older inefficient plants and reduce high transmission and distribution losses.

3.7.2 Barriers

Global warming abatement is low on the priority list of developing countries. Other problems such as food security, health care, housing, employment, and literacy are more pressing. However, these countries also face severe capital constraints that make the cost-effective energy investments described above essential if they are to achieve their economic growth goals. The energy sector capital requirements for developing countries and Eastern Europe to achieve projected rates of GDP and population growth are almost $70 billion 1990 dollars per year through 2000, and $145 billion per year from 2001 through 2025 (Levine et al., 1991). The electric power sector alone would absorb almost half of this capital investment at current rates of GDP and electricity demand growth (Ciliano, 1991). Most non-oil-producing developing countries already have a significant international debt burden, of which a large fraction is energy loans. Current World Bank energy-related lending is less than $4 billion per year, and commercial bank lending has steadily declined during the past decade. As an alternative, investments in energy efficiency would cut the loan requirement by half while meeting the same energy service demand, saving a few trillion dollars over the life of the investments.

International lending policies to date have favored loans for large centralized capital-intensive energy supply projects. Only a small fraction of lending agency funds have gone to energy efficiency projects (only 1% of the World Bank's total lending from 1980–1990 was for end-use efficiency). Life-cycle cost analysis and consideration of environmental and social externalities should be criteria for loan qualification. To reduce greenhouse gas emissions, more capital will be needed for both efficiency improvements and projects to supply conventional electricity and natural gas using more efficient new technology.

Energy pricing policies are another barrier to the dissemination of efficiency in developing countries. Heavily subsidized electricity and fuel prices that do not even cover the costs of production are serious impediments to cost-effectiveness of energy efficiency, and they increase the debt burden of national utilities (Kosmo, 1987; Reddy, 1990; Wilbanks, 1990). Political and social goals currently dominate electricity pricing, especially in the residential sector. As in the industrialized countries, externalities are not included in tariff structures. Commercial and industrial tariffs should be brought up to marginal costs, and time-of-use pricing implemented. Increases in residential prices, traditionally subsidized, must be phased in and combined with social welfare programs and public education.

Another major factor impeding the implementation of energy efficiency in developing countries is the absence of a supporting "delivery infrastructure" linking manufacturers, vendors, distributors, and trained installers, operators and maintenance technicians. Existing buildings and plants tend to be poorly operated and maintained, and funds for renovation and replacement are scarce. Countries are dependent on foreign technology and foreign expertise. If enhanced efficiency is to succeed, a sustainable, indigenous capability must be established that transfers and adapts the technology innovations and market experience of the industrialized nations to the developing nations' unique circumstances.

The structure of international trade also makes regulation of greenhouse gas emissions complex. Treaties and trade policies should be devised that reduce emissions and foster environmental protection, rather than weaken existing regulations. "Emissions trading" is another practice that may be promising if logistics and enforcement can be negotiated.

3.7.3 U.S. and Other OECD Contributions

Developing countries are not necessarily bound to follow the development path that the industrialized countries have pursued. As shown in Figure 3.2, a great discrepancy exists in the ratio of energy consumption to GDP between the U.S. and Japan, for example. Developing countries can "leapfrog" the early inefficient stages caused by investment in energy-intensive infrastructure.

Industrialized countries can contribute to this leap-frogging path in two major ways. The first is by improving their own energy efficiency. Developing countries cannot be expected to curtail greenhouse gas emissions while developed countries continue to consume energy at substantially higher rates and enjoy a much higher standard of living. The allure of "high technology" has been nearly irresistible worldwide, and developing country planners resent the introduction of appropriate technology that they view as obsolete. New technologies such as compact fluorescents or low-emissivity windows will be much more widely used if they have already been commercialized in the US/OECD; lower their prices will be lower.

The second way that the industrialized world can aid the developing world is through technology transfer. For long-term success, domestic talent and infrastructure must be cultivated. Educational programs in efficient building and industrial process design, training programs in preventive maintenance, and access to efficient technologies, are all excellent investments in global warming reduction as well as in international human capability.

3.8 SUMMARY

The technical discussion in this chapter illustrates that we can continue to double the energy efficiency of new products every 20 years for the foreseeable future. This is significant because it results in capping energy and CO_2 growth at today's level and yet allows world energy demand to be met in 2031 using the same amount of energy consumed in 1990.

In Section 3.3, we show that the technical potential for reduction of energy demand in the buildings, industrial, and transportation sectors is large and that this reduction can be accomplished at a net economic benefit. Details of these savings are provided in Figure 4 for the buildings sector, in Figures 3.5 and 3.6 for the industrial sector, and in Figure 9 for the transportation sector. Table 3.13 shows remarkable potential savings of 36 percent of current CO_2 emissions and $82 billion per year from efficiency investments.

In Sections 3.3 and 3.4, we describe barriers that can hinder the full realization of this technical potential and advocate adoption of aggressive energy policies and research, development and demonstration (RD & D) projects to overcome these barriers. In Sections 3.5 and 3.6 we provide examples of energy-efficient technologies that were successfully developed through such projects and discuss directions for future RD & D efforts. Table 3.14 illustrates savings already achieved through three Federal RD & D projects and from implementation of refrigerator standards. These savings, which were generated from a one-time public investment of less than $10 million, are $16.5 billion per year.

Finally, Section 3.7 provides a perspective on energy and CO_2 savings potential for developing countries and discusses the relationship between US/OECD efforts and the needs of developing countries.

REFERENCES

Akbari et al., *Summer Heat Islands, Urban Trees, and White Surfaces*, presented at ASHRAE January 1990 Meeting, Atlanta, GA, (LBL-28308). 1990.

Atkinson, B., J.E. McMahon, E. Mills, P. Chan, T. W. Chan, J. H. Eto, J. D. Jennings, J. G. Koomey, K. W. Lo, M. LeCar, L. Price, F. Rubenstein, O. Sezgen, T. Wenzel, *Analysis of Federal Policy Options for Improving U.S. Lighting Energy Efficiency: Commercial & Residential Buildings,* Lawrence Berkeley Laboratory (Report 31469), Berkeley, CA, December 1992.

Bevington, R., A. H. Rosenfeld, Energy For Buildings and Homes, *Sc. Am.*, Vol. 263, No. 3, 1990, pp. 77–86.

Ciliano, R., Electrotechnologies: Potential for Improving Manufacturing Productivity, (EM-5259), Electric Power Research Institute, Palo Alto, CA, 1987.

Ciliano, R., F. Stern, C. Lang, The Electric and Gas Utility Modeling System, Final Technical Documentation, prepared for U.S. EPA, Office of Planning Policy and Evaluation, prepared by RCG/Hagler Bailly, Inc., 1990.

Ciliano, R., North American Electric DSM Status and Implication for Transferability to Developing Countries, presented at Energy Planning, Production and Utilization in Developing Countries and Emerging Democracies Conference, Washington D.C., May 1991.

Ciliano, R., T. Aleraza, C. Watson, Development of Refinements and Expansions to the BPA Industrial Conservation Supply Curve Model, prepared for the Bonneville Power Administration by ADM Associates, Inc. and RCG/Hagler, Bailly, Inc., 1991

Davis, S., P. Hu, *Transportation Energy Data Book: Edition 11*, Oak Ridge National Laboratory, Oak Ridge, TN, January 1991.

Difiglio et al., Cost Effectiveness of Future Fuel Economy Improvements, *Energy J.*, Volume 11, No. 1, 1990, pp.65–86.

Energy Information Administration, *Monthly Energy Review*, March 1991, p. 25.

Energy Information Administration, U.S. Department of Energy, *Energy Consumption and Conservation Potential: Supporting Analysis for the National Energy Strategy*, December 1990.

Faruqui, A., M. Mauldin, S. Schick, K. Seiden, G. Wikler, Efficient Electricity Use: Estimates of Maximum Energy Savings, prepared by Barakat & Chamberlin, Inc. for the Electric Power Research Institute (CU-6747) Electric Power Research Institute, Palo Alto, CA, 1990.

Felix, C. S. et al., *Environmental Benefits of National Demand-Side Management Program Implementation*, prepared for the Edison Electric Institute by Energy Research Group, 1990.

Gilleard, J.D. *The Status of Building Technology Innovation*, Georgia Institute of Technology, Atlanta, GA, 1989.

Goldemberg, J., T. B. Johansson, A. K. N. Reddy, R. H. Williams, *Energy for a Sustainable World*, Wiley Eastern Limited, New Dehli, 1988.

Gordon, D., L. Levenson, DRIVE+: Promoting Cleaner and More Fuel Efficient Motor Vehicles Through a Self-Financing System of State Sales Tax Incentives, *J. Policy Anal. Manage.*, Vol. 9, No. 3, 1990.

International Energy Agency, Program on Energy Conservation in Buildings and Community Systems, *Buildings of the 21st Century: Developing Innovative Research Agendas*, Workshop Proceedings, May 1989.

International Energy Agency, *Study on Energy Efficiency and its Contribution to Environmental Goals*, 1992.

Intergovernmental Panel on Climate Change, *Formulation of Response Strategies*, Report, prepared for IPCC by Working Group III, June 1990.

Kelly, H., *The Transformation of the Building Construction Industry*, International Energy Agency, Program on Energy Conservation in Buildings and Community Systems, Buildings of the 21st Century: Developing Innovative Research Agendas, Workshop Proceedings, May 1989.

Kosmo, M., Money to Burn? The High Costs of Energy Subsidies, World Resources Institute, 1987.

Ledbetter, M., M. Ross, A Supply Curve of Conserved Energy for Automobiles, 25th IECEC, Reno, NV, August 12–17, 1990.

Levine, M. D., Ashok Gadgil, S. Meyers, J. Sathaye, J. Stafurik, T. Wilbanks, Energy Efficiency, Developing Nations, and Eastern Europe, International Institute for Energy Conservation, A Report to the U.S. Working Group On Global Energy Efficiency, April 1991.

Levine, M., J. Sathaye, A. Ketoff, CO_2 Emissions from Major Developing Countries, *Energy J.,* Vol. 12, No. 1, 1991.

The Levy Partnership, Residential Technical Operating Plan (ResTOP), Advanced Housing Research, NY, May 1990.

McMahon, J. E., et al., Impacts of U.S. Appliance Energy Performance Standards on Consumers, Manufacturers, Electric Utilities, and the Environment, ACEEE 1990 Summer Study on Energy Efficiency in Buildings Proceedings, 1990.

Meier, A., J. Wright, A.H. Rosenfeld, Supplying Energy Through Greater Efficiency: The Potential for Conservation in California's Residential Sector, Berkeley, University of California Press, 1983.

Miller, P. M., et al., The Potential for Electricity Conservation in New York State: Final Report, New York State Energy Research and Development Authority, 1989.

National Academy of Sciences, *Policy Implications of Greenhouse Warming: Mitigation, Adaptation, and the Science Base,* National Academy Press, Washington, D.C., 1992.

National Research Council, Construction Productivity — Proposed Actions by the Federal Government to Promote Increased Efficiency in Construction, Commission on Engineering and Technical Systems, National Academy Press, Washington, D.C., 1986.

National Research Council, Confronting Climate Change: Strategies for Energy Research and Development, National Academy Press, Washington D.C., 1990.

Oak Ridge National Laboratory, Oak Ridge, TN, Energy Efficiency: How Far Can We Go?, 1989.

Philips, M., The Least-Cost Energy Path for Developing Countries, International Institute for Energy Conversation, Washington, D.C., 1991.

Reddy, A. K. N., Barriers to Improvements in Energy Efficiency, Indian Institute of Science, Department of Management Studies, 1990.

Rosenfeld, A., L. Price, Options for Reducing Carbon Dioxide Emissions, in Global Warming: Physics and Facts, American Institute of Physics Conference Series, AIP, NY, 1991.

Rosenfeld, A., The Role of Federal Research and Development in Advancing Energy Efficiency, Hearing on DOE Conservation Budget Request, before James H. Scheuer, Chairman, Subcommittee on Environment, Committee on Science, Space, and Technology, U.S. House of Representatives, Washington, D.C., 1991.

Ross, M., Industrial Sector Energy Analysis and Audits, Volume 2: Final Report to Michigan Electricity Options Study, 1987.

Shaw, H., The Effects of Energy Consumption on Climate Modification, New Jersey Institute of Technology, presented at the CHEMRAWN VII World Conference on The Chemistry of the Atmosphere: Its Impact on Global Change, Baltimore, MD, 1991.

Shepard, M., How to Improve Energy Efficiency, *Issues Sc. Technol.,* 7 (4), 85–91, 1991.

U.S. Congress, Office of Technology Assessment, Construction and Materials Research and Development for the Nation's Public Works, Washington D.C., June 1987.

U. S. Congress, Office of Technology Assessment, Changing By Degrees: Steps to Reduce Greenhouse Gases, USGPO, Washington, D.C., 1991.

U.S. Department of Energy, A Compendium of Options for Government Policy to Encourage Private Sector Responses to Potential Climate Change, (DOE/EH-0103), 1989.

U.S. Department of Energy, Energy Conservation Trends: Understanding the Factors That Affect Conservation Gains in the U.S. Economy, (DOE/PE-0092), 1990.

U.S. Department of Energy, CE Multi-Year Program Plan, Chapter 4, Industrial Technologies, Office of Industrial Technologies, January 31, 1991.

U.S. Department of Transportation, Moving America, New Directions, New Opportunities: A Statement of National Transportation Policy Strategies for Action, February, 1990.

Wilbanks, T. J., Institutional Issues in Energy Research and Development Strategies for Developing Countries, Oak Ridge National Laboratory, Oak Ridge, TN, 1990.

Wilbanks, T. J., The Outlook for Electricity Efficiency Improvements in Developing Countries, Oak Ridge National Laboratory, Oak Ridge, TN, 1990.

Wilson, A., Consumer Guide to Home Energy Savings, *Home Energy Magazine,* Berkeley, CA., 1991.

Chapter 4

ENERGY SUPPLY

Authors: Martin Hoffert, Seth D. Potter
Contributors: Jerry Delene, Peter E. Glaser, Michael Golay,
Harold M. Hubbard, Murali Kadiramangalam, Alfred Perry,
Myer Steinberg, Carl-Jochen Winter

4.1 INTRODUCTION

Global warming projected for the next century by current climate models arises mainly from increased radiative heating by carbon dioxide (CO_2) accumulating in the atmosphere from the combustion of fossil fuels (Houghton et al., 1990; MacCracken et al., 1990). Also contributing to the predicted warming is deforestation (which can increase atmospheric CO_2 by transferring carbon from large standing carbon pools in the land biosphere) and other greenhouse gases whose concentrations have increased along with the recent human population explosion, methane (CH_4), nitrous oxide (N_2O), and chlorofluorocarbons (CFCs). But the core greenhouse problem is carbon dioxide from fossil fuel burning.

The present world consumption rate of commercial energy is (Hammond, 1990, p. 316) 2.95×10^{20} J $y^{-1} = 9.35 \times 10^{12}$ W (9.4 TW). Some 95% of this is the heat of combustion of the fossil fuels (coal, oil and natural gas) burned each year, with the remaining 5% attributed to nonfossil primary electricity production (hydro, nuclear and geothermal — values of nonfossil electrical energy are based on the heat value of their electricity: 1 kW/hr $= 3.6 \times 10^6$J). Though it is not included in many statistical compilation, some 1.3 TW additional is probably consumed as "noncommercial" energy, mainly wood burned for fuel outside cash economies in developing nations (Goldemberg et al., 1988, p. 192). Thus, fossil fuels currently supply some 90% of a total world energy consumption of 10.7 TW.

Our civilization is beneficiary of a massive boon of nature: the fossil fuel reserve. Fossil fuels have accumulated in the Earth's crust over hundreds of millions of years from organic carbon produced by photosynthesis — a small fraction of which escaped oxidation by burial in peat bogs (coal) or marine sediments (gas and oil). Some 5.7×10^{12} kilograms of carbon are transferred from the lithosphere to the atmosphere each year (5.7 Gt C y^{-1}) by fossil fuel combustion as CO_2 gas. At this rate, reserves of economically recoverable natural gas and oil will be gone in the next 50 years or so, though coal could continue to provide energy for several hundred years. Clearly, the convenience, cleanliness and transportability in pipelines of gas and oil favor their use over coal. If depletion of oil and gas were the problem, a synthetic fuel technology in which gaseous and liquid hydrocarbons were derived from more abundant coal (and perhaps shale) would seem a viable solution. Humanity would have more than a century to face the need for nonfossil energy sources. Until recently, such a synfuel strategy was conventional wisdom in the energy industry. Global warming has changed the game by raising the specter of an early transition from fossil fuels, possibly as soon as the beginning of the 21st Century.

How much of this could be produced by nonfossil sources if we really had to? How much of the increased energy demand of the next century could be produced by nonfossil sources? And how rapidly could it be deployed?

4.2 ENERGY AND GLOBAL CHANGE

Global energy supplies are constrained by technological, economic and, increasingly, environmental considerations. In contrast to the technological optimism prevailing in the early part of this century are present-day concerns over undesirable environmental impacts generated by technology — photochemical smog, nuclear reactor accidents, toxic wastes, acid rain, oil spills, deforestation, ozone depletion, and endangered species, to name a few. In addition to engineering feasibility and economic viability, contemporary criteria for evaluating any new technology include its compatibility with natural ecosystems and sustainability over time.

At the United Nations Framework Convention for Climate Change in Rio de Janeiro in June 1992, representatives of the Earth's human populations agreed to "stabilization of greenhouse gas concentrations in the atmosphere that would prevent dangerous anthropogenic interference with the climate system (Bolin, 1994)." There is no consensus yet on what constitutes "dangerous interference." Carbon cycle models indicate that major fossil fuel emission reductions may be needed to stabilize atmospheric CO_2 and climate to acceptable levels (Harvey, 1990; Jain et al., 1994; Wigley, 1995). "Integrated assessments" of global change — which link the causal chain of emissions to climate to environmental impact and perform cost/benefits analyses for different mitigating strategies — are only now being developed (Naki´cenovi´c et al., 1994). Despite uncertainties, it is within the range of outcomes that fundamental changes in worldwide energy consumption and production will be needed to mitigate adverse impacts of climate change from humankind's greenhouse gases.

4.2.1 Fossil Fuel Reserves and the Carbon Cycle

The viability of energy alternatives depends on the amount, cost and environmental impact of fossil fuel reserves. Although fossil energy resources and reserves are fairly well known, real uncertainties exist. And there is sometimes an "apples vs. oranges" confusion in the literature from imprecise definitions of the terms "energy resources" and "energy reserves."

The *total resource base* of an energy source is defined here as the combination of undiscovered and identified, subeconomic as well as economic, concentrations of naturally occurring solid, liquid and gaseous materials in the Earth's crust; the *accessible resource* is that subset of the resource base that can be captured, mined, or extracted by current technology or technology which may be available in the near future (regardless of economics); and the *proven reserves* are that subset of the accessible resource which is identified and can be economically and legally extracted to yield useful energy (SERI, 1990). Oil and gas fuels, for example, are formed at elevated temperatures in the subsurface by the transformation of the organic remains of marine organisms, and coal from undecayed land plants (Tissot and Welte, 1978). The amount of buried organic carbon (kerogen) is thousands of times greater than oil and gas reserves, but it is too diffusely distributed to be economically viable. Similarly, there is much more uranium in the Earth's crust than what is economically extractable as powerplant fuel.

Table 4.1 summarizes energy and carbon contents of proven reserves of coal, oil, and natural gas and uranium-235 (the fuel of conventional fission reactors). The total carbon content of the fossil energy reserve based on these numbers is ~2500 Gt C (1 Gt C = 10^{15} g C), excluding shale. Present estimates of oil shale are 4–5 times the crude oil reserve (Tissot and Welte, 1978, p. 235), or ~500 Gt C of carbon additional. The proven fossil fuel carbon reserve based on present technology is therefore ~3000 Gt C. We estimate that the accessible resource could likely double to ~6000 Gt C given improvements in mining and extraction technologies available in the 21st century. Historical energy consumption records indicate ~220 Gt C have already been transferred from the lithosphere to the atmosphere and oceans since preindustrial times as fossil fuel CO_2 (Sundquist, 1993).

For conservation of mass, the sources, sinks and accumulation of elemental carbon cycling through the Earth system must balance instantaneously and over the long term. Since each 2.13 Gt C remaining in the atmosphere increases atmospheric CO_2 concentration 1 ppm, the observed increase in atmospheric CO_2 concentration from its preindustrial value ~279 ppm (594 Gt C) to its present value ~354 ppm (754 Gt C) (Boden et al., 1991) represents ~160 Gt C added to the atmosphere. Current carbon cycle models indicate the oceans absorbed an additional ~140 Gt C since preindustrial times (Sundquist, 1993). The sum of atmosphere and ocean carbon "sinks" is therefore ~300 Gt C. This is not quite enough to balance the

Table 4.1   Proven energy reserves and their carbon content. (Hammond, 1990, p. 320.)

Energy Source [size]	Energy Content[a] [10^{21} J]	[GW-yr]	Carbon Content[b] [1 Gt C = 10^{15} g C]
Hard coal [2.77 x 10^6 mtce]	77.3	2,450,000	1900
Soft coal [1.27 x 10^6 mtce]	17.7	561,000	440
Oil [0.124 X10^6 mtoe]	5.2	165,000	104
Natural gas [0.109 x 10^6 bcm]	4.2	133,000	57
Uranium [2.34 x 10^6 tu] (recoverable at < $130 kg)	1.4	44,000	0
Totals	105.8	3,353,000	2501

[a] Energy conversion factors: 1 million tonnes oil equivalent = 41.87 PJ; 1 billion cubic meters natural gas = 38.84 PJ, 1 million tonnes coal equivalent (hard coal) = 27.91 PJ, 1 million tonnes coal equivalent (soft coal) = 13.96 Pl; 1 tonne uranium metal = 0.58 PJ (based on 0.72% ^{235}U isotope abundance) (10^6 PJ =10^{21} J = 31,700 GW-yr).

[b] carbon conversion factors: 1 PJ oil = 20.0 Gt C; 1 PJ natural gas = 13.6 Gt C; 1 PJ coal = 24.6 Gt C.

estimated ~380 Gt C that has entered the atmosphere/ocean system from carbon "sources" from 1750–1990: ~220 Gt C from fossil fuel burning plus ~160 Gt C from changes in land use (Sundquist, 1993).

If the ~80 Gt C excess of sources over sinks is real (larger than uncertainties in source and sink estimates), then some "missing sink" is needed to balance the carbon budget. The most likely candidate is increased carbon uptake by terrestrial biomass fertilized by higher CO_2 and nitrate in the environment.

The ratio of CO_2 remaining in the atmosphere to the total CO_2 input from 1750–1990 (the mean airborne fraction) was ~160/380, ~42%. In general, the airborne fraction depends on the time-dependence of the source function. It can go as high as 90% or more if emissions increase faster than the oceans absorb excess CO_2. The asymptotic airborne fraction approached after anthropogenic CO_2 emissions have ceased does not approach zero but a value in the range of 13–30% depending on the amount of carbon injected to the system (Maier-Reimer and Hasselmann, 1987). Still, the 220 Gt C emitted by fossil fuel burning thus far is a small fraction of the carbon in the fossil fuel reserve. Given the pressures of population and economic growth, burning of the global fossil energy reserve to depletion seems a likely outcome in the absence of global energy supply alternatives.

4.2.2 How Much Fossil Fuel Can be Burned?

Scenarios for rapidly burning fossil reserves to depletion typically have CO_2 emission rates increasing from the present ~5.7 Gt C y^{-1} four or five times by the year 2100, and then declining over the next 300 years such that ~6000 Gt C are transferred from the lithosphere to the atmosphere. This carbon represents the present fossil reserves plus a fraction of the accessible reserves that might be economically exploited in the next century. According to current carbon cycle models, such a rapid burning scenario will produce ~1200 ppm atmospheric CO_2 by the year 2100, with a subsequent peak of ~2200 ppm by the year 2200, followed by a slow decline to ~800 ppm by the year 5000 (Walker and Kasting, 1992). These values of CO_2 are large enough to trigger a global greenhouse unprecedented since the Cretaceous era 100 million years ago when the Earth may have been ~10°C warmer (Hoffert, 1990).

Even "freezing" the global fossil fuel burning emission rate at the present rate ~5.7 Gt C y^{-1} would result in CO_2 levels >500 ppm by

2100, and this would continue to rise until the fossil fuel supply is depleted (Krause et al., 1989; Wigley, 1995). And for such a constant emission scenario, only ~570 Gt C would be burnt over the next hundred years — only 10% of the 21st century's accessible reserve. Walker and Kasting (1992) explored as a possible longterm goal whether it might be possible to hold atmospheric CO_2 below 500 ppm while still exhausting fossil fuel reserves. They found that sustainability in this long-term sense would only be possible if emissions were reduced by a factor of 25 — not the factor of two sometimes quoted in the press. The conclusion is inescapable that keeping atmospheric levels below 500 ppm would probably require switching to nonfossil energy supplies long before fossil fuel reserves are depleted.

The environmental impacts of global warming are still controversial, whereas the economic stake in the present energy system is well-established and enormous. Accordingly, energy response strategies proposed thus far have emphasized improved energy efficiency and switching from high CO_2 to lower CO_2 emitting fossil fuels, as opposed to abandoning fossil fuel technology entirely. In the coming decades, as increases in carbon emissions come predominantly from the developing world, the issue of equity is likely to loom importantly.

It is possible that improved end use efficiency and switching from high to low CO_2 emitting fossil fuels could mitigate some of humanity's greenhouse gas emissions and their greenhouse warming. Energy conservation, efficiency increases and fuel switching to lower emitting fossil fuels can also buy time as human populations shift to renewable solar, fission breeder reactors or fusion energy (Harvey, 1990). But it is only prudent to vigorously research nonfossil global energy systems as a hedge against the worst case scenarios.

4.3 ENERGY SUPPLY AND WEALTH

A historic goal of nation-states is to increase, or at least to maintain, their economic productivity, as measured by gross national product (GNP) or (more recently) gross domestic product (GDP). GNP is the value of all goods and services produced by a country measured in a convertible currency; GDP is similar, but includes the effect of trade imbalances with other countries. The present gross world product

(GWP) is ~$21 trillion (21 x 10^{12} $US); with the United States the largest national economy (GDP ~$5.6 trillion) followed by Japan (GDP ~$3.3 trillion) (Hammond, 1994).

Although economic activity is increasingly globalized, it remains useful to consider the existing nation-states as a set of data points from which properties of large-scale (macro) economies can be derived. Among other things, this approach supports macroeconomics theories that take a thermodynamic perspective (emphasizing *production* of goods) over neoclassical economics (emphasizing *exchange* of goods according to subjective human preferences).

Cleveland et al. (1984) rationalize energy-based macroeconomics as follows: "Production is the economic process that upgrades the organizational state of matter into lower entropy goods and services. Those commodities are allocated according to human wants, needs and ability to pay. Upgrading matter during the production process involves a unidirectional, one-time throughput of low entropy fuel that is eventually lost (for economic purposes) as waste heat. Production is explicitly a work process during which materials are concentrated, refined, and otherwise transformed. Like any work process, production uses and depends on the availability of free energy. The laws of energy and matter control the availability, rate, and efficiency of energy and matter use in the economy and therefore are essential to a comprehensive and accurate analysis of economic production." And consequently, the economic productivity of nations should be directly linked to their energy consumption. This is observed (Figure 4.1).

Whereas most economic models are open-ended with regard to economic growth, GDP is clearly constrained by the supply of energy and other resources, and ultimately by environmental degradation (Meadows et al., 1992). The Malthusian question is how these constraints work and whether they can be indefinitely overcome by technology. The limits to growth are not obvious. In a pioneering study, Revelle (1985) estimated that bringing an additional 300 million hectares under cultivation for food production and 1000 million hectares for biomass energy plantations would support global population growth from the present 5.3 billion to an estimated 10 billion by the middle of the next century — citing the main limitation as soil erosion. It is now evident that such expansion of cultivated land into natural ecosystems could result in major loss of biodiversity — particularly since unprecedented species extinctions have already been created by humanity's land use appropriations thus far (Wilson, 1992; Hammond, 1994).

4.3.1 Economic Productivity of Energy

The developed world has essentially stabilized its population and enjoys the highest per capita income in human history. But because of the lopsided distribution of wealth between rich and poor nations — exacerbated by rich vs. poor disparities in population growth rates — the coming decades may see conflicts between economic growth by the developing world and preserving the environment from adverse effects of fossil fuel-induced climate change. These objectives might be at least partly reconciled by increasing the economic productivity of energy (Goldemberg et al., 1988; Redi and Goldemberg, 1990).

Mid-range projections by the Intergovernmental Panel on Climate Change based on United Nations data show global population rising from the present 5.3 billion to 8.4 by 2025 and 11.3 by 2100, and CO_2 emissions rising from the present 5.7 billion tonnes carbon per year to 10.7 by 2025 and 19.8 by 2100 (IPCC scenario IS92a; Leggett et al., 1992). In this scenario ~95% of population growth and ~75% of carbon emissions growth by the year 2025 comes from developing countries. By the middle of the next century the world population will be ~10 billion, of which less than 15% — some 1.4 billion people — will live in presently developed countries and 8.7 billion will live in presently less-developed countries. The relationship between energy supply and economic development is critical for developing a global strategy to reduce greenhouse gas emissions.

For the global population as a whole, present per capita GDP, energy consumption and carbon emission rates are

$$\text{GDP} = \frac{21 \times 10^{12} \, \$US/yr}{5.3 \times 10^9 \, \text{people}} \sim 4000 \, \$US \, yr^{-1} \text{-person}^{-1}$$

(Eq. 4.1)

$$\text{energy consumption} = \frac{10.7 \times 10^{12} \, W}{5.3 \times 10^9 \, \text{people}} \sim 2 \, kW \, \text{person}^{-1}$$

(Eq. 4.2)

$$\text{carbon emissions} = \frac{5.7 \times 10^9 \text{ tonne carbon}}{5.3 \times 10^9 \text{ people}} \sim 1 \text{ tonne C person}^{-1} \text{ yr}^{-1}$$

(Eq. 4.3)

These averages are important reference points. To address the energy supply problem we need to also consider the *distribution* of wealth and energy consumption over the Earth's nation-states.

Figure 4.1 shows per capita GDP (GDP/N) versus per capita energy consumption (E/N) of United Nations member states plotted on a log-log scale.

That logarithmic scales spanning more than two orders of magnitude are needed to display the data is a measure of the inequity in the distribution of wealth of nations. For example, the mean energy consumption rate of the U.S. is about 10,000 W (~10 kW) while that of Bangladesh is only 60 W (~0.06 kW). For scenario analysis, countries are often aggregated into classes based on regions or economic groups; or simply into "developed" (primarily, the U.S., Japan and Western Europe plus the Former Soviet Union) and "developing" (everyone else). Goldemberg et al. (1988, p. 302) cite the present E/N of developed and developing nations as 6.3 and 0.54 kW, respectively.

Whereas global mean CO_2 carbon emissions are ~1 tC per person, the rich, industrial nations presently produce much more than their share of this greenhouse gas (Figure 4.2). Inhabitants of the United States emit five times as much carbon as the global mean. Part of the differences in per capita emission is due to differences in wealth, part is associated with differences in land use (deforestation, agriculture, etc.), and part can be attributed to differences in energy efficiency.

The correlation between per capita gross domestic product (GDP/N) and per capita energy consumption (E/N) on an individual nation basis is evident in Figure 4.1. Let the ratio P = GDP/E = (GDP/N)/(E/N) be the *economic productivity of energy*. P is the wealth created by a unit of raw energy consumption as hypothesized by energy theories of macroeconomics, and is influenced by the end-use energy conversion efficiency. It applies to renewables as well as fossil-fuel-based energy sources. The economic productivity of energy for the world as a whole based on present numbers is

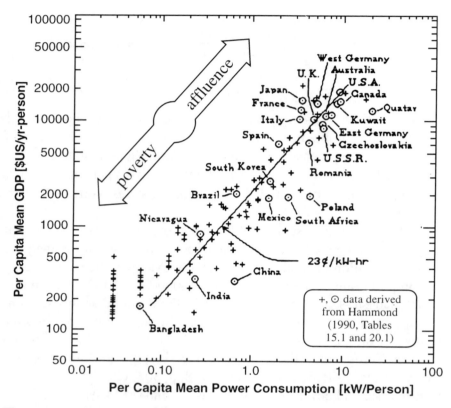

Figure 4.1   Distribution of per capita GDP (GDP/N) vs. per capita raw commercial energy consumption (E/N) for members of the United Nations. Named countries are denoted by circles, unnamed by crosses.

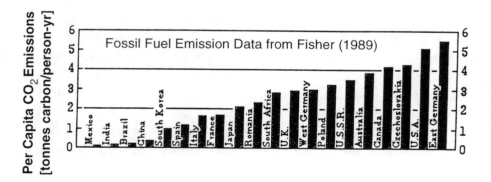

Figure 4.2   Per capita CO_2 emissions in tC person from fossil fuel burning for some nations shown in Figure 4.1.

$$P \sim \frac{\dfrac{4000\ \$US}{person\text{-}yr}}{\dfrac{2\ kW}{person}} \times \frac{1\ yr}{8760\ hr} \times \frac{100\ \cancel{c}}{1\ \$US} \sim 23\ \cancel{c}\,/kW\text{-}hr.$$

(Eq. 4.4)

The solid line in Figure 4.1 is the linear equation (GDP/E) = P x (E/N) for P = 23¢ GDP/kW-hr. Countries above this line are more efficient at converting energy to GDP than the world average (France, Japan, West Germany), countries below the line less efficient (USSR, Poland, China). Most of the variance over three log cycles (factor of a thousand) is accounted for by the GDP-energy correlation with a residual factor-of-two scatter associated with nation-to-nation differences in energy-to-GDP efficiency.

A legacy of our hydrocarbon economy is that energy (and energy per unit GDP) is often expressed as equivalent oil amounts.[1] Crude oil is a mixture of petroleum liquids and gases in various combinations. However, conversion factors between barrels or metric tons of "oil" and energy have become standardized (Häfele et al., 1981):

1 barrel oil equivalent (42 gallons) =
1 boe = 6.12×10^9 J = 1700 kW-hr

1 metric ton oil equivalent =
1 toe = 44.76×10^9 J = 12,430 kW-hr

The energy consumed in producing a unit of GDP — the inverse of the economic productivity of energy P — is the *energy intensity* I = E/GDP = (E/N)/(GDP/N) = P^{-1}. The global mean energy intensity in metric tons of oil equivalent per thousand U.S. dollars for the world economy is

$$I \sim \frac{1\ kW\text{-}hr}{23\ \cancel{c}} \times \frac{1\ mtoe}{1.243 \times 10^4\ kW\text{-}hr} \times \frac{100\ \cancel{c}}{1\ \$US} \times \frac{1000\ \$US}{\$1000\ US} \sim 0.35\ \frac{mtoe}{\$1000\ US}$$

(Eq. 4.5)

[1]In Europe, oil amounts are reported in metric tons (tonnes); in Japan in kiloliters (kl). But in the U.S. and Canada, and colloquially throughout the world, the basic unit is the "barrel" (bbl). In his historical account, Yergin (1991) relates that when oil began flowing from the wells of Western Pennsylvania in the 1860s, "...desperate oil men ransacked farmhouses, barns, cellars, stores and trashyards for any kind

The value of I differs for individual nations as they move through different phases of economic development, and differs among nations at any given time. Such I-variations are often displayed to analyze effects of end-use efficiency. (Note that economies which are more efficient at converting raw energy use to GDP have *lower* values of I.)

Figure 4.3 shows that as Western Europe and North America developed, their energy consumptions grew faster than their GDPs. This is normally attributed to the building of industrial infrastructures. Roads, bridges, homes and heavy industry involved capital investment in materials that did not yield immediate GDP increases. The variation since 1840 of both I and P is shown (on vertical log scales) for the U.S., U.K., West Germany, France and Japan. The present global mean P = 23 ¢/kW-hr is also shown for reference. During their infrastructure-building periods, P dipped as low as 8–10 ¢/kW-hr in England and the U.S., although France and Japan were able to avoid P < 15 ¢/kW-hr. (Note that peaks in I are valleys in P.)

Advocates of efficient energy end-use technology to reduce CO_2 emissions from developing nations observe that materials can now be produced with less energy, and that smaller quantities of modern materials can replace larger amounts of older ones (Reddy and Goldemberg, 1990). Consequently, developing nations may be capable of achieving comparable levels of industrialization at a lower ratio of consumed energy to GDP growth. This is represented by the shaded area projections for developing nations. The critical questions regarding end-use efficiency are quantitative. Consider the optimistic scenario of Figure 4.3 in which the global mean P increases by a factor of two by 2040 to, say, 46¢/kW-hr. This could cut per capita emissions by 50% at constant GDP even with continued (but more efficient) fossil fuel energy technology.

But halving per capita emissions would essentially be canceled by the projected doubling of world population 50 years hence. And this does not address economic growth or "equity" issues. A factor of two

---

of barrel — molasses, beer, whiskey, cider, turpentine, salt, fish, and whatever else was handy. But as coopers began to make barrels specially for the oil trade, one standard size emerged, and that size continues to be the norm to the present. It is 42 gallons. The number was borrowed from England, where a statute in 1482 under King Edward IV established 42 gallons as the standard size barrel for herring in order to end skullduggery and divers deceits in the packing of fish... By 1866, seven years after Colonel [Edwin L.] Drake drilled his well, Pennsylvania producers confirmed the 42-gallon barrel as their standard..." Few petroleum workers have ever seen a "barrel," except in a museum.

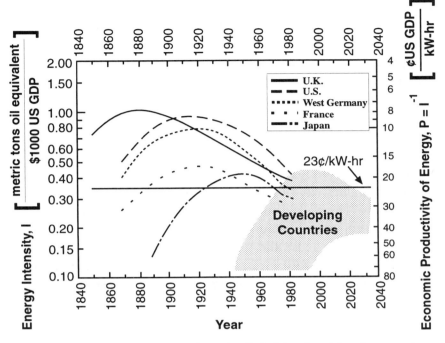

Figure 4.3   Evolution of energy intensity, I, (left hand scale) and economic productivity of energy, P, (right hand scale) for selected nations.
Because of improvements in materials science and energy efficiency, the maxima in I (minima in P) reached by developed countries
during industrialization have decreased over time. An objective of
energy end-use strategies for developing countries is to avoid
repeating the history of the industrialized world by using energy
more efficiently. (Adapted from Reddy, A.K. and Goldemberg, J.
Sci. Am., 263, 11, 1990. With permission.)

from energy efficiency cannot by itself bridge the gap between rich
(developed) and poor (developing) nations. To do this new global
scale nonfossil fuel energy supplies will be needed.

To address equity issues, some redistribution of per capita energy
from rich to poor countries is implied. Presumably, GDP could be
maintained in rich nations with declining per capita energy consumption by improving end-use efficiency. Goldemberg et al. (1988) advocate a "no-global-energy-growth" scenario in which per capita energy
consumption falls in developed countries from 6.3 to 3.5 kW by the
year 2020, and rises in developing countries from 0.54 to 1.2 kW.
These authors also consider — and discount, as requiring unavailable
capital investment funds — a scenario with global energy growth of
2% per year at constant per capita energy use by developed nations.
At a 2% overall energy growth rate, energy consumed per capita in the

developed world can remain at 6.3 kW and rise in the developing world to 2.2 kW. But global energy consumption must increase to 19.4 TW by the year 2020.

There is no way to set goals for the future economic and environmental state of the Earth without invoking human values. The morality of a world in which energy resources are shared more equitably by nations seems positive enough to move in that direction. But a zero sum game in which the poor get richer at the expense of the rich getting poorer is politically naive and may be akin to re-shuffling deck chairs on the Titanic. A better strategy is to combine increases in the economic productivity of energy with new energy supplies and modest per capita energy redistributions. Still, there is no optimum path independent of subjective human values.

We therefore propose the goal of a global economy by the year 2040 (20 years after Goldemberg et al.'s projections) with a population of 10 billion consuming 20 TW in which CO_2 emissions are cut 50% relative to the present. Global per capita energy consumption would stay the same as today's 2 kW. If P rose by a factor of two or more from improvements in energy efficiency, the Gross World Product could more than double, most of which could be distributed to less-developed nations. Five TW could still be generated by fossil fuel combustion; but global nonfossil energy technologies would have to be in place fifty years from now capable of producing 15 TW of thermal power free of greenhouse gas emissions — 150% the present global consumption rate.

4.3.2 Fossil Fuel Costs, Carbon Taxes, and the Cost of Electricity

To assess the global scale power supply it is necessary to consider engineering feasibility and cost. We can identify at this time only a few energy systems with the capacity to power human civilization. Will economics determine whether any of these will replace fossil fuels?

Classical economics holds that technologies are in a kind of Darwinian competition with each other in which the "invisible hand" of the market acts as natural selection. But the winners are not preordained. A particular energy technology like the light water nuclear reactor is no more inevitable than the elephant or hummingbird. More recent theories view the global economy as a complex adaptive system which is constantly making predictions based on its internal models of the world. If the greenhouse model of climate change is accepted by governments, that model would itself influence the development of energy supply technologies, regardless of whether the climatic changes predicted for fossil fuel burning would have occurred. This is no different from the way the economy may work in other regards.

John Holland of the Santa Fe Institute put it this way (Waldrop, 1992, p. 147): "…Complex adaptive systems have many *niches*, each one of which can be exploited by an agent adapted to fill that niche. Thus, the economic world has a place for computer programmers, plumbers, steel mills, and pet stores, just as the rain forest has a place for tree sloths and butterflies. Moreover, the very act of filling one niche opens up more niches — for new parasites, for new predators and prey, for new symbiotic partners. So new opportunities are always being created by the system. And that, in turn, means that it's essentially meaningless to talk about a complex adaptive system being in equilibrium: the system can never get there. It is always unfolding, always in transition… And by the same token, there's no point in imagining that the agents in the system can ever 'optimize' their fitness, or their utility, or whatever. The space of possibilities is too vast; they have no practical way of finding the optimum. The most they can ever do is change and improve themselves relative to what the other agents are doing."

Practically speaking, an optimum energy strategy for the world may not exist. Even if one could define a theoretical optimum utility function it would be impossible to predict. But there are better and worse solutions. And just as in evolutionary biology, where not all organisms are possible, there are physical and economic constraints on energy supplies. A practical upper limit on the cost of a global-scale energy source is when it becomes more expensive than the wealth it produces. Below this limit — and one seeks to stay as much below this limit as possible — incentives like carbon taxes can make nonfossil energy sources competitive earlier and thereby help limit CO_2 emissions.

Convenience, high energy density, relative cleanliness and ease of transportation have led oil to dominate as the leading fossil energy source in recent years. Since production began in 1861, crude oil prices have fluctuated with market pressures from < \$5/bbl in the mid-1960s to ~\$20/bbl after the Arab oil embargo of 1973 to as much as

~$40/bbl in the 1979–1981 oil panic (inflation-corrected prices in 1990 $US/bbl from Yergin, 1991, p. 785). The price of oil also varies day-to-day, and depends on the source, which is why it is traded as a commodity. At present, oil is relatively inexpensive. Crude prices quoted in the *New York Times* of Thursday, June 13, 1996 are:

	Wed.	Tue.
Saudi Arabian light $ per bbl fob	16.60	16.57
North Sea Brent $ per bbl fob	18.25	18.22
West Texas Intermed $ per bbl	20.08	20.13
Alaska No. Slope del. US Gulf Coast	18.60	18.71

For illustrative purposes, assume the cost of unprocessed fossil fuel is CF ~20 $US/boe. In cents per kilowatt-hour of chemical (or thermal) energy this representative fossil fuel cost works out to

$$CF \sim \frac{20\ \$US}{1\ boe} \times \frac{1\ boe}{1700\ kW\text{-}hr} \times \frac{100\ \cent}{1\ \$US} \sim 1.18\ \cent/kW\text{-}hr\ .$$

(Eq. 4.6)

The ratio of the global mean productivity of energy to the cost of fossil fuel energy is then P/CF ~ (23 $\cent$/kW-hr)/(1.18 $\cent$/kW-hr) ~ 20. In other words, the present productivity of energy implies a twenty-fold return on investment to the global economy from fossil fuel consumption. Conversely, the world's consumption of fossil fuels at the rate of ~ 5% of the GWP is >$1 trillion/yr — a prodigious spending rate. It is not surprising that the relatively few multinational corporations dealing in oil (and often other forms of fossil energy) have annual incomes exceeding the GDPs of most developing nations.

Fossil fuels are the most cost-effective global-scale energy sources available at present if "externalities" like the adverse environmental impact of global warming are ignored. The science of climate impact analysis is too uncertain at this point to provide reliable environmental costs as a basis for carbon taxes. Indeed, some argue that higher CO_2 levels should be assigned a credit associated with enhanced agricultural yields. But as we have discussed previously, it is within the range of outcomes that serious disruption to global ecosystems could occur if more than a small fraction of the fossil fuel reserve is burned.

In any event, nonfossil fuel energy technologies will significantly penetrate global energy markets only if their costs come down and/or fossil fuel costs go up. A policy to make alternative energy technologies more competitive is the imposition of carbon taxes (Edmonds and Reilly, 1985, 1986; Nordhous, 1991; National Academy of Sciences, 1992). Taking oil as a representative fossil fuel (1 mtoe = 0.89 tC = 12430 kW-hr), each $10/tCO$_2$ carbon tax increases the fuel cost per unit thermal energy by

$$\$10/\ tCO_2 \sim \frac{10\ \$US}{tCO_2} \times \frac{44\ tCO_2}{12\ tC} \times \frac{0.89\ tC}{12430\ kW\text{-}hr} \times \frac{100\ \cancel{c}}{1\ \$US} \sim 0.26\ \cancel{c}\ /kW\text{-}hr.$$

(Eq. 4.7)

carbon taxes would fall differently on different fossil fuels. Relative to oil (= 1.00), carbon emissions per unit energy are 1.23 for coal and 0.68 for natural gas. A tax of $10 per tonne of CO_2 emitted would therefore be ~ 0.32 ¢ for coal, ~ 0.26¢ for oil and ~ 0.18¢ for gas for the same kilowatt-hour of thermal energy. Among other things, carbon taxes tend to deplete oil and gas reserves earlier than a tax-free policy.

Based on the oil standard, a tax of ~$10/tCO$_2$ would raise the "fossil fuel" price ~20% (from ~1.2 to ~1.4¢/kW-hr) whereas ~$100/tCO$_2$ would more than triple it (from 1.2 to 3.8¢/kW-hr). Of course, these costs are representative, and a comprehensive analyses should include additional market variables; e.g., coal is cheaper than oil, but has a higher emission factor; synthetic fuels derived from coal are more expensive; and so on. Econometric models including these details indicate that a 50% reduction in fossil fuel carbon emissions would require taxes at the margin in the range 27–164 $/tCO$_2$ (100-600 $/tC) (Nordhous, 1991; Manne and Richels, 1992; Scheraga and Leary, 1992). An average carbon tax of $46/tCO$_2$ would double fossil fuel costs (CF + CTAX ~ 2.4¢/kW-hr), half the ratio of the productivity of energy to-the fuel cost (P/(CF + CTAX) ~ 10) and provide revenue of over a trillion $US worldwide (~5% of GWP).

It is not clear who would collect carbon taxes or to what use they would be put. If governments were the agents, the balancing of national budgets, health care, etc., are likely candidates. Alternately, multinational energy corporations might be more than willing to act as *de facto* tax collectors by raising prices, as OPEC did in 1973 —

particularly if antitrust litigation could be avoided. A more rational policy from our point of view would be to assign carbon tax revenue to research, development and demonstration of nonfossil fuel energy sources.

There are precedents for subsidizing alternative energy technologies. Grubb and Meyer (1993) analyzed some lessons from the last energy crisis: "The rapid development of wind power in California in the early 1980s was a product of 'sledgehammer' promotion. Generous federal tax credits, combined with state incentives, produced large subsidies for wind installations. In addition, the Public Utilities Policy Act mandated that utilities buy energy at its full avoided cost from independent generators, thus ensuring a market for wind power... Installation rates in California increased six-fold in one year, rising from 10 MW in 1981 to 60 MW in 1982, and by 1984 had reached 400 MW per year. By 1986, the cumulative investment in wind energy totaled about $2 billion, with the value of energy generated put at $100 million per year."

But rapid development of wind energy has its cautionary tales. Grubb and Meyer relate that, "...Many of the machines were of poor quality and broke down during the first season of operation. As one manufacturer complained, some of the early companies knew more about tax minimization than they did about engineering. Machines were often sited carelessly, and some were sold based on blatantly fraudulent promises... Such slipshod technology, as well as visions of thousands of motionless machines threatened to destroy wind power in the United States. However, several companies, aided by favorable state-financing packages, invested heavily in wind energy technology and helped advance the field significantly. The total net cost of the program to the California economy (minus fuel savings) is estimated to be about $500 million, but without such a program, it is unlikely that wind energy would have reached its current stage." The message to anyone viewing cost-effective wind farms today — for example, at the Altamont Pass near the Lawrence Livermore National Laboratory in California — is that tax incentives to alternative energy technologies can work if proper safeguards are in place.

A recent, if modest, threat to the fossil fuel infrastructure is the requirement in certain states that automobile manufacturers begin selling zero emission vehicles (ZEVs). Again, California is taking the lead. It has mandated that by 1998, 2% of all vehicles sold be ZEVs, rising to 10% by 2003. ZEVs are being promoted not to reduce CO_2

emissions, but to reduce photochemical smog precursors (hydrocarbons and nitrogen oxides). If electricity used to charge an electric car's batteries were produced by fossil fuel burning, the net effect would be an *increase* in CO_2 emissions. It is normally more efficient to use internal combustion engines in motor vehicles than to drive electric generators in remote fossil fuel-fired powerplants, transmit the electricity over power lines, charge onboard vehicle batteries, and deliver rotational energy to wheels via electric motors.

If the goal is smog reduction with reasonable performance, a compromise could be hybrid electric-internal combustion. Such cars would generate electricity onboard from a combination of internal combustion and regenerative braking. They require smaller batteries than electric cars while eliminating transmission line losses. In the future, hydrogen-powered vehicles might do even better. But with today's technology, "zero emissions" means battery-powered vehicles with low stored energy-to-mass ratios and high costs. To meet ZEV quotas, manufacturers would presumably have to spread these costs over their nonelectric fleets. This has fostered protests by auto makers and oil companies that electric cars are being subsidized contrary to normal marketplace competition.

Historically, many key technologies in the commercial sector have been stimulated by government policies. The Interstate Highway System initiated by President Dwight D. Eisenhower is a continuing federally-supported investment in the internal combustion engine. Commercial jet aircraft, computer chips and the Internet are spin-offs of military research. The tax-subsidized development of wind energy in California begun in the energy crisis years of the late 1970s helped to make wind the most cost-effective renewable energy source today. More recently, the Strategic Defense Initiative (SDI) of the Reagan and Bush administrations fostered remote sensing and space technologies, which has been tested in a joint DoD/NASA mission to orbit the moon by the Clementine space probe. It has been argued that a Strategic Environment Initiative (SEI) including tax incentives for new technologies and disincentives for old could likewise focus technological talent on environmentally appropriate energy technologies (Gore, 1992).

In any event, a transportation system aimed at slowing the growth of atmospheric CO_2 should employ a primary power source that minimizes CO_2 emissions. It is better, for example, to make H_2 for hydrogen-powered cars by water electrolysis using nonfossil fuel electricity than to use H_2 from steam-reformed methane — 99% of present-day industrial hydrogen is derived from natural gas, essentially methane. From a carbon emissions point of view, the goal of ZEVs should be to build an energy infrastructure in which nonfossil electricity (and/or hydrogen) is a universal energy carrier. Methodologically, cost comparisons of fossil fuel energy with or without carbon taxes versus renewables, fission, fusion and solar power satellites imply such a universal energy carrier.

Electricity is a thermodynamically higher quality of energy than heat. Most of the world's electricity is produced by fossil or nuclear heat engines with H_2O (water and steam) as the working fluid. In the classical Rankine cycle, superheated steam generated in a boiler passes through a turbine (driving an electrical generator) to a condensor, and is returned to the boiler as water. The advantage relative to a gas turbine is that it is cheaper to pump liquid from a low-pressure condensor to a high-pressure boiler than to compress a gas. However, the efficiency of a vapor/liquid cycle is lower than Carnot because cold fluid plunged into the hot boiler vapor mixes irreversibly.

Consider a typical boiler providing steam to a turbine at ~200 bars and T_{max} ~ 540°C (813 K), after which it enters a condenser at ~2 bars and T_{min} ~ 60°C (333 K). The Carnot efficiency is $(T_{max} - T_{min})/T_{max}$ ~ 59%. But even with regenerative heaters to improve performance, the theoretical efficiency of this steam turbine cycle is ~45% (Faires and Simmang, 1978, p. 258). Assuming efficiencies of 95% each for combustion-to-thermal and turbine-to-electrical power, the overall electrical conversion efficiency of a fossil fuel powerplant is thus only η_e ~ 40%. In that case, 1 kW/hr (thermal) ~ 0.4 kWe/hr (electrical). A typical generating plant fired by oil at $20/boe (1.18 ¢/kW/hr) has a fuel cost per unit electrical power of (1.18/0.4) ~ 3.0 ¢/kWe/hr exclusive of capital costs.

The levelized cost of electricity (COE) is the weighted average cost of energy production in ¢ per kWe/hr including capital investment recovery, fuel and nonfuel operating costs projected over the facility's economic lifetime,

$$COE = IC\left(\frac{FCR + OMR}{DUTY}\right) + \left(\frac{CF + CTAX}{\eta_e}\right).$$

(Eq. 4.8)

Here IC is the initial capital cost of the plant ($ per kWe installed x 100 ¢/$), FCR is the fixed charge rate and OMR the operation and maintenance rate (% of IC per yr), DUTY is hrs per year on-line (% time operational x 8760 hr.), CF is the cost of fuel and CTAX the carbon tax (¢ per kW/hr), and η_e the electrical conversion efficiency (the dimensionless ratio). The first term contains all plant costs except fuel and carbon taxes. We chose values for a reference oil-fired powerplant such that the first term is ~ 3.6 ¢/kWe/hr and COE (no carbon taxes) ~ 3.6 + 3.0 ~ 6.6 ¢/kWe-hr — close to the 7¢/kWe/hr cited as the average cost of electricity in the U.S.

Table 4.2 shows the effect of carbon taxes on the cost of electricity computed for our hypothetical average fossil-fuel powerplant. The Low (0–10 $/tCO$_2$), Med (10–100 $/tCO$_2$) and High (100–200 $/tCO$_2$) categories are similar to those employed by the National Academy of Sciences (1992) in their study of mitigation of greenhouse warming. Apart from their effect on reducing CO$_2$ emissions by encouraging energy conservation, carbon taxes at the 100 $/tCO$_2$ level would essentially double the cost of electricity to ~ 13¢/kWe/hr. Compare, for example, the cost of fossil fuel electricity in our Table 4.2 example with the cost of energy from renewable sources estimated by the Department of Energy in their report to the U.S. Congress in Table 4.3.

Based on these numbers, geothermal, wood-fired electric, hydroelectric and wind are all nearly cost-effective today, and would be significantly cheaper than fossil fuel electricity at 13¢/kWe/hr. Notice in Table 4.3 that terrestrial solar thermal and photovoltaic electricity

Table 4.2 Effect of carbon tax on cost of electricity (COE). Calculations assume an oil-fired powerplant @ $20/boe, initial capital cost = $1500/kWe, fixed charge rate = 15%/yr, operation & maint. = 3%/yr, duty cycle = 85%; hence, generating cost per electrical energy output (less fuel cost) = 3.6¢/kWe/hr. Also assumed are electricity/fuel conversion efficiency = 40% and carbon emission factor = 0.122 tC/boe = 0.89 tC/toe.

	Carbon Tax		**fuel+Ctax**		**electricity**
Tax Category	**$/tCO$_2$**	**$/tC**	**¢/kW/hr**	**¢/kWe/hr**	**¢/kWe/hr**
None	0	0	1.2	3.0	6.6
Low	0–10	0–37	1.2–1.4	3.0–3.5	6.6–7.2
Med	10–100	37–370	1.4–3.8	3.5–9.5	7.2–13.1
High	100–200	370–730	3.8–6.4	9.5–16.0	13.1–19.6

Table 4.3   Projected levelized cost of electricity (COE) for terrestrial renewable energy technologies in current U.S. ¢/kWe/hr. (After Bradley et al., 1991, Appendix E.)

Technology	1990	2030
Geothermal	5.5	4.4
Wood-fired electric	7.2	7.2
Hydroelectric	8.3	8.3
Wind	9.5	5.0
Solar thermal electric	18.7	9.4
Photovoltaics	54.3	5.1

cannot provide cost-effective electricity today even with high carbon taxes, though they might be by the year 2030. However, the potential magnitude of an energy source is important. Even carbon taxes in the medium to high range would be insufficient to drive fossil fuels from the market unless their replacements could collectively supply the ~10 TW of global demand.

To summarize: low cost, high energy density, and availability will tend to continue humanity's reliance on fossil fuels in the near term unless there are incentives to change. Although there is no "optimum" energy supply, a nonfossil energy infrastructure capable of powering human civilization at >20 TW will be needed sooner or later — perhaps as early as the middle of the next century.

Displacing fossil fuels by alternate energy sources prior to depletion of recoverable oil, gas and coal would be fostered by carbon taxes. A worldwide investment as large as ~5% of GWP might be needed to cut CO_2 emissions 50% — on the order of $1 trillion per year. Carbon taxes could work only if they were international — particularly if revenue were targeted to research, development and demonstration of new technologies. Developed nations could probably reduce emissions by a factor of two by well-understood improvements in energy conversion. But new ideas, technology and financing are needed to provide greenhouse-free energy to the developing world where 95% of global population growth is projected by the year 2040. An appealing vision is the implementation of a "green energy" infrastructure compatible with the biological species diversity of the planet, and sustainable over time scales comparable to recorded human history.

4.4 TOWARD A NONFOSSIL FUEL CIVILIZATION

It is well-known that cost estimates of innovative energy technologies (for example, those in Table 4.3) are problematical because of the difficulty of estimating economies of scale and future manufacturing efficiencies. A technological optimist could, for example, cite dramatic decreases in unit costs of very large scale integrated circuits (VLSI) in personal computers and consumer electronics. Few would have predicted in 1970, for example, that transistors would be virtually free today — you can now buy a transistor for 4,000-millionths of a cent (Gilder, 1993). A pessimist could cite erroneous predictions by the Atomic Energy Commission of the 1950s that nuclear power would be "too cheap to meter." Is there a methodology that can help policy makers make choices regarding alternate energy technologies?

Eminent scientists have made notoriously wrong predictions of technological change. John von Neumann, father of the modern electronic computer and one of the 20th century's great mathematicians made two such predictions in the early 1950s: first, that in the future computers would become so complex that only governments could afford them, and second, that computers would be able to make accurate long-range weather forecasts (Kaku, 1994). He based his predictions on linearly extrapolating trends at the time, and he was wrong on both counts: computers are now inexpensive enough to have become home appliances, whereas the weather is known to be statistically unpredictable more than three to five days in the future. In the first instance, von Neumann missed the advent of quantum electronic devices like VLSI; in the second, he missed the significance of nonlinearities on the predictability of the fluid dynamical equations governing atmospheric motion (chaos).

In the short run — a few years — it is reasonable to quantitatively extrapolate existing technologies linearly. On the decadal to century time scales of global warming it is more likely that qualitative breakthroughs will become dominant, and that new industries will open up in unexpected places. One reason why the system behaves nonlinearly is that technologies form "clusters" that reinforce each other and create whole new capabilities. Jesse Ausubel (1994), who studies the impact of technology change on global warming, realizes this: "...Imagining how the clusters will affect lifestyles and restructure the economy, and thus affect emissions and vulnerability to climate, is a tremendous intellectual challenge."

Science fiction writers may have more intuitive feeling for nonlinear technology forecasting than establishment scientists. A recently discovered manuscript by Jules Verne written in the 1860s, *Paris in the 20th Century*, was rejected for publication in its time because it pictured a future too strange to be credible (Riding, 1994): In this work, Verne imagined a future in the 1960s where people traveled by subway and in gas-driven cars, where they communicated by fax and telephone, where they used computers, and where "electric concerts" provided entertainment. In this world, everyone could read but no one read books. It was a society dominated by money where destitute homeless people roamed the streets. Strange indeed.

In contrast to what is often assumed in studies of global warming mitigation, energy technologies of the 21st century are likely to differ from those of the 20th century as much as 20th century technology differed from 19th. If the past is any guide, global warming mitigation studies that assume progressive decarbonization of fossil fuels by switching from coal to oil to methane, increasing end use efficiency of energy conversion devices, and even the more widespread use of existing renewable energy technologies (wind, terrestrial solar and biomass) will prove too conservative. Polls taken in the 1920s predicted that we would have, within a few decades, huge fleets of blimps taking passengers across the Atlantic (Kaku, 1994). The reality was that Lindbergh crossed the Atlantic in 1929 in the heavier-than-air *Spirit of St. Louis,* while the *Hindenberg* explosion of 1938 effectively ended air travel by blimps. The progress of theoretical aerodynamics and propulsion technology during World War II, particularly the invention of the aviation gas turbine, led to the present era of commercial jet aviation as the dominant long-distance transport mode. Given their high frontal area and drag, there is no way that lighter-than-air craft can compete for high-speed passenger service. Again, the technology of transportation turned on the physics.

What we have seen in the past century is that technological progress was determined more in the long run by limitations and opportunities stemming from fundamental scientific principles than by naive extrapolations of existing technology. It seems fruitful, therefore, to assess 21st Century nonfossil fuel energy technologies as imaginatively as possible, but always constrained by our understanding of the physics. The laws of physics determine what *cannot* be achieved — nonconservative energy sources, perpetual motion machines, efficiencies higher than those allowed by thermodynamics, and (probably) "cold fusion" — but they do not guarantee what *can*. They are necessary but not sufficient conditions. What can be done is likely to be constrained by available energy fluxes and reservoirs, innovations in energy transmission and storage, and the impact of harvesting and redistributing energy flows on the Earth's environment and ecosystems.

4.4.1 Renewable Energy

Many (but not all) sustainable energy sources are forms of solar energy. The sun is a normal main sequence star (spectral class G2) powering the Earth's atmospheric and oceanic circulations, hydrological cycle and biosphere. It radiates energy derived from hydrogen to helium fusion in its core at a slowly increasing rate — the sun is more than 25% brighter now than when the solar system formed ~4.6 x 10^9 yr ago. Its present radius, $R_{sun} \approx 6.96$ x 10^8 m, and luminosity, $L_{sun} = 3.90$ x 10^{26} W, correspond to an energy flux at the solar surface of $F_{sun} = L/(4\pi R_{sun}^2) = 64.1$ x 10^6 W m^{-2} (Allen, 1976). It follows from the blackbody radiation law ($F = \sigma T_{eff}^4$ where $\sigma \approx 5.67$ x 10^{-8} Wm^{-2} K^{-4} is the Stefan-Boltzmann constant) that the effective temperature of the sun's visible surface (the photosphere) is $T_{sun} = (F_{sun}/\sigma)^{1/4} \approx$ 5800 K.

Most of the energy radiated from sun to the universe as a whole is, from our point of view, wasted. Solar flux on a surface perpendicular to the sun's rays decreases inversely with the square of the distance from its center. By the time solar photons have spread to the Earth's mean orbital distance, a = 1 A.U. = 1 astronomical unit ≈ 1.5 x 10^{11}m, the flux per unit area (solar constant) has dropped to $S_0 = F_{sun}$ x $(R_{sun}/a)^2 \approx 1380$ W m^{-2}.

4.4.1.1 Terrestrial Solar Energy

We are mainly concerned in this section with energy fluxes available at the surface of the Earth, of area $4\pi R_E^2 = 5.1$ x 10^{14} m^2, where $R_E \approx 6.37$ x 10^6 m is the Earth's radius. The *hectare* is a convenient SI unit employed in scientific literature for areal measure (1 ha = 10^4 m^2). Land, ocean, and total surface areas in 10^6 ha are (Sverdrup et al., 1970):

$$A_l = 15,000 \text{ x } 10^6 \text{ ha}$$
$$A_w = 36,000 \text{ x } 10^6 \text{ ha}$$
$$A_E = 51,000 \text{ x } 10^6 \text{ ha.}$$

Changing land use patterns, as the human population grew from 600 million in 1700 to 4,430 million in 1980, based on recent estimates, and assuming that "badlands" (deserts, ice sheets, etc.) remained constant at 11% of A_1, are shown in Table 4.4.

Table 4.4   Global land use in 10^6 ha from 1700 to 1980. (Richards, J.F., in *The Earth as Transformed by Human Action*, Turner III, B.L. et al., Eds., Cambridge University Press, New York, 1990.)

Year	Forests & Woodlands	Grassland & Pasture	Deserts & Icesheets	Agricultural Croplands
1700	6,215	6,860	1,661	264
1850	5,965	6,837	1,661	537
1920	5,678	6,748	1,661	913
1950	5,389	6,780	1,661	1,170
1980	5,050	6,788	1,661	1,501

Despite a "Green Revolution" in crop yields subsidized by energy in the form of fertilizers and farm machinery, humankind's agriculture has grown from a few percent of global land use preindustrially to ~ 10% today, primarily at the expense of temperate and tropical forests.

Per unit surface area of the spinning Earth, solar flux incident at the top of the atmosphere is $S_0/4 = 345$ W m^{-2}, of which 70% is absorbed by the atmosphere, clouds and surface: That is, $(S_0/4)(1 - \alpha) \approx 242$ W m^{-2} is absorbed by the atmosphere/earth system. The factor of 4 is the ratio of the Earth's surface to disk area, and $\alpha = 0.30$ is the Earth's visible reflectivity (albedo). Energy conservation requires solar energy absorbed to be balanced in the steady state by radiative cooling to space at the blackbody temperature, $T_{eff} = [(S_0(1 - \alpha)/(4\sigma)]^{1/4} \approx 255$ K. Radiative cooling is affected by infrared absorbing atmospheric gases (H_2O, CO_2, O_3,...). These warm the surface ~ 33 K above T_{eff} (the greenhouse effect). For the Earth as a whole, $T_s \approx 288$ K. In addition, differential absorption of solar energy at high and low latitudes creates an equator-to-pole surface temperature gradient which drives atmosphere and ocean circulations (Peixoto and Oort, 1992).

Of the 345 W m^{-2} solar flux at the top of the atmosphere, only 51% reaches the surface. The rest is absorbed by the clear sky and clouds or reflected back to space (Wallace and Hobbes, 1977, p. 321). Thus $0.51 \times (S_0/4) = 180$ W m^{-2} is a typical solar flux at the surface. Because of cloudiness, seasonality and the rotation of the Earth, there is substantial intermittency in the local values of surface solar insolation. $F_s \approx 180$ W m^{-2} should be considered a long-term average over diurnal and seasonal cycles and over the Earth's entire surface area. A fraction of this flux is available, in principle, for humankind's energy end uses: space heating, transportation, industrial processes and electricity.

Apart from land, the direct conversion of solar flux to electricity or fuel places stringent requirements on materials. These must absorb the solar flux to produce electronic excitations, separate the electron-hole pairs making up the exciton, and convert the separated charge either into electrical energy in an external circuit or into chemical fuel. These steps occur in nature (in photosynthesis) as well as human technology (in photovoltaic, or PV, solar cells). There is a common body of theory linking these processes (APS, 1979).

A key material property for solid state solar energy conversion is the *bandgap energy*, $\varepsilon_o = h v_o$, where $h \sim 6.63$ x 10^{-34} J-s ~ 4.14 x 10^{-15} eV-s is Planck's constant and v_o the bandgap frequency. The monochromatic efficiency of a single bandgap absorber is (Schockley and Queisser, 1961; Thorndike, 1976):

$$\eta(v,v_o) = \left\{ \begin{array}{ll} (v_o/v); & v \geq v_o \\ 0 & v < v_o \end{array} \right\},$$

(Eq 4.9)

where v is the frequency of light. Light from a monochromatic laser at the bandgap frequency v_o is absorbed with an efficiency of 100%. But more energetic photons can only transfer a fraction of their energy, and less energetic photons are not absorbed at all.

The high efficiency of monochromatic light absorption by PV cells might be exploited in special applications where it is desirable to transmit a narrow beam of energy over long line-of-sight distances. For example, a recent proposal to power a lunar outpost during exploratory missions to the moon is to beam the energy from Earth-based sites using high-powered lasers tuned to the bandgap frequency of a PV receiver on the lunar surface (Matthews et al., 1994).

For the (multicolored) spectral distribution of sunlight per unit frequency $I_v(v)$ available in the natural environment, the efficiency of photon energy transfer is

$$\overline{\eta}(\nu_0) = \frac{\displaystyle\int_0^\infty \eta(\nu,\nu_0) I_\nu\,(\nu)d\nu}{I},$$

(Eq. 4.10)

where $I = \int_0^\infty I_\nu\,(\nu)d\nu$ is the frequency-integrated sunlight.

Power engineers (and natural selection) favor materials with bandgap energies (ε_0 or ν_0) that optimize the integrated energy absorption efficiency, $\overline{\eta}(\varepsilon_0)$. Typical bandgap energies are $\varepsilon_0 \sim 1.1$–1.6 eV for single and polycrystalline silicon (Si), gallium arsenide (GaAs-GaAlAs) and related PV materials, and $\varepsilon_0 \sim 1.8$ eV for green plant chlorophyll photosynthesis (APS, 1979). The theoretical peak efficiency of photon absorption for normal sunlight at the Earth's surface is $\sim 35\%$ (Thorndike, 1976, pp. 33–35), while the realized peak efficiencies of PV cells is $\sim17\%$ for Si crystals and $\sim22\%$ for GaAs-GaAlAs crystals (APS, 1979). In principle, more efficient cells could be created using multiple light-absorbing layers of different bandgap energies (Vos, 1980).

In practice, commercialization of PV power is limited more by manufacturing costs than by energy efficiency, as such. In the case of silicon, the raw material (sand) is virtually free and yet the current cost of PV electricity is >50 ¢/kWe/hr (Table 4.3). Recent efforts to lower PV costs emphasize amorphous (as opposed to crystalline) semiconductors, which have slightly lower efficiencies but are cheaper to fabricate (Ovshinsky and Madan, 1983).

Liquid hydrocarbon fuels are perhaps the greatest challenge to replace with solar energy. Because of its high energy per unit mass, hydrogen derived from water electrolysis has many advantages as an energy carrier, although a prime power source is, of course, needed, and problems of onboard vehicle storage remain (Winter and Nitsch, 1988). As early as 1923, J.B.S. Haldane suggested electrolytic hydrogen production from wind power; in 1927 A.J. Stuart proposed hydrogen from hydroelectricity; and in 1960, J. O'M. Bockris proposed solar power as an energy source for hydrogen electrolysis (Bockris, 1980). Advantages of solar hydrogen are that it buffers intermittent and spatially nonuniform sunlight. Its main disadvantage is high cost (on top of high PV costs).

Ogden and Williams (1989) advance an ingenious argument for "mining" the Earth's crust for amorphous silicon solar power versus mining it for uranium and thorium to fuel breeder reactors: "Even though the energy released in the fissioning of a single uranium nucleus is 100 million times greater than the 'energy released' when a photon is absorbed in amorphous silicon, a uranium atom can fission only once, whereas a silicon solar cell can repeatedly absorb photons and convert solar energy into electricity. An amorphous solar cell contains an amorphous silicon layer about 1 micron thick, amounting to some 3 grams of silicon per square meter of cell area. A 15-percent efficient PV system operated in the southwestern United States, where the insolation averages about 250 watts per square meter, would thus produce about 3300 kW-hr per gram of silicon over the expected thirty-year PV system life — about the same as the amount of electricity from a gram of nuclear feedstock using breeder reactors."

Assuming cost and materials problems can be solved, the fundamental problem of terrestrial solar energy is low power density. If we assume that it will eventually be possible to divert 5% of solar flux at the surface without adversely affecting the global environment, then ~10 W m^{-2} could be available for human use by solar thermal power-plants and PV cells. The heat rejected would in any event be radiated to space, but could be redistributed by global scale solar energy systems. Radiative forcing by greenhouse gases emitted by humankind is already ~2.5 W m^{-2} (Shine et al., 1990, Table 2.6). Note also that solar flux available for human use as electricity is further reduced by low efficiencies of solar collectors and converters and by the area devoted to storage and transmission of this intermittent resource. In the case of PV cells, F_{pv} ~ 1 W m^{-2} is a reasonable electricity output flux, not counting storage.

4.4.1.2 Biomass Energy

Ecosystems of the terrestrial and marine biospheres are energized by solar photons (hν) converted to chemical energy in green plants, algae and cyanobacteria. The photosynthesis reaction is hν + CO_2 + H_2O → CH_2O + O_2, where CH_2O is organic carbon (shorthand for 1/6 of a glucose molecule, $C_6H_{12}O_6$) and ΔH = 467 kJ mol^{-1} = 3.89 x 10^4 J gc^{-1} is the heat of reaction (Bolton and Hall, 1991). Net primary productivities (NPP, photosynthesis less respiration) per unit area of real ecosystems are in the range (NC/A) = 200–2000 gC m^{-2}

y^{-1} (Schlesinger, 1991, Table 5.2) corresponding to NPP efficiencies in the range $\eta_p = [(N_C/A)\Delta H]/F_s \sim 0.1-1.3\%$.

The theoretical maximum efficiency of photosynthesis has been estimated by Bolton and Hall (1991) at $\eta_{p,max} \approx 9\%$ (independent of the cell or ecosystem in which it occurs). This is virtually impossible to achieve in real organisms because it ignores plant respiration, reflected light, limitations due to inadequate water and nutrients, and functional requirements of plant growth and physiology.

What is impressive is that ecosystems with sufficient water and nutrients like swamps and rainforests have evolved NPPs within an order-of-magnitude of the theoretical peak. Tropical rainforests are vertically organized to waste as few photons as possible. Broad-crowned, widely scattered emergent trees tower 50 m or more above the main canopy. The densest foliage, containing many coexisting tree species, is frequently at heights of 20–30 m. Below this second tier, small plants energized by scattered, diffuse sunlight take their places in vertical sequence — treelets, shrubs, and finally, herbs (Terborgh, 1992). These are complex and fragile ecosystems, home to a great diversity of biological species — how many is unknown, but it is of the order of millions. Tropical rainforests can impact nontropical environments through the hydrological cycle, atmospheric chemistry and in other important but poorly understood ways (Dickinson, 1987). But unchecked, the expansion of human populations through land use conversion could effectively destroy the tropical rainforests and make countless species extinct in the next century (Table 4.4).

In the process of photosynthetic organic carbon production, trees remove CO_2 from the atmosphere, virtually all of which is regenerated by respiration when the trees decay. The primary productivity of the terrestrial biosphere collectively is of the order of 100 Gt C y^{-1} (Schlesinger, 1991). Global biomass is close to a steady state over the long term, but if the standing carbon crop in trees decreases or increases, the result is a transient carbon source or sink to the air/sea system. Current rates of tropical deforestation are believed to be releasing ~ 2.4 Gt C y^{-1} on top of ~ 5.6 Gt C y^{-1} from fossil fuel burning (Houghton et al., 1990).

Meyers and Goreau (1991), and others, have proposed CO_2 absorption by rapidly growing tropical tree plantations to mitigate carbon dioxide buildup by fossil fuel combustion and deforestation. But carbon sequestration is a one-shot use of land whose scarcity is likely to

increase in the 21st century. Such land use also runs counter to perceived economic aspirations of developing nations in the tropics
(Marland, 1991) — not to mention the threat to the biodiversity of
indigenous ecosystems by monoculture farms.

Hall et al. (1991) and Marland (1991) observed that biomass farms
are preferable to CO_2 sequestration by unharvested trees because the
land use is renewable. This is a valid argument up to a point. Although
the combustion of wood (and its alcohol fuel derivatives) produces
CO_2, biomass farms are an essentially greenhouse-free form of solar
energy because virtually all the CO_2 recycles rapidly regardless of
whether the tree is burned for fuel or decays naturally. But the threat
to biodiversity and competition with agriculture remains; and biomass
energy plantations have the additional need to restore soil nutrients
depleted by the harvest. For a realistic 1% energy efficiency by biomass plantations, we estimate $F_{BIO} \sim 2$ W m^{-2} which drops to ~ 1 W
m^{-2} after conversion to electricity in (say) wood-fired power plants.
This does not include the energy needed to farm the trees (nutrients,
extraction, transportation, etc.).

4.4.1.3 Windpower

Atmospheric winds driven by temperature gradients from differential heating by the sun are another form of solar energy. The kinetic
energy flux per unit *vertical* cross sectional area is $(1/2)\rho_a U^3$, where
ρ_a is air density and U horizontal wind speed. The large-scale circulation is driven by the equator-to-pole temperature gradient $\Delta T \approx 45$
K interacting with Coriolis forces produced by the Earth's spin ($\Omega =$
2π radians/day ≈ 7.3 x 10^{-5} s^{-1}) embodied in the Coriolis parameter, f
$= 2\Omega\sin(\text{latitude}) \sim 10^{-4}$ s^{-1}. The result is strong high-altitude winds
in the East-West (zonal) direction at midlatitudes (the jet stream).

The vertical gradient of the zonal wind is (Peixoto and Oort, 1992,
p. 156), $\partial{<}u{>}/\partial z \sim (g/fT_s)\ \partial T/\partial y \sim g\Delta T/(fT_s R_E)$, where z is altitude,
y the meridional coordinate and g ≈ 10 m s^{-2} the gravitational accel-
eration. Zonal winds in the jet stream near the tropopause altitude, H_t
~ 8 x 10^3 m, are of order $<u_{jet}> \sim gH_t\Delta T/(fT_s R_E) \sim 20$ m s^{-1}.
Unfortunately, such high wind speeds (and very high kinetic energy
fluxes) eight kilometers up are inaccessible to current energy technology, whereas mean surface winds are much less intense (Oort, 1983),
$<u_s> \sim 3$ m s^{-1}. Surface air density, $\rho_a \approx 1.2$ kg m^{-3} and $<u_s> \sim 3$ m s^{-1}
give a kinetic energy flux $F_w \approx (1/2)(1.2)(3)^3 \sim 16$ W m^{-2}. But surface

windspeed is highly variable, and the mean kinetic energy flux (proportional to $<u_s^3>$) at a site is usually much higher than the flux at the mean windspeed $<u_s>$. Typically (Grubb and Meyer, 1993), $<u_s^3> \sim 2<u_s>^3$. Also, current wind farm practice suggests a ratio of wind turbine disk area to horizontal area $A_F/A_H \sim 1/10$.

Combining these estimates gives a mean wind energy flux in the atmospheric boundary layer per unit *surface* area $F_w \sim (A_F/A_H)\rho_a <u_s>^3 \sim 3$ W m^{-2}. The major environmental impacts of wind turbines are their land and materials requirements and their need to be integrated into the power production infrastructure.

4.4.1.4 Ocean Thermal Energy

The oceans are a source of renewable energy. A most promising technology is ocean thermal energy conversion (OTEC) — floating heat engines grazing the surface-to-deep temperature gradient. The vertical mean ocean temperature gradient occurs mainly in the first kilometer of ocean depth (the thermocline). A typical horizontally-averaged temperature difference is $\Delta T_w = T_{surface} - T_{deep} \sim 15$ K. OTECs date to 1881, when d'Arsonval first proposed a closed-Rankine-cycle device in which a working fluid is vaporized by heat exchange with cold seawater drawn through a vertical pipe from 700–1200 meter depth (Dugger et al., 1983).

Because of the small temperature gradient, and the need to divert power to the cold water pump, the energy conversion efficiency of OTECs is only a few percent. On the positive side, OTECs bring nutrients, as well as cold water, to the surface, offering along with energy the possibility of aquiculture (food) (Savage, 1994). An upper limit on power available is the rate heat is transferred to the surface from deepwater by upwelling. Nature pumps cold high-latitude seawater to the deep ocean by polar downwelling at a volumetric rate of (Tolmazin, 1985, Table 7.2) $Q_w \sim 34$ x 10^6 m^3 s^{-1}, after which the cold water upwells into the world's oceans. The thermal power of this flow is $\rho_w C_p Q_w \Delta T_w$, where $\rho_w \sim 1030$ kg m^{-3} and $C_p \sim 4000$ J kg^{-1} K^{-1} are the density and heat capacity of seawater, so the upward power flux per unit ocean area is $(\rho_w C_p Q_w \Delta T_w)/A_w \sim$ (1030)(4000)(34 x 10^6)(15)/(3.6 x 10^{14}) ~ 5.8 W m^{-2}. This upward flux of cold water is balanced in the steady state by downward flux of warm surface water by vertical mixing. The balance determines thermocline structure.

Were the entire 5.8 W m^{-2} thermal flux exploited by OTECs, it could destroy (make isothermal) the thermocline structure. Marine ecosystems might not survive. But diverting a fraction, say 10%, of the thermocline energy flux might be sustainable. A 10% diversion would make ~ 0.6 W m^{-2} available as OTEC input power, with fertilization of surface ocean plankton and fisheries a bonus. For a realistic energy conversion efficiency of 3%, electric power output flux harvestable over the oceans is F_{OTEC} ~ 0.02 W m^{-2}.

4.4.1.5 Hydroelectricity

The largest present-day form of commercial renewable energy is hydroelectricity. Driven by differential solar heating, water in liquid, vapor and solid phases cycles through clear air, clouds, rain, snow, rivers, seawater, sea ice and glaciers. The hydrological cycle, among other things, produces an excess of precipitation over evaporation over elevated land surfaces, which runs off to the seas. Volumetric runoff by all by the world's rivers is (Budyko, 1974, p. 228; Tolmazin, 1985, Table 7.3) Q_r ~ 1.3 x 10^6 m^3 s^{-1}.

Apart from flowrate, hydropower depends on the "total head," H = h + $V^2/(2g)$, where the first term on the right is gravitational potential energy and the second kinetic energy (both in units of elevation). Mean elevation above sea level of the Earth's land surfaces is (Sverdrup et al., 1970, p. 19) H_1 ≈ 840 m. An approximate first-principles estimate of global hydropower per unit land area is therefore $(\rho_w g H_1 Q_r/A_1)$ ~ (1000)(10)(840)(1.3 x 10^6)/ (1.5 x 10^{14}) ~ 0.07 W m^{-2}. Assuming 20% can actually be converted to electricity gives F_{HDRO} ~ 0.014 W m^{-2}, smaller in potential than most renewables but the one most exploited today. Although hydropower is concentrated in rivers (ideally with large changes in elevation, or head), the proper conditions do not often occur naturally. Niagara Falls is an exception.

It is typically necessary to create artificial reservoirs with dams by diverting rivers and flooding existing lands to tap hydropower, with potentially adverse environmental impacts. A recent example is the Three Gorges hydroelectric project on the Chang Jiang (Yangtze) River in China, approved by the National People's Congress in April, 1992. Chinese officials estimate that the reservoir will partially or completely inundate 2 cities, 11 counties, 326 townships, and 1,351 villages. About 23,800 hectares of cultivated land will be submerged, and 1.1 million people will have to be resettled. The Yangtze dolphin, one of the world's most endangered species, would be further threatened, as would other aquatic life — not to mention the destruction by the dam and reservoir of some of China's finest scenery. On the positive side, the project's 9.6 GW output is expected to displace 40–50 million tons of coal per year (Hammond, 1994, p. 68).

4.4.1.6 Geothermal Energy

It is now feasible to tap some of the Earth's interior heat with steam turbines at "hot spots" where underground steam is produced by the proper geologic conditions. Steam geysers like "Old Faithful" in Yellowstone Natural Park are indicative of such locales. In general, upwelling geothermal heat comes mainly from radioactive decay of uranium, thorium and potassium isotopes (^{238}U, ^{235}U, ^{232}Th, ^{40}K) in the Earth's mantle — 80% of the heat flux is from radioactive decay, 20% from cooling of the primordial Earth. On a global mean basis, the geothermal flux is substantially lower than that associated with solar energy. Measured fluxes are 0.0505 W m^{-2} in the continental crust and 0.0782 W m^{-2} in the (thinner) oceanic crust, with a global mean of ~0.07 W m^{-2} (Turcotte and Schubert, 1982, Table 4.1).

Geothermal heat is accessible to present-day technology only when there is an existing hydothermal-convection system that transfers heat to the surface by underground circulating water and steam. Ideally, one seeks geothermal steam fields such as those in New Zealand associated with the Tonga deep ocean trench in the western Pacific. These are normally found in volcanic zones near geologic faults occupying a few percent at of the Earth's area (Shepard et al., 1977). Moreover, some 80% of underground hydrothermal reservoirs are not hot enough to drive a turbine.

Our estimate for accessible geothermal energy is ~2% of the global mean heat flux; or F_{GEO} ~ 0.0014 W m^{-2}. This could conceivably increase by drilling deeply into dry rock. In the most optimistic case one could recover only <0.1% of the solar flux at the surface (averaged over large areas). Apart from the technological challenge of deep drilling through the crust to underlying magma, the impact on plate tectonics of large-scale mining of the Earth's interior heat by humans is unknown.

4.4.1.7 Can Renewable Energy Fluxes be Captured?

Advocacy of renewable energy as the goal of human energy supply can be found in many articulate analyses (Lovins, 1977; Williams, 1979; Goldemberg et al., 1988; Ogden and Williams, 1989; Brower,

1990). Why have these technologies had so little impact? The latest global commercial energy consumption statistics show fossil fuels remain, by far, the dominant source (90.4%), followed by nuclear (6.8%), hydro (2.4%) and geothermal plus wind (0.4%) (Hammond, 1994, Table 21.1). Biomass burning, primarily by developing countries off the cash economy, does not even show up as commercial energy.

Undoubtedly, part of the resistance comes from the entrenched infrastructure and investment in fossil energy systems. But a more fundamental problem, and one following from the underlying physics, is low power density. Naturally available energy flows are thin gruel. They impose severe constraints on land use, materials, energy transmission and energy storage, all of which have to be addressed before they can become practical. Our analyses indicate energy fluxes potentially exploitable by renewable energy technology in the range of 0.01 to 10 W m^{-2}. These are orders of magnitude below power extraction rates per unit area from coal and oil fields, and fossil and nuclear thermal power plant heat fluxes. Although significant variations can be found in estimates of harvestable renewable energy fluxes, our numbers are generally comparable in order-of-magnitude to those derived by others.

Figure 4.4, for example, shows power flux versus area regimes for various renewable and fossil energy sources estimated by Smil (1991). As he observes, "Graphic comparison of power densities of various energy supply modes with those of final uses reveals a wealth of spatial implications... The most important conclusions concern the spatial consequences of the eventual transition to a solar civilization. *These contrasts bypass the dubious cost comparisons to demonstrate the fundamental physical difficulties of such a transformation* (our italics)." Large power densities of present-day fossil and nuclear steam turbine powerplants derive from the concentrated nature of their energy sources and the consequently large heat fluxes attainable in the boilers. For example, a typical heat absorption rate per unit area of the walls of a pulverized coal furnace is 160,000 W m^{-2} (~50,000 BTU hr^{-1} ft^{-2}, from Kessler, 1978, Fig. 31, p. 9.21). Present-day thermal engineering can handle hundreds of kW m^{-2} to as much as 10 MW m^{-2} in blast furnaces, Such energy fluxes are available primarily from fossil, nuclear and (presumably) fusion power sources. These inherently require less materials and land than terrestrial renewable energy.

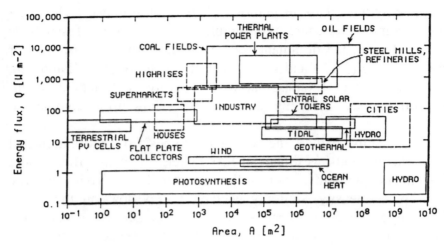

Figure 4.4   Flux-area domains of energy sources and sinks. This comparison highlights differences — often several orders of magnitude— and overlaps between the harnessing of renewable energy fluxes and extraction of fossil fuels on the one hand (solid rectangles), and energy end use fluxes on the other (dashed rectangles). Since the present global energy infrastructure is based on high energy density fossil fuels, a transition to low power density renewable energy is likely to require socio-economic, as well as technological, transformations. (Adapted from Smil, V., *General Energetics: Energy in the Biosphere and Civilization,* John Wiley & Sons, New York, 1991. With permission.)

The flux of solar energy on a given spot can be increased with optical concentrators, where the area ratio of the concentrator aperture to the solar image spot size is the concentration ratio, CR. However, the collection area for a given radiant power to an energy conversion device is as large as without concentration, and the materials and land use implications are similar. Solar concentrators are worth pursuing — as in "power tower" or heliostat configurations — because higher temperatures in a working fluid near the focal point can drive more thermodynamically efficient heat engines. Two-axis mirror tracking is normally required to keep axisymmetric concentrators pointed at a fixed focal spot on a heat exchanger.

The thermodynamic limit on the concentration ratio of axisymmetric solar concentrators is (Welford and Winston, 1981, p. 3.11) $CR_{max} = 1/(\sin\theta)^2$, where θ is the semi-angle subtended by the sun ~ (R_{sun}/a) ~ 4.64×10^{-3} radians ~ $0.27°$. Hence, $CR_{max} \sim \theta^{-2} \sim 46,000$. In principle the solar constant flux, 1380 W m^{-2}, could be focused on a spot with a flux of $(1380)/(4.64 \times 10^{-3})^2 \sim 64$ MW m^{-2} — the radiative intensity at the solar photosphere. But because of the low mean power flux available, and the intermittent nature of the source, even if fluxes of order $\sim$ 1–10 MW m^{-2} could be produced by the sun at the focal spot of a central power station, land and material requirements would be orders of magnitude larger than fossil or nuclear plants of the same capacity.

The land use problem is fundamental. A renewable energy yield of 1 W m^{-2} requires 10% of the earth's land area (1.5 x 10^{13} m^2) to produce 15 TW (15 x 10^{12} W), the level of nonfossil energy we estimated earlier might be required in the next 50 years to mitigate global warming by greenhouse gases.

Boyle (1994) has reported an assessment by Greenpeace International and the Stockholm Environmental Institute–Boston (SEI–B) of a Fossil Free Energy Scenario (FFES) in which energy efficiency is combined with the phaseout of fossil fuel by the year 2100 and the phasein of renewable technologies. Their scenario assumes that biomass technologies, windpower and a limited expansion of hydropower provide the bulk of the renewables in the short run, while over the long term a solar energy system (solar PV, solar thermal power and solar water and space heat) linked to hydrogen predominates, providing 80% of global energy by 2100. Boyle's (1994) FFES scenario requires 8% of global arable/pasture/forest and woodland (6% of total global land area). These land use results are comparable to ours. Boyle argues that the use of non-productive land (deserts and degraded areas) plus roof tops can alleviate potential problems. Note that the technology employed in the FFES scenario assumes renewable energy technologies of the 21st century are those already "on the shelf" in the latter part of the 20th century.

As for land requirements, rooftops are convenient sites for solar collectors, particularly for solar space heating of well-insulated structures, but are inadequate to power a solar civilization. Moreover, the use of remote areas for renewable energy poses special problems of long-distance energy transport. Deserts are well endowed with sunlight, but usually distant from the intended points of energy use. Hydrogen has been proposed as a potentially important energy carrier (Winter and Nitsch, 1988), but needs to be considered in a total energy analysis as many energy conversion steps are needed (water electrolysis, transmission of H$_2$ by pipeline, combustion or electric power generation in a fuel cell).

Given the low power density and intermittency of natural energy fluxes, transmission and storage are critical, and relatively unexplored, components of sustainable global renewable energy systems. We recommend that innovative technologies be explored which can qualitatively improve global energy infrastructure through new modes of transmission and storage.

4.4.1.8 Electric Power Transmission

Electric power cannot efficiently be transmitted today more distantly than several hundreds of kilometers owing to the finite resistance of ambient temperature conductors (aluminum and copper). This reach is insufficient for a global solar energy system to bridge day/night cycles or to bypass overcast sunless regions. A major problem for any global-scale renewable energy system is "load matching" — connecting (or wheeling) supply and demand over long distances between consumers and producers.

Ohm's law ($V = IR$) indicates that power losses from the resistance, R, of an electrical transmission line are minimized when electricity is transmitted at the highest possible voltage, V, for a given transmitted power ($P = IV$), $P_{loss} = I^2R = P^2R/V^2$. The volumetric resistivity of transmission lines is a material property, $\rho = RA/l$ ($\sim 8.4\ \mu\Omega$-cm for aluminum-clad steel), where l is the length of the conductor and A its cross sectional area; so power loss per unit length is $P_{loss}/l = P^2\rho/(AV^2)$. Per unit length of wire, the material cost increases nearly linearly with A, whereas the cost of the power loss decreases inversely with A. The sum of these (the total cost) vs. A thus has a minimum which determines the most cost-effective wire cross section. Maximum line voltage is normally limited by the breakdown of insulation or air separating the conductors (~ 1000 kV in current practice).

Nikola Tesla reasoned in the early years of this century that the highest possible voltages should be used to minimize I^2R losses in electric power transmission. He invented a series of alternating current (AC) generators and transformers to accomplish this. Undaunted by Ohm's law, Telsa's former employer (and later commercial opponent), Thomas Edison, backed low voltage direct current (DC). Edison and Tesla fought a bitter battle for public acceptance called the "war of the currents" (Hackmann, 1993). Tesla had to win (and did), because physics favors high voltage, and because the technology of the time for controlling voltage was AC. Three-phase AC power lines became standard in much of the world for transmitting electricity the few hundred kilometers or less from fossil and nuclear central power stations to consumers. But existing grids are inadequate for global renewable energy systems where electrical sources and sinks would necessarily be much more widely dispersed.

Technological breakthroughs could make long-distance electricity transmission a reality in the 21st century. Remarkably, the resistivity of certain substances vanishes entirely ($\rho \rightarrow 0$) at temperatures close to absolute zero. Typically, these "Type I" superconductors require cooling by liquid helium (He) which, at atmospheric pressure, boils at 4.2 K — an expensive proposition because of the large refrigeration power and the scarcity of He gas. In a pivotal discovery, Bednorz and Muller (1986) found ceramic oxides exhibiting a new form of super-conductivity ("Type II") at temperatures as high as 30 K. Chu and Wu, and others, later found materials exhibiting Type II superconductivity above the 77 K boiling point of liquid nitrogen (Cambel and Koomanoff, 1989). This development opens up the possibility of cost-effective superconducting power transmission because N_2 is an abundant atmospheric constituent, and the power needed to liquefy it is orders of magnitude less than for He.

Superconducting transmission lines would most likely be buried cryogenically cooled cables (Bockris, 1980). Preliminary studies emphasize AC lines compatible with electric utility practice in the U.S. (Giese et al., 1988). However, a potentially fatal power loss of AC superconductors is the "skin effect" caused by vortex motion into and out of the cable (Dale et al., 1990).

A solution to the AC loss problem may simply be to use DC in the superconducting link, with state-of-the art semiconductor AC alternators and inserters at the ends (Engström, 1978). Edison did not have these. But there is no theoretical reason why they should not work. A superconducting power line 100 km long could be >99.8% efficient if the only power loss were that needed to replenish the liquid N_2 (Giese et al., 1988). Maintaining this power loss per unit length over a 10,000 km transmission link gives a point-to-point transmission efficiency of $\eta_T = (.998)^{100} \times 100\% \sim 81\%$, quite acceptable for wheeling renewable energy planetwide.

The prospect of global superconducting transmission lines was raised by Buckminster Fuller even before the discovery of Type II superconductivity (Fuller, 1981). His interest was motivated partly by geometric thinking and partly by the realization that population growth is inversely correlated with per capita GDP and energy consumption. The Fullerian approach to the latter problem was to foster economic development worldwide by a planetwide electricity grid (Figure 4.5).

Fuller's thinking is best expressed in his own inimitable language (Fuller, 1991): "…We must be able to continually integrate the progressive night-into-day and day-into-night hemispheres of our revolving planet. With all the world's electric energy needs being supplied by a twenty-four-hour around, omni-integrated network, all of yesterday's, one-half-the-time-unemployed, standby generators will be usable all the time, thus swiftly doubling the operating capacity of the world's electrical energy grid." Regarding his icosahedral map projection, he explains, "… A half century ago I discovered with my nonvisibly distorted, one-world-island-in-one-world-ocean, 90° longitude-meridian-backbone, north-south-oriented, sky-ocean world map that a world energy network grid would be possible if we could develop the delivery reach…"

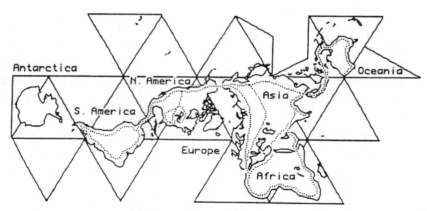

Figure 4.5   A proposed global superconducting electric grid (dotted lines). An icosahedron is a polyhedron of twenty equilateral triangles approximating a sphere. The world's land masses and oceans are shown in an icosahedral, as opposed to a Mercator, map projection to more realistically represent the size of continents, and to emphasize their near contact points where superconducting transmission lines could be run. Buckminster Fuller's ideal was a common electrical currency for the global economy of 1¢/kWe/hr. (Adapted from Fuller, R.B., *A Critical Path,* St. Martin's Press, New York, 1981. With permission)

Fuller's grid is an example of how important ideas can come from unanticipated directions when a creative mind is at work. Combined with plausible developments in superconductivity, his vision provides an electricity distribution approach that could enable terrestrial renewables to be effective global energy sources by the 21st century.

4.4.1.9 Energy Storage

In addition to new power transmission systems, more effective energy storage will be needed by a global nonfossil fuel civilization. Physics today defines matter in terms of elementary quarks and leptons and their antiparticles. Energy is stored in matter by the four fundamental forces through which these particles can interact — the strong and weak nuclear, electromagnetic and gravitational forces.

It follows from special relativity that the energy that can be stored per unit mass is limited to $\Delta E/m \leq c^2$, where m is the rest mass and c $\sim 3.0 \times 10^8$ m s^{-1} the speed of light. The theoretical maximum (for matter-antimatter annihilation) is much larger than what has been obtained by any present-day storage system. Table 4.5 lists values of the specific energy, $\Delta E/m$, and the rest mass energy fraction, $\Delta E/(mc^2)$, for various materials and systems. The variation over so many orders of magnitude arises from the different strengths of the fundamental forces and the distances over which they act. In general, the higher specific energy systems require less material and land.

The most technologically ready large-scale electrical energy storage system is pumped-storage. In operation, a reversible hydroelectric generator-pump pumps water from a river or lake into a reservoir at higher elevation during periods of low power demand from utilities. During demand peaks, the system is reversed to recover the stored energy. The round-trip efficiency of pumped storage is ~65% (Culp, 1978, p. 428). However, the specific energy is a very low ~1.0×10^3 J kg^{-1} for a 100 meter head, leading to substantial land requirements. Pumped-storage also requires a specific type of topography which limits its applicability. In any event, the diversion of rivers and lakes for this purpose can lead to the same kinds of ecosystem disruption as hydroelectric reservoirs and dams discussed previously.

Batteries have more than 100 times the $\Delta E/m$ of pumped storage, ~1.2×10^5 J kg^{-1} for the conventional lead-acid automotive type, but are low in specific energy compared with fossil fuels. (This is the main problem with battery-powered vehicles.) The U.S. Department

Table 4.5   Energy stored in matter. (Culp, *Principles of Energy Conversion,* McGraw-Hill, 1979.)

Material or System	$\Delta E/m$ [J kg^{-1}]	$\Delta E/(mc^2)$ [-]
Antimatter (M/A annihilation reaction)	9.0×10^{16}	1.0
Deuterium (D/D fusion reaction)	3.5×10^{14}	3.9×10^{-3}
Uranium 235 (fission reaction)	7.0×10^{13}	7.8×10^{-4}
Reactor fuel (2.5% enriched UO_2)	1.5×10^{12}	1.7×10^{-5}
80% plutonium-238 (radioactive decay)	1.8×10^{9}	2.0×10^{-8}
Hydrogen (lower heating value)	1.2×10^{8}	1.3×10^{-9}
Gasoline (lower heating value)	4.4×10^{7}	4.9×10^{-10}
Lead-acid battery (automotive)	1.2×10^{5}	1.3×10^{-12}
Flywheel (uniformly-stressed disk)	7.9×10^{4}	8.7×10^{-13}
Organic elastomer ("rubber band")	2.0×10^{4}	2.2×10^{-13}
Falling water (H = 100 m)	1.0×10^{3}	1.1×10^{-14}
Torsion spring (steel)	2.4×10^{2}	2.6×10^{-15}
Capacitor (communication electronics)	1.6×10^{1}	1.8×10^{-16}

of Energy (DoE) is supporting research on improved batteries for both automotive and stationary energy applications. Sodium-sulfur batteries are considered the most promising (Cohen, 1990, p. 265). A goal of the DoE is 80% efficiency over 2,500 charge-discharge cycles at a cost of $90 per kWe/hr capacity. This may be overly optimistic, but if it were attained the cost per charge-discharge cycle would be ($90/2,500) x 0.8 ~ 2.8 ¢/kWe-hr, not counting installation and maintenance. Materials and land use implications of a global electrical energy battery storage system have not been analyzed.

A global energy economy employing H_2 generated by electrolysis [$2H_2O \rightarrow 2H_2 + O_2$] as "currency" has been much discussed in recent years (Bockris, 1980; Winter and Nitsch, 1988; Ogden and Williams, 1989). Advantages of hydrogen as an energy carrier are storability at high specific energy and the absence of greenhouse gas emissions. Deriving H_2 from water-splitting is crucial to CO_2 emission reduction. Commercial H_2, as made today by steam reforming of natural gas [$2H_2O + CH_4 \rightarrow 4H_2 + CO_2$], is more likely to *increase* CO_2 emissions. Of course, the primary power source also needs to be non-fossil. Conventional water electrolysis with aqueous electrolytes employs an asbestos diaphragm and alkaline electrolytes to prevent corrosion of the metal components. H_2 production with alkaline electrolytes is well-established, but has been used thus far mainly for small to medium-sized plants ($< 5MW_e$; Winter and Nitsch, 1988, p. 179).

The specific energy of H_2 fuel based on the mass of hydrogen is $\Delta E/m \sim 1.2 \times 10^8$ J kg^{-1} — three times higher than gasoline (Table 4.5). This doesn't count the mass of the storage tank, and the containment of gaseous hydrogen remains a major challenge for applications. Present-day technologies include cryogenic liquid hydrogen (LH$_2$) tanks and adsorption by metal hydrides such as iron-titanium (FeTiH$_x$). Hydrogen to total (tank plus hydrogen) mass fractions for H_2-powered vehicles were estimated by Carpetis (1988, Table 9.2) as ~ 0.10 kg(H$_2$)/kg for liquid hydrogen (LH$_2$) and ~ 0.018 kg(H$_2$)/kg for a metal hydride combination tank. Including these tank-age factors decreases the $\Delta E/m$ of hydrogen to 1.2×10^7 J kg^{-1} (LH$_2$) and 2.2×10^6 J kg^{-1} (metal hydride)— less than gasoline but far better than storage batteries.

The main problems with hydrogen storage at this point are high cost and possible limitations from toxic (asbestos) and rare (titanium) materials. Relatively low power densities of electrolyzers and issues of water availability in remote locales may also pose problems for a global hydrogen economy.

The optimum energy storage system in a land-limited world may be that with the highest specific energy. There would, for example, be advantages to an economy in which energy was stored in antimatter and recovered through matter-antimatter reactions (Table 4.5). The technological challenge is major, but so is the payoff. When antiprotons annihilate with protons, the products are, on average, five elementary particles — three charged and two neutral pions. The charged pions live 26 ns and contain 60% of the rest mass energy (Forward, 1987). Thus - 5.4×10^{16} J kg^{-1} in kinetic energy of charged pions could be available for conversion to (say) electricity — eight orders of magnitude more than for hydrogen!

Antimatter is virtually nonexistent in the natural environment for reasons of interest mainly to cosmologists. But antiprotons are produced routinely today in particle accelerators, albeit in small quantities. When relativistic protons strike a dense metal target, their kinetic energy, which is many times their rest mass energy, is converted into a spray of particles, some of which are antiparticles. A magnetic field focuses and a selector separates antiprotons from the resulting debris and directs the antiprotons into a storage ring.

Storing antimatter is more difficult than storing hydrogen — the containment must avoid mutual annihilation on contact with normal matter. We do not know how to do it yet, but given its high $\Delta E/m$, the energy stored as antimatter would make an excellent space propulsion energy source.

Again, science fiction may lead reality. The starship *Enterprise,* of the *Star Trek* series, is powered by matter/antimatter (M/A) reactions. According to the latest technical manual (Sternbach and Okuda, 1991, p. 60), "The key element in the efficient use of M/A reactions is the dilithium crystal... Dilithium permits the antihydrogen to pass directly though its crystalline structure without actually touching it, owing to the field dynamo effect created in the added iron atoms..." Although Arthur C. Clarke has reminded us that a sufficiently advanced technology is indistinguishable from magic, we cannot count on dilithium crystals or other "unobtainium" to provide antimatter containment in the near future.

Fortunately, there are options consistent with known physics (Forward, 1987). It is already possible to slow antiprotons in a particle accelerator storage ring to almost zero velocity and capture them in a small electromagnetic ion trap no larger than a thermos bottle, although the technique is limited to low ion densities. One possibility is to add positrons in the ion traps to slowly build up "cluster ions" of antihydrogen — large agglomerations of neutral hydrogen atoms clustered around a single antiproton. Forward (1987) proposes that the net electric charge of the cluster ion would permit it to be held in an ion trap until the mass is increased to the point where it could be electrostatically levitated without touching the walls of a cryogenically-cooled trap. This antimatter could then be transported to applications.

There is no guarantee that technologies speculated on here will become cost-effective in the next century. But it is likely that 21st century energy supplies will differ from today's more than what might expected by linear extrapolation of current trends. Given the explosive technology changes of the past century, it is at least possible that the low power fluxes of terrestrial renewables could be harvested with the help of superconducting global power grids and antimatter energy storage.

4.4.1.10 Land Use, Population Growth, and Energy

Any effective global energy system should minimize the impact on land use. Humankind has already appropriated ~10% of the land surface for agriculture (Table 4.4) — a fraction that will almost certainly increase by the middle of the next century. At the same time, there will be increasing pressure on natural ecosystems for renewable energy conversion. We (and others) estimate this will require another 10% of the Earth's land surface. Humankind has already appropriated ~40% of the world's biological productivity, if one includes NPP used for food, the wood used for timber, paper and firewood, the diversion of organic matter harvested but not eaten, and modification of the land for human habitats intentionally (asphalt, pastures and cropland) and unintentionally (desertification). This proportion of terrestrial NPP diversion is the largest by any one species (and its servant species) since life colonized the land (Diamond, 1987).

Tropical rainforests are being cut back for agriculture and habitats at rates that could destroy them by the middle of the 21st century. Some 34 mostly poor countries with high fertility rates account for ~97.5% of tropical rainforests — Brazil, Zaire, and Indonesia contain more than 50%. Meyers (1991, Table 1) estimates that only ~530 x 10^6 ha remain of ~1,360 x 10^6 ha of primary tropical rainforest. Thus $[(1,360-530)/1,360]$ x 100% ~ 61% of the primary rainforests have already been destroyed. If the remaining 39% is deforested at 2% per year, then only $(.98)^{50}$ x 39% ~ 14% will survive 50 years hence. There is worldwide concern over the implications of this biomass and species destruction, but studies indicate that much tropical deforestation is the result of "slash and burn" shifting cultivation by the very poor in developing nations, who see no alternative to encroachment on the marginal environments of the tropical forests (Meyers, 1991).

As a society becomes more wealthy (as measured by its per capita income, GDP/N, or per capita energy consumption, E/N) the birth rate of its people tends to decline (Meadows et al., 1992, Fig. 2-7). All of the poorest nations experience birth rates between 20 and 50 per thousand people per year, while the rich nations generally have birth rates below 20 per thousand. There are some exceptions like China and Sri Lanka which have anomolously low birth rates for their level of income, and some Middle Eastern countries have anomolously high birth rates for theirs. But in general the inverse correlation between birth rate and GDP/N holds. The present-day population dynamics of tropical developing nations is characterized by a positive feedback loop in which poverty leads to overpopulation and overpopulation leads to further poverty, with concomitant impacts on natural ecosystems.

Protection of the global environment against adverse effects of deforestation, loss of biodiversity and climatic change is thus intimately linked to the economic level and energy use of developing countries. As Meadows et al. (1992, p. 40) observe, "Any positive feedback loop that grinds a system down, however, can be turned around to work the other way. More prosperity, widely distributed, can lead to slower population growth, which can lead to more prosperity." It was for precisely those reasons that Buckminster Fuller was motivated to find a more equitable global energy distribution system.

The development of sustainable nonfossil energy systems is a multifaceted problem that must be assessed in an integrated way. However desirable for other reasons, viable solutions must provide greenhouse-free energy supplies for developing nations which foster reasonable economic development along with population control. In the long run, that is the only way to prevent negative feedbacks from destroying the natural environment. In this section we analyzed advantages and limitations of renewable energy sources, and of transmission and storage systems, that could reduce greenhouse gas emissions. The low power fluxes of renewables can be most efficiently exploited by yet-to-be-invented energy distribution and storage systems. Such systems are going to be major challenges for 21st century technology, and should be major research priorities now.

4.4.2 Nuclear Fission

Nuclear fission fuels are derived from heavy element radionuclides in the Earth's crust originally synthesized by exploding supernovae. The detritus of such an explosion, including nuclear fuel, was incorporated into the matter cloud from which our solar system formed. Of immediate relevance to global energy supply are the uranium and thorium resources residing in the Earth's crust which decay with half-lives ($t_{1/2}$) of order the solar system age ($\sim 5 \times 10^9$ yr):

$$^{238}U \rightarrow {}^{206}Pb + 8{}^4He, \quad t_{1/2} \sim 4.5 \times 10^9 \text{yr}$$
$$^{235}U \rightarrow {}^{207}Pb + 7{}^4He, \quad t_{1/2} \sim 0.7 \times 10^9 \text{ yr}$$
$$^{232}Th \rightarrow {}^{208}Pb + 6{}^4He, \quad t_{1/2} \sim 14.0 \times 10^9 \text{ yr}$$

In the late 1930s, Hahn and Strassmann discovered that bombarding natural uranium with neutrons moving at energies of only a few electron volts (eV) would split the ^{235}U nucleus and release about 170 million electron volts (MeV) — a ferociously exothermic reaction with output exceeding input by five orders of magnitude.

At the time of a supernova explosion, ^{235}U is produced more abundantly than ^{238}U, but the more rapid decay of the lighter (fissionable) isotope over the age of the solar system and the time prior to its formation has resulted in a present $^{235}U/^{238}U$ ratio in uranium ore of only ~0.72%. Economically recoverable fissionable uranium reserves available to present-day light water nuclear reactors (LWRs) is quite modest, less than oil or natural gas in energy equivalents at present energy prices (Table 4.1).

The fuel of LWRs is mostly ^{238}U enriched with ~3% readily fissionable ^{235}U — about 4 times the natural abundance. Bombs require much more enrichment (close to 100%). Stimulated by a perceived Nazi threat to develop the first atomic bomb, U.S. Manhattan Project scientists in Los Alamos during World War II made weapons from both highly-enriched ^{235}U and ^{239}Pu — a man-made plutonium isotope bred from uranium. The first atomic explosion on July 16, 1945 at the Trinity site in the New Mexico desert had an imploding plutonium core. The "Little Boy" bomb dropped on Hiroshima at the end of WW II incorporated ~50 kg of highly purified ^{235}U (Serber, 1992).[2] The possible proliferation of nuclear weapons from nuclear power plants has been of concern ever since. Subsequent to the Gulf War of 1990, Iraq informed the International Atomic Energy Agency (IAEA) that it had on hand ~14 kg of 80% enriched and 11.3 kg of 93% enriched ^{235}U from its "peaceful" nuclear energy programs (Norman, 1991) — of the order to produce a nuclear weapon. Concern over nuclear proliferation has now shifted to North Korea. The so-called n-country problem in which progressively more nations acquire weapons-grade materials from their nuclear energy programs is clearly upon us, and getting worse. Some analysts argue that global nuclear energy should be developed only by politically "stable" societies, and that an electricity or electrolytic hydrogen be product marketed to others, but this is clearly a value-laden position.

---

[2]Three critical masses <20 kg each were distributed between a ring fixed around the muzzle of a small cannon and a "bullet" inserted into the cannon breech ahead of the bags of cordite and fired at the appropriate time up the barrel to complete the supercritical assembly. Los Alamos considered Little Boy to be sufficiently conservative that it delivered the weapon without further testing. Smaller plutonium bombs are possible: A solid sphere of ~5 kg ^{239}Pu surrounded by a thick shield (tamper) explodes immediately on assembly (Serber, 1992).

While environmentalists tend to favor renewable energy sources, there are good engineering arguments for the high power densities of nuclear reactors over the generally much lower energy fluxes available from solar, wind and biomass energy. On the anti-nuclear side is also a public concern about nuclear safety. A severe accident occurred in March 1979 at the Three Mile Island (TMI) power plant, an LWR of the type that now generates 85% of the world's nuclear-generated electricity. Meltdown of about half of the fuel occurred though the public was protected from hazardous radiation releases by the containment building. A much more serious accident occurred at the Chernobyl reactor complex in the Soviet Union in April 1986 resulting in some 30 prompt fatalities and a potentially much larger number of adverse health effects from the radiative contaminant released. There are also unresolved problems in disposing of radioactive wastes and spent fuel from power plants when they are decommissioned. Advocates of nuclear power argue that inherently safe nuclear power can be developed, and cite global warming by fossil fuel carbon dioxide as a motivation for doing so (Weinberg, 1992).

An approach that overcomes low ^{235}U abundance is *breeder reactors* — reactors that "breed" fissionable fuels from nonfissile uranium and/or thorium. A critical parameter for both nuclear fission and nuclear fuel breeding is η — the ratio of neutrons produced by a nuclear reaction to neutrons absorbed by the fuel (Table 4.6). In order to sustain a chain reaction, we need $\eta > 1$. In order to "breed" nuclear fuel we need $\eta > 2$; one neutron to continue the fission chain, and one neutron to be absorbed by the fertile nucleus to produce a new fuel atom. It follows from the values in Table 4.6 that for thermal neutrons ^{233}U is the best thermal breeder reactor fuel, and that it may be possible to build a thermal breeder reactor operating on the thorium-uranium-233 fuel cycle. The value of η for fast neutrons is significantly

Table 4.6   Neutrons produced per neutron absorbed by the fuel for common nuclear fuel isotopes, η. (Culp, *Principles of Energy Conversion*, McGraw-Hill Companies, 1979.)

Fuel Isotope	Thermal Neutrons, (KE ~ 0.25 eV)	Fast Neutrons (KE > 1.5 MeV)
^{233}U	2.28	2.60
^{235}U	2.10	2.65
^{239}Pu	2.09	3.04

> 2 for all three fuel isotopes, which has motivated so-called fast breeder reactors.

France is singular in its commitment to breeder technology, having constructing the Rapsodie, Phenix, and Super Phenix breeder reactors to explore its potential. These are important experiments. But a long time may be needed to breed enough plutonium and to build plutonium-burning reactors to replace fossil energy on a global scale — perhaps 50 years. It may already be too late to replace fossil energy with breeder reactors at the rates needed to mitigate global warming.

Alvin Weinberg, pioneer of "inherently safe" worldwide breeder reactors as a solution to the fossil fuel greenhouse, expresses his case profoundly to nuclear engineering colleagues — but sidesteps the thorny issues of nuclear proliferation and disposal of radioactive wastes (Weinberg, 1992):

"...There was no heavenly-ordained requirement that the age of fossil fuel be replaced by the age of fission, nor a cosmically-ordained anthropic cosmological principle that required η to exceed 1 so that a chain reaction be possible, or exceed 2 so that breeding is possible. Had fission not been developed we would, willy-nilly, be conserving energy at a much faster rate than now, we would be pushing fusion even harder — and as a last resort, we would turn to solar energy. A solar world in which primary energy is 3 times as expensive as it is today is hardly an impossible world, especially since in such a world, energy would be much more strongly conserved than now. We would not be relegated to Malthusian poverty, were the only reactor upon which man depended for energy located 150×10^6 kilometers away from the Earth.

"But Hahn and Strassmann's discovery of 50 years ago and God's providence in adjusting the nuclear constants so as to make a power breeder practical have given us another option. We nuclear engineers of the first nuclear era have had successes, yes, with our 500 commercial nuclear reactors, and our practical breeders. But the job is only half finished. The generation that follows us must resolve the profound technical and social questions that are convulsing nuclear energy. The challenge is clear, even the technical paths to meet the challenge are clear. All of us old-timers wish we will be here to see how these challenges are met; but even if we shall not be here, we wish the new generation well in fashioning an acceptable Second Nuclear Era!"

A reactor design developed by DoE that addresses nuclear safety and waste disposal but not breeding issues is the Integral Fast Reactor (IFR) being readied for test near Idaho Falls by Argonne National Laboratory *(Discover,* 1994). The IFR uses high-energy fast neutrons to trigger the chain reaction, in contrast to conventional light-water reactors that slow their neurons with graphite rods. Because fast neutrons can cause many more types of elements to undergo fission, the IFR is not limited to the uranium and plutonium that conventional reactors use as fuel. It can burn highly radioactive elements with half-lives of tens of thousands of years that are waste by-products of uranium and plutonium fission. In contrast to existing breeders, the IFR burns plutonium rather than producing it, thus precluding, in principle, the possibility that a cache of weapons-grade plutonium will fall into the hands of unstable regimes or terrorists. IFR project manager, Yoon Chang, claims that the reactor safely system is designed to be invulnerable to catastrophic loss-of-coolant accidents that crippled the Chernobyl and Three Mile Island reactors.

Even if the new generation of breeder reactors prove themselves inherently safe, a coordinated worldwide strategy for conversion of nuclear reactors would be needed to deal with the prospect of ^{235}U running out before plutonium was bred. The classical reactor strategy would have required that prior to the year 2000 breeders be installed to compliment the less efficient burner reactors. For this to have occurred, there would have to have been worldwide fast breeder programs already in place comparable to that of the French (Hafele et al., 1981, p. 55). An alternate tack is a converter-breeder strategy which can be fueled with ^{233}U. It remains to be seen whether the IFR or a comparable innovative reactive design can fit such a strategy, and provide safe nonfossil energy supplies on a global scale in the 21st century and beyond.

The authors (MIH and SDP) acknowledge the contributions of the Energy Supply (Chapter 4) panel members during the Engineering Response to Global Warming Workshop at the Sheraton Palm Coast resort, June 1–6, 1991, and thereafter. However, the opinions expressed in this chapter, as well as any errors of omission or commission, are the sole responsibility of the authors. Research and manuscript preparation for this chapter was supported in part by grants from the U.S. Department of Energy (Global Change Assessment Research) and National Institutes of Global Environmental Change to New York University.

REFERENCES

Allen, C.W. (1976) *Astrophysical Quantities*. Athlone Press, London, 310 pp.

APS (1979) *Principle Conclusions of the American Physical Society Study Group on Photovoltaic Energy Conversion,* H. Ehrenreich, Chair, American Physical Society, New York, 190 pp.

Ausubel, J.H. (1994) Technical progress and climatic change. In Nakicenovic, N., W.D. Nordhaus, R. Richels, F.L. Toth (eds.), *Integrative Assessment of Mitigation, Impacts, and Adaptation to Climate Change,* PC-94-9, pp. 501–512, International Institute for Applied Systems Analysis, Laxenburg, Austria, 669 pp.

Bednorz, J.G., K.A. Muller (1986) *Z. Phys.,* B64, 189.

Bockris, J.O'M. (1980) *Energy Options: Real Economics and the Solar Hydrogen System.* John Wiley & Sons, New York, pp. 212–233.

Boden, T.A., R.J. Sepanski, F.W. Stoss (1991) *Trends '91: A Compendium of Data on Global Change.* Carbon Dioxide Information Analysis Center, Oak Ridge National Laboratory, Oak Ridge, Tennessee, 665 pp.

Bolin, B. (1994) Next step for climate-change analysis. *Nature,* 368, 94.

Boulton, J.R., D.O. Hall (1991) The maximum efficiency of photosynthesis. *Photochem. Photobiol.* 53, pp. 545–548.

Boyle, S. (1994) Toward a fossil free future: the technical and economic feasibility of phasing out global fossil fuel use. In Naki´cenovi´c, N., W.D. Nordhaus, R. Richels, F.L. Toth (eds.), *Integrative Assessment of Mitigation, Impacts, and Adaptation to Climate Change,* CP-9S9, pp. 353–378, International Institute for Applied Systems Analysis, Laxenburg, Austria, 669 pp.

Bradley, R.A., E.C. Watts, E.R. Williams (1991) *Limiting Net Greenhouse Gas Emissions in the United States, Vol. 1: Energy Technologies.* Report DoE/PE-0101, U.S. Department of Energy, Office of Environmental Analysis, Washington, DC.

Brower, M.C. (1990) *Cool Energy: The Renewable Solution to Global Warming.* Union of Concerned Scientists, Cambridge, MA, 89 pp.

Budyko, M.I. (1974) *Climate and Life.* Academic Press, New York, 508 pp.

Cambel, A.B., F.A. Koomanoff (1989) High-temperature superconductors and CO_2 emissions. *Energy,* 14, pp. 309–322.

Carpetis, C. (1988) Storage, transport and distribution of hydrogen. In Winter, C-J., J. Nitsch (eds.), *Hydrogen as an Energy Carrier: Technologies, Systems, Economy.* pp. 249–289, Springer-Verlag, New York, 377 pp.

Cohen, B. (1990) *The Nuclear Energy Option: An Alternative for the 90s.* Plenum Press, New York, 338 pp.

Cleveland, C.J., R. Costanza, C.A.S. Hall, R. Kaufmann (1984) Energy and the U.S. economy: A biophysical perspective. *Science,* 225, pp. 890–897.

Culp, A.W., Jr. (1979) *Principles of Energy Conversion.* McGraw-Hill, New York, 199 pp.

Dale, S.J., S.M. Wolf, T.R. Schneider (1990) *Energy Applications of High Temperature Superconductors, Vol. 1.* Electric Power Research Institute, Palo Alto, CA, 89 pp.

Diamond, J.M. (1987) Human use of world resources. *Nature,* 328, pp. 479–480.

Dickinson, R.E. (ed.) (1987) *The Geophysiology of Amazonia: Vegetation and Climate Interactions.* John Wiley & Sons, New York, 526 pp.

Discover (1994) Breakthroughs: fast neutrons in Idaho. *Discover,* April 1994, p. 23.

Dugger, G.L., D. Richards, E.J. Francis, W.H. Avery (1983) Ocean thermal energy conversion: Historical highlights, status and forecast. *J. Energy,* 7 (4), pp. 293–303.

Early, J.T. (1989) Space-based solar shield to offset greenhouse effect. *J. Brit. Interplan. Soc.,* 42, pp. 567–569.

Edmonds, J., J.M. Reilly (1985) *Global Energy: Assessing the Future.* Oxford University Press, New York.

Edmonds, J., J.M. Reilly (1986) *The IEA/ORAU Long-Term Global Energy CO_2 Model: Personal Computer Version A84PC.* Institute for Energy Analysis, Oak Ridge Associated Universities, Oak Ridge, Tennessee (Available from National Technical Information Service, Springfield, Virginia).

Engström, P.C., et al. (1978) Direct-current power transmission. In Fink, D.G., H.W. Beaty (eds.), *Standard Handbook for Electrical Engineers.* McGraw-Hill, New York, Chapter 15, 2,448 pp.

Faires, V., C.M. Simmang (1978) *Thermodynamics,* Sixth edition. Macmillan Publishing Company, New York, 646 pp.

Forward, R.L. (1987) *Advanced Space Propulsion Study: Antiproton and Beamed Power Propulsion.* AFAL TR-87-070, Air Force Astronautics Laboratory, Air Force Systems Command, Edwards Air Force Base, CA, 29 pp. plus seven appendices.

Fuller, R.B. (1981) *Critical Path.* St. Martins Press, New York, 471 pp.

Giese, R.F., R.A. Thomas, E.B. Forsyth (1988) AC transmission. In Wolsky, A.M. et al. (eds.), *Advances in Applied Superconductivity: A Preliminary Evaluation of Goals and Impacts,* ANL/CNSV-64, pp. 69–100, Argonne National Laboratory, Argonne, IL; available from National Technical Information Service, U.S. Department of Commerce, Springfield, VA, 192 pp.

Gilder, G. (1993) When bandwidth is free. Interview in *Wired,* September/October 1993, pp. 38–41.

Glaser, P.E. (1968) Power from the sun: Its future. *Science,* Vol. 162, pp. 857–861.

Glaser, P.E., F.P. Davidson, K.I. Csigi (1993) *Solar Power Satellites: The Emerging Energy Option.* Ellis Horwood, New York, 300 pp.

Goldemberg, J., T. B. Johansson, A.K.A. Reddy, R.A.H. Williams (1988) *Energy for a Sustainable World,* Wiley Eastern, New Delhi, India, 517 pp.

Gore, A. (1992) *Earth in the Balance: Ecology and the Human Spirit.* Houghton Mifflin Co., New York, 408 pp.

Grubb, M.J., N.I. Meyer (1993) Wind energy: resources, systems, and regional strategies. In Johansson, T.B., H. Kelly, A.K.N. Reddy, R.H. Williams (eds.), *Renewable Energy: Sources for Fuels and Electricity,* pp. 157–212, Island Press, Washington, DC, 1160 pp.

Hackmann, W. (1993) Tesla's sparks of imagination. *Nature,* 363, p. 592.

Häfele, W., et al. (1981) *Energy in a Finite World: Paths to a Sustainable Future.* Vol. 1, Ballinger Publishing Company, Cambridge, Massachusetts, 225 pp.

Hall, D.O., H.E. Mynick, R.H. Williams (1991) Cooling the greenhouse with bioenergy. *Nature,* 353, pp. 11–12.

Hammond, A.L. (ed.) (1990) *World Resources 1990–91*. Oxford University Press, New York, 383 pp.

Hammond, A.L. (ed.) (1994) *World Resources 1993–94*. Oxford University Press, New York, 403 pp.

Harvey, L.D. (1990) Managing atmospheric CO_2: Policy implications. *Energy,* 15, pp. 91–104.

Hoffert, M.I. (1990) Climate sensitivity, climate feedbacks and policy implications. In Mintzer, I.M. (ed.), *Confronting Climate Change: Risks, Implications and Responses,* pp. 33–54, Cambridge University Press, New York, 382 pp.

Houghton, J.T., G. Jenkins, J.J. Ephraums (eds.) (1990) *Climate Change: The IPCC Scientific Assessment.* Cambridge University Press, New York, 365 pp.

Jain, A.K., H.S. Kheshgi, M.I. Hoffert, D.J. Wuebbles (1994) Distribution of radiocarbon as a test of global carbon cycle models. *Global Biogeochem. Cycles,* 9(1), 153–166.

Kaku, M. (1994) *Hyperspace.* Oxford University Press, New York, 359 pp.

Kessler, G.W. (1978) Steam boilers. In Baumeister, T., E.A. Avalone, T. Baumeister III (eds.), *Mark's Standard Handbook for Mechanical Engineers, Eight Edition,* pp. 9.7-9.35, McGraw-Hill, New York, 1,864 pp.

Krause, F., W. Bach, J. Koomey (1989) *Energy Policy in the Greenhouse,* Vol. 1, International Project for Sustainable Energy Paths (IPSEP), El Cerrito, California.

Leggett, J., W.T. Pepper, R.J. Stewart (1992) Emission scenarios for the IPCC. In Houghton, J.T., B.A. Callander, S.K. Varney (eds.) *Climatic Change 1992: The Supplementary Report to the IPCC Scientific Assessment,* pp. 69–95, Cambridge University Press, New York, 200 pp.

Lovins, A.B. (1977) *Soft Energy Paths: Toward a Durable Peace.* Harper Colophon Books, New York, 239 pp.

MacCracken, M.C.(ed.) (1990) *Energy and Climate Change: Report of the DoE Multi-Laboratory Climate Change Committee.* Lewis Publishers, Chelsea, MI, 161 pp.

Maier-Reimer, E., K. Hasselmann (1987) Transport and storage of CO_2 in the ocean — An inorganic ocean-circulation carbon cycle model. *Clim. Dyn.,* 2, pp. 63–90.

Marland, G. (1991) A commentary on: "Tropical forests and the greenhouse effect: A management response." *Climatic Change* 19, pp. 227–232.

Matthews, R.B., E.P. Coomes, E.U. Khan (1994) Hierarchical analysis of options for lunar-surface power. *J. Prop. Power,* 10, pp. 425–440.

Manne, A.S., R.G. Richels (1992) *Buying Greenhouse Insurance: The Economic Costs of Carbon Dioxide Emission Limits.* MIT Press, Cambridge, MA, 182 pp.

Meadows, D.H., D.L. Meadows, J. Randers (1992) *Beyond the Limits,* Chelsea Green Publishing Co., Post Mills, VT, 300 pp.

Meyers, N. (1991) Tropical forests: Present status and future outlook. *Climatic Change* 19, pp. 3–32.

Meyers, N., T.J. Goreau (1991) Tropical forests and the greenhouse effect: A management response. *Climatic Change* 19, pp. 215–225.

Naki´cenovi´c, N., W.D. Nordhaus, R. Richels, F.L. Toth (eds.) (1994), *Integrative Assessment of Mitigation, Impacts, and Adaptation to Climate Change,* CP-94-9, pp. 501–512, International Institute for Applied Systems Analysis, Laxenburg, Austria, 669 pp.

NAS (1992) *Policy Implications of Greenhouse Warming: Mitigation, Adaptation, and the Science Base.* Panel on Policy Implications of Greenhouse Warming, National Academy of Sciences, National Academy Press, Washington, DC, 918 pp.

Nordhaus, W.D. (1991) The cost of slowing climate change: a survey. *Energy J.,* 12 (1), pp. 37–65.

Nordhaus, W.D. (1992) An optimal transition path for controlling greenhouse gases. *Science,* 258, pp. 1315–1319.

Norman, C. (1991) Iraq's bomb program: A smoking gun emerges. *Science,* 254, pp. 644–645.

Ogden, J.M., R.H. Williams (1989) *Solar Hydrogen: Moving Beyond Fossil Fuels.* World Resources Institute, Washington, DC, 123 pp.

Oort, A.H. (1983) *Global Atmospheric Circulation Statistics.* NOAA Professional Paper 14, p. 157, F40-F43, National Oceanic and Atmospheric Administration, Geophysical Fluid Dynamics Laboratory, Princeton, NJ, 180 pp., includes microfiche data tables (F01-F47).

Ovshinsky, S.R., A. Madan (1983) Amorphous semiconductors equivalent to crystalline semiconductors. U.S. Patent 4,409,605, Oct.11, 1983.

Peixoto, J.P., A.H. Oort (1992) *Physics of Climate.* American Institute of Physics, New York, 520 pp.

Reddy, A.K., J. Goldemberg (1990) Energy for the developing world. *Scientific American,* 263 (Sept. 1991), pp. 111–118.

Revelle, R. (1985) Soil dynamics and sustainable carrying capacity of the Earth. In Malone, T.F., J.G. Roederer (eds.), *Global Change,* Cambridge University Press, New York, pp. 465–473.

Richards, J.F. (1990) Land transformation. In Turner III, B.L. et al. (eds.), *The Earth As Transformed by Human Action: Global and Regional Changes in the Biosphere over the Past 300 Years,* pp. 163–201, Cambridge University Press, New York, 713 pp.

Riding, A. (1994) Back to the present in a long-lost novel by Verne. *The New York Times,* September 27, 1994, p. C15.

Savage, M. (1994) *The Millennial Project.* Little, Brown and Co., New York, 508 pp.

Scheraga, J.D., N.A. Leary (1992) Improving the efficiency of policies to reduce CO_2 emissions. *Energy Policy,* May 1992, pp. 394–403.

Schlesinger, W.H. (1991) *Biogeochemistry: An Analysis of Global Change.* Academic Press, New York, 443 pp.

Schockley, W., H.J. Queisser (1961) Detailed balance limit of efficiency of p-n junction solar cells. *J. Appl. Phys.,* 32, pp. 510–519.

Serber, R. (1992) *The Los Alamos Primer: The First Lectures on How to Build An Atomic Bomb.* University of California Press, Berkeley, CA, 98 pp.

SERI (1990) *The Potential of Renewable Energy: An Interlaboratory White Paper.* SERI/TP-260-3674, Solar Energy Research Institute, Golden, CO.

Shepard, M., J.B. Chaddock, F.H. Cocks, C.M. Harmon (1977) *Introduction to Energy Technology.* Ann Arbor Science, Ann Arbor, MI, 300 pp.

Shine, K.P., R.G. Derwent, D.J. Weubbles, J-J. Morcrette (1990), Radiative forcing of climate. In Houghton, J.T., G.J. Jenkins, J.J. Ephraums (eds.) *Climate change: The IPCC Scientific Assessment,* pp. 41–68, Cambridge University Press, New York, 364 pp.

Smil, V. (1991) *General Energetics: Energy in the Biosphere and Civilization.* John Wiley & Sons, New York, 370 pp.

Sternbach, R., M. Okuda (1991) *Star Trek: The Next Generation Technical Manual.* Pocket Books, New York, 184 pp.

Sundquist, E.T. (1993) The global carbon dioxide budget. *Science,* 259, pp. 934–940.

Sverdrup, H.U., M.W. Johnson, R.H. Fleming (1970) *The Oceans, Their Physics, Chemistry and General Biology.* Prentice-Hall, Englewood Cliffs, New Jersey, 1087 pp.

Terborgh, J. (1992) *Diversity and the Tropical Rain Forest.* W.H. Freeman & Co., New York, 242 pp.

Thorndike, E.H. (1976) *Energy and Environment: A Primer for Scientists and Engineers,* Addison-Wesley, Reading, MA, 286 pp.

Tissot, B.P., D.H. Welte (1978) *Petroleum Formation and Occurrence.* Springer-Verlag, New York, 538 pp.

Tolmazin, D. (1985) *Elements of Dynamic Oceanography.* Allen & Unwin, Boston, MA, 181 pp.

Turcotte, D.L., G. Schubert (1982) *Geodynamics: Application of Continuum Physics to Geological Problems.* John Wiley & Sons, New York, 450 pp.

Vos, A.D. (1980) Detailed balance limit of the efficiency of tandem solar cells. *J. Phys. D: Appl. Phys.,* 13, pp. 839–846.

Waldrop, M.M. (1992) *Complexity: The Emerging Science of Order and Chaos,* Touchtone, New York, 380 pp.

Walker, J.C.G., J.F. Kasting (1992) Effects of fuel and forest conservation on future levels of atmospheric carbon dioxide. *Paleogeogr., Paleoclimatol., Paleoecol.* (Global and Planetary Change Section), 97, pp. 151–189.

Wallace, J.M., P.V. Hobbs (1977) *Atmospheric Science: An Introductory Survey.* Academic Press, New York, 467 pp.

Weinberg, A.M. (1992) The first and second fifty years of nuclear fission. In Kuliasha, M.A., A. Zucker, K.J. Ballew (eds.), *Technologies for a Greenhouse-Constrained Society,* pp. 227–237, Lewis Publishers, Boca Raton, FL, 835 pp.

Welford, W.T., R. Winston (1981) Principles of optics applied to solar energy concentrators. In Kreider, J.F., F. Kreith (eds.), *Solar Energy Handbook,* McGraw-Hill, New York, Chapter 3, 1,120 pp.

Wigley, T.M.L. (1995) Global-mean temperature and sea level consequences of greenhouse gas stabilization. *Geophys. Res. Lett,* 22, pp. 45–48.

Williams, R.H. (ed.) (1980) *Toward a Solar Civilization,* MIT Press, Cambridge, MA, 250 pp.

Wilson, E.O. (1992) *The Diversity of Life.* Harvard University Press, Cambridge, Massachusetts, 424 pp.

Winter, C.J., J. Nitsch (eds.) (1988) *Hydrogen as an Energy Carrier: Technologies, Systems, Economy.* Springer-Verlag, New York, 377 pp.

Yergin, D. (1991) *The Prize: The Epic Quest for Oil, Money and Power,* Simon & Schuster, New York, 877 pp.

Chapter 5

WATER RESOURCES

Authors: William H. McAnally
Co-Authors: Phillip H. Burgi, Darryl Calkins, Richard H. French,
Jeffery P. Holland, Bernard Hsieh, Barbara Miller, Jim Thomas
Contributors: William D. Martin, James R. Tuttle

5.1 INTRODUCTION

Current estimates of climatic change have significant implications
for the hydrologic cycle and water resource systems. This chapter provides an overview of the problems and issues likely to result from
potential changes in the magnitude, and distribution of water as a
result of global climate change. Potential response strategies to cope
with current hydrologic variability, as well as projected changes, are
also examined. Finally, research and development needs are outlined
to improve the flexibility, resiliency, and robustness of water resource
systems to deal with projected, but uncertain, climatic changes and
present variability that stresses existing systems.

5.2 PROBLEM OVERVIEW

Global climate change is projected to alter temporal and spatial
patterns and ranges of temperature, precipitation, and evaporation.
Changes in other important climate variables, such as solar radiation,
cloud cover, wind, and humidity are also predicted. These changes
and their interactions could significantly alter key components of the
hydrologic cycle, including runoff, soil moisture, groundwater
recharge, and snowmelt patterns. The magnitude, distribution, and frequency of extreme climatic events, such as droughts and floods, may
also be modified.

Currently, water resources projects are planned, designed, and
operated based on historical patterns of water availability, quality, and

demand assuming climate is constant. Changes outside the expected normal range of variation in these parameters can stress the ability of hydraulic structures to perform safely as designed and the ability of water resource systems to meet designated, and often competing, uses. Some examples of the potential effects of climatic change on water resources are described in the following sections.

5.2.1 Water Supply

Under many climate change scenarios, both the supply and demand of water will change, creating potential imbalances between the two or exacerbating existing imbalances. Reservoirs that store surface runoff from snowmelt or precipitation input may be either too small to accommodate the increase in demand or oversized for the change in requirements. Recharge and withdrawals from groundwater basins may be altered. Transmission and distribution systems (canals and conduits) that convey the water to the end user may be similarly impacted.

5.2.2 Navigation

Navigation channels and ports have been designed with channel depths and widths based on historical water levels, flows and sedimentation rates. Climate-induced changes could potentially alter these physical conditions, and thus affect channel and port navigational capacity. The result would be alteration of operation and maintenance costs and reduction of the confident use of the port, as well as making the identification and justification of new facilities difficult.

5.2.3 Flood Control

Flood control structures (channels, dams and spillways, levees, retention and detention basins, and floodwalls) are designed to provide protection against events with an occurrence frequency based on statistics derived from historical data. Potential climate changes may alter the frequency distribution of storms that produce flood events and the magnitude of probable maximum floods; and thus existing projects may have future protection requirements that are quite different from the initial design requirements.

5.2.4 Groundwater

Existing conjunctive use of ground and surface water may be altered. Groundwater withdrawal in many coastal areas has been shown to cause dramatic effects on ground subsidence. High water withdrawal rates near the coastal regions often induce salinity intrusion, and any sea level rise combined with increased groundwater pumping will magnify these problems. Groundwater recharge areas will also be affected by changes in the amount of precipitation, its form, and distribution.

5.2.5 Power Production

Hydroelectric power facilities would be directly impacted by changes in flow regimes. Decreased flows would reduce hydrogeneration. Higher flows would increase generation, but the facilities might be undersized to effectively utilize the excess capacity. Thermal power plants that depend on water supply for cooling would be impacted by higher river temperatures. Multiple use demands could curtail optimum hydropower scheduling in areas experiencing reduced flows.

5.2.6 Water Quality and Aquatic Biota

Climate induced changes in the magnitude, timing and quality of runoff will impact waste assimilative capacity, nutrient levels, water temperature, salinity, turbidity, and dissolved oxygen levels, which in turn affect riverine, reservoir, estuarine, wetland, and lake environments. The balance in the environmental conditions for some sensitive biota is very delicate. Users such as waterfowl, raptors and other higher trophic species are affected by water quality conditions.

5.2.7 Recreation

In recent years, public demands have required those agencies with water control authority to allocate water resources for boating, fishing, rafting, swimming and other recreational uses in addition to the original water use requirement of the designed facility. With climate changes, additional pressures for continued or increased recreational uses will create more demand for sustainable water supplies.

5.2.8 Agriculture and Irrigation

The projected increase in world population combined with a change in the climate will stress the existing food production regions to increase their output. This may result in increased demand for irrigation in some regions. If the climate shifts or changes, existing irrigation systems may not be adequate to accommodate the increased demand. Resource allocations to these systems will have to be adjusted, which will be difficult without accurate hydrologic predictions. In addition, agricultural areas that have traditionally not required irrigation may, under changed climatic conditions, require irrigation.

5.2.9 Water and Wastewater Treatment

An increase or decrease in the demand for water from surface supplies will impact both the water supply facilities and wastewater treatment facilities. Changes in water quantity and quality could require modified or new strategies for treatment.

5.2.10 Institutional and Legal Matters

The competitive use of water is increasing and with a change in climatic conditions, institutions and the legal and social framework for allocating water and resolving conflict may be forced to change. The pressure to meet competing interests and supply water for multiple uses can increase as a result of any climate change.

5.3 IMPACTS ON THE HYDROMETEOROLOGICAL CYCLE

5.3.1 Range of Predicted Changes

Current estimates of equilibrium warming range from 1.9 to 5.2°C for a doubling in CO_2 concentrations, or its radiative equivalent, from preindustrial levels (National Academy of Sciences, 1991). Scientists predict that as the atmosphere warms, its ability to hold water vapor will increase at an approximate rate of 5 to 6% per °C (Rosenberg et al., 1989). Higher air temperatures will also increase surface evaporation. To maintain equilibrium in the moisture budget, the precipitation rate must also increase (MacCracken and Luther, 1985).

Global average increases in precipitation and evaporation are presently estimated at 7 to 15% for a doubling in CO_2 concentrations. These increases, however, will vary both spatially and temporally. Individual regions might actually experience reduced precipitation. Changes in the seasonal distribution of precipitation will also vary from region to region (Waggoner, 1990; Rind and Lebedeff, 1984). Moreover, recent studies indicate that when complex plant-climate interactions are considered, changes in evapotranspiration can vary from –20 to +40% in specific river basins (Martin et al., 1989).

Predicted changes in precipitation and evaporation are likely to be coupled with changes in other climatic variables such as solar radiation, cloud cover, wind, and humidity. The range of these changes on a global and regional basis are outlined in Table 5.1, while a comparison of general circulation models' (GCM) results for a specific region (the Tennessee River Basin in the Southeastern U.S.) with global projections are presented in Table 5.2.

Projected changes in these meteorological variables could significantly alter key components of the hydrologic cycle, including runoff, soil moisture, groundwater recharge and discharge, and snowmelt patterns. Due to the nonlinear relationship between precipitation and runoff, small changes in rainfall and evaporation can produce significant changes in runoff and regional water availability (Gleick, 1986, 1989). Nemec and Schaake (1982) used hydrologic watershed models to show that a 1°C temperature increase, combined with a 10% decrease in precipitation, can cause a 25% reduction in average annual runoff in a humid basin and a 50% reduction in an arid basin. Using statistical correlations in the Colorado River Basin, Revelle and Waggoner (1983) estimated that a 2°C rise in temperature, even when coupled with a 10% increase in average annual precipitation, could still produce an 18% reduction in annual runoff due to increased evaporatranspiration. Although more complex studies and further refinements in hydrologic analysis procedures may alter the magnitude of these results, the sensitivity of runoff to changes in meteorology remains an important consideration.

5.3.2 Spatial and Temporal Variations

In addition to changes in the average annual values of hydrometeorological variables, shifts in regional distribution, seasonality, extremes, variability, and recurrence frequency are also likely. Most GCMs predict distinct changes in the regional distribution and seasonality of key hydrologic variables (IPCC, 1990; MacCracken and Luther, 1985; Rind and Lebedeff, 1984). Although GCMs are currently incapable of predicting detailed regional hydrologic effects, some general patterns are apparent. Surface air is expected to warm faster over the land than over oceans, with a minimum of warming predicted to occur around Antarctica and in the northern North

Table 5.1 Range of climate changes. From: Mearns, L., P.H. Gleick, and S.H. Schneider, 1990, "Prospects for Climate Change," in Waggoner, P., Ed., *Climate Change and U.S. Water Resources*, (John Wiley & Sons, New York for AAAS. With permission.)

Phenomenon	Projection of Probable Global Annual Average Change[a]	Regional Average	Distribution of Change — Change in Seasonality	Distribution of Change — Interannual[b] Variability	Significant Transients	Confidence of Projection — Global Average	Confidence of Projection — Regional Average	Estimated Time for Research That Leads To Consensus (years)
Temperature	+2 to +5°C	-3 to +10∘C	Yes	Down?	Yes	High	Medium	0-5
Sea Level	+10 to +100cm[c]		No	?	Unlikely	High	Medium	5-20
Precipitation	+7 to 15%	-20 to 20%	Yes	Up	Yes	High	Low	10-50
Direct Solar Radiation	-10 to +10%	-30 to 30%	Yes	?	Possible	Low	Low	10-50
Evapotranspiration	+5 to 10%	-10 to +10%	Yes	?	Possible	High	Low	10-50
Soil Moisture	???[d]	-50 to 50%	Yes	?	Yes	?	Medium	10-50
Runoff	Increase	-50 to +50%	Yes	?	Yes	Medium	Low	10-50
Severe Storms[e]	?	?	?	?	Yes	?	?	10-50

[a] For an "equivalent doubling" of atmospheric CO2 from the pre-industrial level. These are equilibrium values, neglecting transient delays and adjustments.

[b] Interference based on preliminary results for the United States (Rind et al., 1989).

[c] Increases in sea level at approximately the global rate except where local geological activity prevails.

[d] No basis for quantitative or qualitative forecast.

[e] Emanuel (1987) suggests that increased ocean temperatures could increase the intensity of tropical cyclones.

Table 5.2   Projection of climatic changes likely to occur as a result of an equivalent doubling in atmospheric CO_2. From: Miller, B. A., 1990, TVA Engineering Laboratory, Norris, TN.

Parameter	Reported Average Annual Change[1]		GCM Average Annual Projections for the Tennessee Valley[2]		
	Global	Regional	Minimum[3]	Maximum[4]	Average[5]
Temperature (°C)	2+ to +5°C	-3 to +10°C	+3.3°C	+5.8°C	+4.4°
Precipitation	+7 to +15%	-20 to +20%	-30%	+33%	+2%
Runoff	Increase	-50 to +50%	-77%	+94%	-1%
Wind Speed			-31%	+28%	-3%
Solar Radiation	-10 to +10%	-30 to +30%	-4%	+7%	+2%
Evapotranspiration	+5 to +10%	-10 to +10%			
Cloud Cover			-22%	+12%	-6%
Humidity[6]			+10%	+75%	+33%

[1] Reported by L. Mearns. P. H. Gleick, and S. H. Schneider, "Prospects for Climate Change," in Paul Waggoner. Ed. , *Climate Change and U.S. Water Resources* (New York, Wiley for AAAS, forthcoming).

[2] Based on 1987 General Circulation Model (GCM) projections for the Tennessee Valley Authority region, values averaged for grids covering TVA region; values averaged for: NASA's Goddard Institute for Space Studies (GISS) Model Princeton's Geophysical Fluid Dynamics Laboratory (GFDL) Model Oregon State University (OSU) Model.

[3] Minimum is annual average minimum based on monthly values for 3 GCMs.

[4] Maximum is annual average maximum based on monthly values for 3 GCMs.

[5] Average is annual average based on monthly values for 3 GCMs.

[6] Humidity ranges taken from GISS only.

Atlantic region. In other high northern latitudes, however, warming in the winter is projected to be 50 to 100% greater than the global mean. Average winter precipitation is also predicted to increase in middle and high latitude continents, including areas over central North America and southern Europe. Predicted changes in summer precipitation are variable with recent models showing decreases in central North America and southern Europe and increases over areas such as Australia and southeastern Asia (IPCC, 1990). Other investigators suggest that elevated temperatures and evaporation may be accompanied by decreased summer precipitation in the lower latitudes (Shiklamanov, 1987; World Meteorological Organization, 1988).

In the northern and western U.S., where runoff is largely derived from snowmelt, distinct shifts in the relative amount of rain and snow, as well as earlier snowmelt resulting from warmer winters, are likely. The resulting changes in runoff patterns would alter the magnitude, timing and probability of flooding patterns. The availability of water during peak demand periods such as the irrigation season would also be impacted. In the southeastern U.S., where runoff is largely precipitation driven, some GCMs also project shifts in current seasonal patterns (Smith and Tirpak, 1988; Hains and Henry, 1989; Frederick and Gleick, 1988).

Changes in variability, or inter- and intra-annual deviations from mean conditions, can impact the adequacy and reliability of water resource projects. IPCC (1990) suggests that with the possible exception of the increase in the number of intense rainfall events, there is no clear evidence that weather variability will change in the future. For example, in the case of temperature, this assumption implies that changes in mean conditions would shift the entire temperature distribution upward, resulting in more days above some critical temperature on the high end and less days with temperatures on the low end of the distribution; but the deviation from the mean would remain constant. Other investigators have also found that, despite considerable noise in GCM results, the standard deviations of surface temperatures appear to be more likely to decrease than increase under global warming scenarios (Rind et al., 1989). These impacts on variability, however, are surmised from equilibrium double CO_2 conditions after new mean conditions have been established. During the transition from current to potentially changed conditions, shifts in the mean time could appear.

Based on statistical reasoning, several investigators argue that small shifts in mean values can imply large changes in the frequencies of extreme events such as droughts and floods (National Academy of Sciences, 1987; Waggoner, 1990; Mearns et al., 1984). Analysis of tropical cyclones and their relationship to warm low latitude oceans suggests that severe flood frequency could increase in a greenhouse enriched environment (Michaels, 1989). Some modeling evidence suggests that hurricane intensities would also increase with climatic warming (Emanuel,1987). Recent reports, however, argue that the climate models do not give a consistent indication of whether tropical storms will increase or decrease in either frequency or intensity as the climate changes. At present, tropical storms, such as typhoons and hurricanes, develop over seas that are warmer than approximately 26°C. Although the areal extent of seas exceeding this critical temperature would increase in a warming scenario, the critical temperature itself might increase under warmer conditions (IPCC, 1990).

5.3.3 Related Factors

In the long term, alterations in the hydrologic cycle could be complicated by climate change-induced effects on plant growth and land use. Elevated CO_2 concentrations have been shown to increase photosynthesis and plant growth potential in laboratory studies. In these CO_2 enriched environments, plant transpiration is also generally reduced due to increased stomatal resistance (Waggoner, 1990). Under natural conditions, the net effect of these plant responses to increased CO_2 levels will vary depending on plant type, relative changes in a range of climatic variables, and individual ecosystems. Complex plant-climate interactions may moderate or augment predicted increases in evapotranspiration with consequences for plant growth and development (Rosenberg et al., 1989; Martin et al., 1989). These effects, coupled with changes in the hydrologic cycle, could ultimately alter vegetative patterns. Changes in the extent and type of vegetative cover could modify the water content of soil layers, thereby further influencing runoff and water availability (Abramopoulos et al., 1988).

Land use could be further impacted by changes in agriculture, forestry, population distributions, and industrial activity with associated feedback implications for the hydrologic cycle. Higher temperatures, modified rainfall patterns, and increased CO_2 concentrations would affect agricultural practices, the type and distribution of crops, and irrigation needs (see Chapter 7). Similarly, the distribution and composition of forests, wildlife, and other natural ecosytems could be altered. Smith and Tirpak (1988) project potential demographic shifts, as well as possible changes in the location of industrial and agricultural centers. Such conversions in land use would not only affect the relative balance between water supply and demand, but would also impact soil and land surface characteristics, thereby influencing important components of the hydrologic cycle such as infiltration, interception, evapotranspiration and groundwater recharge.

5.3.4 Limitations

It is important to recognize that many of these impacts represent the potential consequences of an equilibrium-double CO_2 scenario. Although abrupt changes and large transients are possible (Schneider, 1989), actual changes could occur more slowly over time or be moderated by other climatic factors. Changes in the hydrologic cycle may also work to the benefit or detriment of individual regions. Further, the projected changes in hydrometerological variables outlined in Tables 5.1 and 5.2 are based on GCM runs conducted prior to 1987. Subsequent model runs, based on recent advancements in climate modeling, may provide different rates and magnitudes of predicted changes. However, the interrelationship among meteorological and hydrological variables, their sensitivity to changes in global average temperature, and the potential for changes in these variables to significantly alter available water resources remain key climate change issues.

5.4 WATER RESOURCES IMPACTS

Water resource projects are generally planned, designed, operated and maintained to accommodate historical ranges and patterns of climatic variability based on the assumption of a statistically constant climate. Current, as well as reasonably projected (based on past experience), patterns of relative water supply and demand are also incorporated in system designs. Projected changes in the magnitude, timing, and distribution of hydrometeorological parameters, particularly if coupled with demographic shifts and changes in industrial and agricultural activity, could impact the safety of hydraulic structures, as well as the ability of water resource systems to effectively balance available supplies against competing water uses. This section provides representative examples of potential beneficial and adverse impacts of climatic changes on water resources.

5.4.1 Water Supply and Demand

Global warming is likely to impact both water supply and water demand. Predicted changes in the hydrologic cycle — changes in precipitation patterns, evapotranspiration rates, temporal and spatial distributions in magnitude of runoff, and frequency and intensity of severe storms—will affect the quantity and quality of water supplies (Frederick and Gleick, 1988). Long-term climatic impacts on demographics, industrial development, agricultural production, irrigation needs, natural systems, and energy use would ultimately impact water demand patterns (Smith and Tirpak, 1988).

Table 5.3 illustrates the vulnerability of major American river basins to changes in supply and demand (U.S. Water Resources Council, 1978; U.S. Geological Survey, 1984). Relative storage capacity, or the ratio of existing maximum storage volume to mean annual renewable supply within a basin, is one indicator of a region's ability to withstand prolonged periods of drought or flood. Basins with a large relative storage (ratio greater than 0.6) provide protection during times of drought and a buffer during floods. Conversely, regions with low ratios (below 0.6) have small relative storage capacities and are more sensitive to changes in hydrology (Frederick and Gleick, 1988). To address basin vulnerability to climate change with regard to water supply however, one must also consider relative demand, or the ratio of demand (consumptive depletions including consumptive use, water transfers, evaporation and groundwater overdraft) to annual mean renewable supply. Water is considered a critical factor in economic development when the relative demand exceeds a ratio of 0.20 (Szesztay, 1970). High relative demand ratios indicate that existing supply is susceptible to stress from growing populations, increased industrial and commercial demand, and climatic fluctuations. The Alaskan basin, for example, has essentially no storage, but as relative demand is also extremely low, the basin is not susceptible to drought for domestic or industrial consumption. Conversely, the Lower Colorado River Basin has the highest relative

Table 5.3 Indicators of vulnerability to climate conditions. From: U. S. Water Resources Council (1978) and U.S. Geological Survey (1986). Taken from: Frederick and Gleick (1988).

River Basin or Region	Measure of Storage[a]	Measure of Demand[b]
New England	.15	.01
Mid-Atlantic	.10	.02
South Atlantic-Gulf	.16	.03
Great Lakes	.08	.02
Ohio	.12	.02
Tennessee	.23	.01
Upper Mississippi	.14	.03
Lower Mississippi	.31	.09
Souris-Red-Rainy	.93	.07
Missouri	1.12	.29
Arkansas-White-Red	.45	.17
Texas-Gulf	.61	.23
Rio Grande	1.89	.64
Upper Colorado	2.61	.33
Lower Colorado	4.22	.96
Great Basin	.35	.49
Pacific Northwest	.19	.04
California	.42	.29
Alaska	.00	<.01
Hawaii	.01	.05
Caribbean	.05	.06

[a] Ratio of maximum basin storage volume to total basin annual mean renewable supply (1985). Regions with values below 0.6 have small relative reservoir storage volumes. Large reservoir storage volumes provide protection from floods and act as a buffer against shortages.

[b] Ratio basin consumptive depletions (including consumptive use, water transfers, evaporation, and groundwater overdraft) to total basin annual mean renewal supply (1985). Water is considered a decisive factor for economic development in regions with values above 0.20 (Szesztay, 1970).

storage capacity, but as consumptive use is 96% of renewable supplies, the basin users of runoff are vulnerable to droughts or reductions in streamflow. Basins such as the Upper and Lower Colorado, Rio Grande, Great Basin, and Missouri, where the relative demand exceeds 0.25, are vulnerable to climate change induced reductions in supply (Frederick and Gleick, 1988).

While relative storage and demand give some indication of the vulnerability of large river basins to climatic change, the range of potential impacts is dependent on other local factors including legal regulations, compacts, and treaties. The relative importance of instream water uses such as hydropower production, recreation, navigation, wildlife, and aquatic habitat can influence the flexibility a water resource system has to meet competitive water use demands and to reallocate water uses during times of scarcity. Climatic impacts on snowmelt and the timing of runoff would also influence the magnitude of impacts. For example, in the Sacramento-San Joaquin River Basin, Lettenmaier et al. (1989) found that the increased temperatures in four GCM scenarios produced major reductions in snow accumulation, resulting in increased winter runoff but reduced spring and summer runoff. Although the total volume of water increased in these scenarios, Sheer and Randall (1989) found that the shift in seasonality resulted in increased probability of spring flooding and substantially reduced water deliveries to consumers in the California Central Valley in the summer. Consequently, under current operating constraints and water allocation policies, water supplies during peak summer demand periods could become a problem.

5.4.2 Power Production

Shifts in the magnitude and seasonality of streamflows would directly impact hydropower generation and capacity, as well as system flexibility and reliability. In a study of the Tennessee Valley Authority (TVA) reservoir system (Miller and Brock, 1988), a warm and wet climate change scenario (4°C increase in average annual temperature and 31% increase in average annual runoff, with monthly variations in runoff ranging from +73% in March to –28% in November) increased average annual hydropower generation by 16%.

Miller and Brock (1988) also evaluated the impacts of a dry climate change scenario (31% reduction in average annual stream flow) on reservoir system behavior and power production. Average annual hydrogeneration was decreased by 24%. Reduced streamflows and operating heads also resulted in substantial losses in dependable hydro system capacity, particularly during the summer months when power demands for residential and commercial cooling are high under current conditions.

Crissman (1988) assessed the impact of climate change on the power production capability of the New York Power Authority system on the Niagara River. Decreased flows from the Great Lakes drainage area coupled with increased lake evaporation could have dramatic negative effects on scheduling of power based on historical records.

Warmer temperatures and hydrologic changes could also impact the operation of fossil and nuclear plants. Elevated water temperatures adversely impact thermal efficiency and the power output of steam turbines. Higher air and water temperatures also reduce cooling tower effectiveness. Elevated water temperatures, which can be exacerbated by high air temperatures and reduced streamflows, can increase the potential for violating environmentally based thermal discharge standards at fossil and nuclear plants. Nuclear power plants also have limits on the maximum intake water temperatures for auxiliary safety systems. When this limit is reached, the Nuclear Regulatory Commission requires plants shut down for safety reasons. Increased water temperatures, therefore, can constrain nuclear power plant operations (Miller et al., 1992).

In coastal areas, the prospects of rising sea levels, increased storm surges, and salt water intrusion could also impact power production. Power plant siting, the location of cooling water intakes, transmission efficiency, and system reliability become important issues (EPRI, 1989). In inland areas, water intakes may have to be relocated during extremely dry conditions.

5.4.3 Flood Control and Dam Safety

Floods result from a combination of heavy precipitation from severe storms, snowmelt, or combinations of rain and snowmelt; the physical characteristics of drainage basins; and modifications to drainage basin characteristics. The data from previous flooding events and the derived statistics are essential for designing cost effective flood protection systems. Climate change will require changes in flood frequency analysis and estimates of probable maximum precipitation (PMP) and hence probable maximum flood. Changes in the hydrometeorology resulting from global climate change could affect the reliability of flood protection systems. (Flood control projects are designed with safety factors. Up to a point, climate change trends toward higher flows will only reduce the margin of safety, not eliminate it.) New, more flexible design and operation procedures will be needed. The use of new design concepts that include overtopping embankments, fuse plugs, and labyrinth spillway designs are good examples of flexible and robust designs for climate variability.

Downstream flood protection can be maintained by incorporating expected climate change scenarios into reservoir operating rules, such as keeping reservoir water levels lower. Changing reservoir operating procedures, however, can have detrimental effects on multi-purpose projects such as those incorporating hydropower production, recreation, water supply, or navigation. Conversely, reduced runoff and flood risk could facilitate the reallocation of flood storage space for other uses, such as raising reservoir levels to increase hydropower generation or recreational opportunities.

5.4.4 Navigation

Frequency and duration of extremes, if increased or decreased due to climate change, could adversely impact navigation industry. Extreme high flows could create problems with bridge clearances, produce wave damage to riverside structures, increase project maintenance (dredging, bank stabilization, structures, etc.), potentially terminate navigation for short periods of time, create shoaling problems in channels due to increased sediment inflow, and may require a modification of dam capacity for lock and dam structures to handle higher than anticipated flows effectively, or will lead to more periods of suspended navigation. High flows increase fuel consumption of vessels.

Extreme low flows will result in possible long-term channel closures resulting in negative economic impacts. More hazardous navigation conditions resulting from low flows cause delays, and subsequent increased travel time from port to port. Increased maintenance of navigation aids will be required to insure safe operation within navigable waters. Shoaling patterns and shoaling rates can change, requiring an increase in maintenance dredging. Low flows in the winter increases the likelihood of an earlier ice cover and, depending on the temperature regime, a thicker ice sheet could be expected in cold regions.

Costs to the shipping industry for the Great Lakes due to a climate warming scenario has been investigated by Crissman(1988). The conclusions drawn from that study indicated that the duration of the lake and connecting river ice covers would decrease which would permit a longer shipping season without the hindrance of the ice. This results in an economic gain for this particular transportation mode in this region because of lower operating costs.

5.4.5 Recreation

Reduced flows with lowered lake levels and releases will diminish recreational use of lakes and their tailwaters. In most cases, an increase in water supply will increase recreational use of waterways (Miller and Block, 1988). However, the timing of these increases in water supply may conflict with recreational needs.

5.4.6 Irrigation and Drainage

Changes in temperature, runoff, snowmelt, evapotranspiration, and other hydrological and meteorological factors could have major effects on irrigated agriculture (see Chapter 7). Peterson and Keller (1989) examined the potential effects of global climate change on irrigation in both the Western and Eastern parts of the U.S. They computed a potential Net Irrigation Requirement (NIR) for four scenarios: (1) present conditions; (2) climate change of +3°C; (3) climate change of +3°C and +10% precipitation; and (4) climate change of +3°C and -10% precipitation. The NIR increased under the latter three scenarios compared to present conditions. The implication is that the percentage of cultivated land requiring irrigation will increase. Western states will find it increasingly difficult to maintain the present level of irrigation without developing "new" water sources through conservation and improved delivery system efficiencies. However, irrigation may well continue to increase in the East. Similar scenarios are likely in other parts of the world.

5.4.7 Water Quality

The warming of the Earth's atmosphere will result in significant modifications in the hydrologic cycle and the quality of surface and ground waters. Water quality of surface water bodies such as rivers and reservoirs could be significantly stressed by temperature increases, reduced dissolved oxygen levels, and potentially higher nutrient loads. Presently stressed systems will be affected the most. Aquatic ecosystems supported by these water bodies, such as fisheries, may be drastically altered or fail catastrophically under current climate change predictions. The survival of sensitive species may be threatened.

5.4.8 Effect on Land Transportation at Water Crossings

Increased rainfall amounts or changes in storm intensities will affect design parameters at highway, pipeline, and rail crossings. The designs typically provide minimum clearance of the bridge low chord above a 25-, 50-, or 100-year flood profile. Changes in rainfall patterns and intensities may result in some structures being over designed. This is of no particular safety concern, but replacement structures should reflect changed conditions. Conversely, other structures may be underdesigned and experience flows which encroach on the bridge. This will lead to higher flowlines upstream, increased scour at the bridge piers and abutments, and possible failure/overtopping, threatening traffic and endangering lives. Additionally, utility crossings such as water, gas, oil and telecommunications lines may be similarly adversely affected. Failure of these transportation and communication systems due to scour undermining them or removing protective overburden and allowing secondary breaks will lead to disruption of services and potential environmental pollution. River and estuarine ports may experience increased periods of time where facility loading/unloading operations are shut down due to higher or lower water levels than were originally anticipated when they were designed. There may be a need to design bridges (highway and rail) to withstand the higher runoff and sediment loads to which they will be exposed. Better data sets of present scour and channel capacities will be needed to estimate future conditions properly.

5.4.9 Region-Specific Impacts

Some regions will experience different or exaggerated climatic impacts because they exhibit extremes of natural climate. These include regions of deserts, permafrost, glaciers, and tropical rain forests.

5.4.10 Secondary Impacts

Areas of increased rainfall will have increased flow downstream of water control structures such as dams. This may cause increased flooding and limit agriculture activities in low lying areas. In most areas, increased water supply will increase agricultural production and/or decrease its cost. Other areas may experience decreased flows, directly impacting activities depending on surface water for irrigation, aquaculture, municipal, or industrial use. This could lead to ground water overdraft. For the case of long term navigation shutdown in which goods are moved by alternate land transportation modes, a higher unit transportation cost will result.

5.4.11 Public Works Policy, Planning, and Management

Extreme shifts in rainfall and temperature could force changes in public works policies and planning. Project management may require reanalysis and alteration to reflect different future conditions. Projects currently in the planning stage may be adversely affected by conditions existing upon their completion in 10–15 years and throughout their economic life expectancy of an additional 50–100 years.

Public works activities tend to be reactive, in that public funds are often unavailable until a disaster strikes. For example, federal flood control activities had been sought for the lower Mississippi valley for almost a century before the great 1927 flood. Two years later, a comprehensive project was formulated and approved for the valley. Public policy should be active, anticipating potential climate change and working to adapt to climate changes during the early stages of project planning/design, rather than reacting to a disaster.

Where public works missions are divided among agencies and political units, fragmentation of responsibility can lead to less effective use of water resources. In reacting to potential climate change, comprehensive planning that crosses political and institutional boundaries will reduce the impact of adverse changes.

5.5 ISSUES

The prospect of climatic change poses serious challenges to water resources managers concerning the availability and quality of future water supplies. Understanding the implications of these changes and preparing for an uncertain future raises several important issues regarding the development, maintenance, and management of water resources in the coming years. These issues are posed as a series of questions that must be adequately addressed to effectively cope with a changing world.

5.5.1 Reliable Predictions

- What are the prospects for climate change in the near and long-term?
- Can the uncertainties regarding current climate change predictions be narrowed?

- Can global climate change projections be translated into reliable regional climate scenarios on temporal and spatial scales useful for water resources analysis and management?

5.5.2 Impacts and Vulnerabilities

- Can more reliable regional climate scenarios be used effectively to assess the sensitivity, range of potential impacts, and critical vulnerabilities of water resources systems?
- Are the analysis tools and data sets currently available adequate to conduct reliable impact analyses?
- Can integrated systems analysis procedures be used effectively to incorporate inter-related disciplines, such as forestry, agriculture, and environmental sciences, into water resources analysis.

5.5.3 Flexibility to Cope with an Uncertain Future

- Are current water resources systems designed for current estimated levels of variability sufficient to accommodate predicted changes in climate?
- Can information regarding future predictions and impact assessments be effectively incorporated into water resources planning, design, construction, and operation such that systems are rendered more robust and flexible to cope with an uncertain future?
- Can water resources planners and agencies move to a more proactive role in recognition and resolution of potential problems?

5.5.4 Institutional and Legal Considerations

- Are current institutional and legal frameworks capable of effectively dealing with increased stresses on available water supplies?
- Are procedures available to effectively address issues of equity, water allocation, and competitive uses if the relative balance between available supplies and demand is significantly changed?
- Can mechanisms be developed to increase inter-agency and inter-governmental cooperation in the sustainable development of water resources?
- Should water be treated as a commodity subject to market forces after basic needs have been provided for?

5.5.5 Mitigation
- Can water resources systems be used to help mitigate the production of greenhouse gases?

5.6 RESEARCH AND DEVELOPMENT

While predictions of regional climate change remain poorly defined and uncertain, the key climate change issue becomes how to prepare in the intervening years. The planning and implementation time horizon for major water resources projects is on the order of 10 to 30 years, and operational lifetimes often exceed 50 years. It is wiser and, ultimately, more cost-effective to consider the prospects of climate change than to ignore its possibility. Thus, in the near-term, response efforts should be directed at education, assessment, and research, accompanied by the development of longer-term adaptation and mitigation strategies. It is important to note that there will be significant benefits from these efforts even if global climate change does not occur. These efforts are discussed in the next section.

5.6.1 Education
Education of water resource managers and planners, as well as the public, to the long-term nature and seriousness of the climate change issue is essential if we are to be prepared for the future. The assumption of a stationary climate, or at least the existence of a climate with predictable and limited variability, is ingrained in engineering training, design and operation. Hydrologists and water resource planners may find it difficult to consider climatic shifts outside traditional expectations. Recent advances in the use of risk/reliability analysis should be endorsed and more widely taught. The utilization of multi-disciplinary teams in problem solving, with all specialities involved adequately represented, will improve the transfer of knowledge among professionals from diverse disciplines and will improve problem solving capabilities.

Information and awareness programs should be instituted in both industrialized and developing countries to educate the public. The importance of water supply, wastewater treatment, water conservation, and the quality of the aquatic environment should also be taught in primary and secondary schools. Information on water and energy conservation measures imparted in such programs could result in substantial resource savings.

5.6.2 Assessment Methodology and Techniques

Increased understanding of the potential impacts of climate change and of climate sensitive activities can be used to improve our flexibility to deal with future change. Sensitivity analyses should be used in the near-term to determine critical thresholds for individual components of water resource systems, while scenario analyses can be effective in identifying the vulnerability of the integrated system to potential climatic changes and to climatic variability.

Increasing competition for a less predictable and more variable water supply will require improved data collection capabilities to manage water resource systems. In addition, improved data sets are needed to provide complete data records for impact analyses. Actions should include increased gathering of climate, streamflow, and water quality data, improved instrumentation and data transmission equipment, better methods of managing and analyzing data, increased use of remote sensing technology, improved real-time operations and flood warning systems. To increase the quantity and quality of collected data, it is essential to find less costly ways of collecting and analyzing data.

Impact assessments, including sensitivity studies as well as scenario analyses, should be initiated quickly. Sensitivity studies can determine critical thresholds for individual components of the water resource systems by evaluating the impacts of incremental changes in temperature, flow, and other pertinent meteorological variables such as humidity or wind speed. Scenario analyses can be used to identify the vulnerability of integrated systems to changes in climatic variables. Changes in seasonality, the frequency and magnitude of extreme events, and the magnitude and distribution of key hydrologic variables, should be explicitly addressed in the formulation of scenarios. Impacts on multiple system uses and purposes, including those associated with environmental and recreational impacts, should be assessed.

5.6.3 Adaption Methodology and Techniques

Strategies should be developed to incorporate climate change uncertainty into water resources planning with the ultimate aim of creating robust, flexible water resource systems. Such strategies include improved hydrologic analyses procedures; monitoring of current trends; basin-wide integrated water management; flexible institutions and enhanced inter-agency and international cooperation; improved mediation procedures to resolve competitive water use issues; improved water conservation strategies; realistic water pricing policies; and more responsive legal and institutional frameworks to deal with future change.

Fragmentation of water resource missions and budgets among various agencies can lead to decisions and plans that are optimum for one agency's mission, but not for the nation. Centralized planning and control should not be the goal, but centralized policies emphasizing a systems approach that transcends agency mission boundaries should be. Similarly, water resources planning and development by one nation can adversely impact other nations and, in cases of very large-scale projects, the global community. Integrated resource management across national boundaries should be encouraged.

Given the uncertainty of existing GCMs and regional models, their best near-term use may be in defining potential limits of the important factors. This information could be used by project personnel and designers in bracketing possible changes in the future. They can then review current procedures to determine if changes are desirable. While this process specifies neither the magnitude nor direction of climate changes that may occur, it does allow more formal consideration of the impacts of existing climate variability than now takes place.

Adaption strategies should be geared toward the development of techniques that incorporate climate change uncertainty into long-range planning with the aim of creating robust, flexible water resource systems. As historical records may no longer adequately predict future trends, hydrologic analyses procedures, such as flood forecasting and water supply determinations, need to account for the possibility of increased variability and changes in the frequency and intensity of extreme events.

As water supply variability is reduced with increased storage in larger basins, the need for basin-wide integrated water management, supported by flexible institutions, will increase in significance. Similarly, enhanced institutional arrangements for interagency cooperation, as well as improved mediation procedures among water use stakeholders, may be needed to resolve competitive use conflicts.

Improved water conservation strategies, combined with realistic legal and water pricing policies, will become important in areas where dry conditions are likely to prevail.

Prospects of climate change offer the opportunity to increase the resiliency, efficiency and productivity of existing water resource systems. Risk analysis should be included in project design and operation to account for variability. Projects that are influenced by possible radical climatological changes may require design lives of shorter duration. More responsive and flexible operating plans for existing and planned reservoirs will be required, and drought contingency plans should be developed. Measures to improve and encourage water and energy conservation will require development. The procedures that national or international water resource agencies must go through to permit modifications to projects will most likely require streamlining.

While a variety of meaningful studies can be conducted using existing technology in conjunction with sensitivity and scenario analyses, more definitive quantification of potential global climate change impacts on water resources systems/infrastructure will require improved GCM predictions, increased understanding of the various hydrometeorological and ecological processes impacted by potential climatic changes, and improved analytical tools. These long-term research and development requirements are discussed below.

5.6.4 Future Research and Development Needs: Robust, Viable Water Resource System

Climatic change has the potential to impact many interrelated water users, purposes of water resources systems, and the many technical disciplines associated with them. With regard to water resources systems/infrastructure, significant research and development investments will be required to quantify the impacts of climatic change and variability, and to create methodologies/designs to mitigate those impacts. We believe that the two broad goals listed below must be accomplished if the impacts of climatic change and variability on water resources systems/infrastructures are to be dealt with effectively. These goals are:

Goal 1: Robust, viable water resources systems that can accommodate present climate variability and potential climatic changes.

Goal 2: Mitigation of greenhouse forcing by increased use of alternative water-related power sources and increased efficiency and conservation in water resources activities.

Amplification of these goals, and the basic components of the research and development required to accomplish them, are discussed below.

- In the context of this discussion, the terms robust, viable, flexible, and resilent have specific meanings as they relate to water resource systems. Robust refers to the ability of a system to perform in a predictable, controlled fashion when conditions approach and exceed its design limits.
- Viable systems are those whose design, construction, and operation have adequate flexibility and resiliency to meet performance, economic, social, cultural, and environmental quality objectives in the face of climatic variability and potential climatic changes.
- Flexible systems can be used in ways not originally intended if conditions require a change.
- Resilient systems maintain their flexibility and operating effectiveness without frequent repair and rehabilitation.
- Vulnerability refers to the degree to which a water resource system fails to provide robustness, flexibility, and resiliency.

Accomplishment of Goal 1, which would result in the design, construction/modification, and operation of water resources systems capable of adequately responding to climatic change and variability, will require completion of four basic technical objectives associated with climatic variability and potential climatic changes as described below.

(1a) Creation of an adequate understanding of hydrometeorological and environmental responses to climatic and anthropogenic influences.

This will require the development of more refined global climate models and increased use of mesoscale modeling in the development of regional climate change scenarios. To produce hydrologic and climatic data for use in water resources development and management, the data from these models should be provided at a spatial scale of 250 km^2 and a daily temporal scale. Improved data collection efforts, for analysis of historical climatic trends and verification of models, should be initiated. The coupling of process-based ground and surface water models, with other models that focus on natural ecosystems, environmental quality, economics, climate, and land use, is required to provide the basic modeling framework to assess the impacts of potential climatic change and climate variability on water resources systems. However, several of the underlying descriptions of the physical, chemical, and biological processes within even the best of the current component models are known to require additional investigation. For example, many aspects of groundwater flow and transport, especially within the unsaturated zone, are poorly understood. Even more well-understood processes, such as watershed rainfall-runoff or turbulent three-dimensional hydrodynamic flow fields, have known inadequacies that will require significant research and development.

(1b) Assessment of the resiliency and vulnerability (including from an environmental quality, performance, and structural perspective) of present water resources systems and infrastructures.

This will require development of engineering tools that provide the capablities to perform integrated systems analyses. These analyses are essential because they provide a mechanism for assessing the river basin-wide, or even inter-basin, responses to climatic inputs. Without these tools, the potential for current water resources infrastructures to flex, but not break, in the face of climatic change cannot be assessed. Use of these tools, however, requires the establishment of evaluation criteria and indices of economic value that appropriately reflect environmental, economic, social/cultural, and performance requirements and public perceptions. Development of these criteria and indices, in forms ranging from reservoir release temperature objectives to hydropower efficiency targets to measures of the relative worth of meeting one of these objectives at the expense of the other (as examples), should be coupled with integrated systems analysis tools. Development of such criteria will allow evaluation and identification of the resiliency and vulnerability of existing systems and infrastructure to climatic changes. This assessment of resiliency and vulnerability must then be documented along with prioritization of efforts to ameliorate areas with significant levels of vulnerability.

(1c) Development of planning and design procedures for water resources infrastructures that meet multiple project uses and objectives.

The development of water resources planning and design procedures that meet multiple uses and purposes requires incorporation of risk/reliability and uncertainity analyses in those procedures. Current planning and design procedures used by many water resources development agencies worldwide, as discussed in previous sections of this chapter, provide for limited consideration of potential climatic changes beyond those previously observed and recorded. Incorporation of risk and uncertainity concepts will allow more straightforward consideration of climatic change and variability in system design and operation.

Methods should also be developed for identifying new water supplies. Weather modification, increased use of desalination, improved conservation, and reuse of storm and wastewater are among those concepts meriting investigation. Implementation of institutional changes that make water a commodity responding to market forces, while providing for basic needs, should be explored within various international socio-economic and cultural frameworks as a potential means of improving water conservation.

(1d) Development, modification, and operation of water resources projects that meet multiple uses and stated objectives in the face of an uncertain climatic future.

Having assessed the need for more robust water resources systems, and having developed the procedures required to plan and design these systems for the future, one would naturally implement these more "climate change-proof" systems. While construction activities are implicit in this statement, there are a number of additional themes that should be investigated. Modification of the legal and institutional constraints that limit water resources systems' flexibility should be considered. This may prove a monumental effort in that multi-decades of agency inertia and legislation will have to be overcome. In concert with this, the scope of participants who input decisions to water resources project operations should be broadened. Given the multiple, and often conflicting, uses these projects generally have, it is imperative that trust and cooperation between all interests be voiced openly and equally. This is extremely important in the design of a project, when changes in project features can often be easily made. Such coordination is equally important during consideration of project operational changes, so that modifications to improve or enhance the accomplishment of certain project objectives are not done to the detriment of differing, and perhaps less obvious, concerns.

New construction materials and methods for rehabilitation/retrofitting of existing water resources infrastructure are also needed. Assuming that climatic changes and variability will require future modifications to water resources systems, the need for cost-effective rehabilitation measures will undoubtedly increase.

The development and use of knowledge-based tools for implementation of flexible operations will improve the responses of water resources systems to climatic changes and variability. These tools, to be most effective, must be driven by real-time data.

5.6.5 Mitigation of Greenhouse Forcing via Water Resources Development

Partial mitigation of greenhouse forcing may be affected through accomplishment of two basic objectives presented below.

(2a) Increased use of alternative water-related power sources.

Approaches to accomplish this objective include development and use of innovative hydropower sources such as low-head (i.e., at navigation dams or irrigation projects), tidal power, and increased use of pumped-storage capabilities or additional off-peak alternatives. Each of these are known to represent largely untapped sources of potential power that heretofore have not been deemed cost-effective.

Within this same topic area, the heat exchange capabilities of various ground and surface water sources should be developed and used. As an example, the use of groundwater aquifer as heat sinks/sources should be explored. The potential of geothermal resources, including hot springs, geysers, etc, should also be investigated.

(2b) Increased efficiency and conservation in water resources activities.

The accomplishment of this objective could be promoted through improvement of hydropower generator and turbine efficiencies. Application of petroleum industry extraction techniques to extend groundwater development should also be explored. Wastewater treatment methodologies that minimize methane and carbon dioxide production must be examined. Additional conservation measures and more efficient water use practices, including the lining of irrigation canals, best management practices on agricultural lands, minimization of evaporation from arid-region transmission channels and reservoirs, reduction of water use in individual households, use of wetlands to trap runoff and assimilate pollutants, and increased recycling in industrial processes appear promising. These and other measures and practices should receive increased emphasis in the near-future.

REFERENCES

Abramopoulos, F., C. Rosenzweig, B. Choudhury, Improved Ground Hydrology Calculations for Global Climate Models (GCMs): Soil Water Movement and Evapotranspiration, *J. Climate*, Vol. 1 (September 1988), pp 921–941.

Brown, Lester R., John E. Young, 1988, *Growing Food in a Warmer World*, World Watch, November/December, Worldwatch Institute, Washington, D.C.

Crissman, C.A., 1988, Impacts of Electricity Generation in New York State, in Impacts of Climate Change in the Great Lakes Basin, Joint Report No. 1 of the U.S. National Climate Change Program Office and the Canadian Climate Centre, Sept. 27–29, Oak Brook, IL.

Electric Power Research Institute, 1989, Development of a Scoping Document to Assess the Potential Effects of Climate Change on Electric Utilities, prepared by ICF, Inc., for EPRI Risk Analysis Program, Palo Alto, CA.

Emanuel, K. A., The Dependence of Hurricane Intensity on Climate, *Nature*, Vol. 326, (1987), pp. 483–485.

Frederick, K. D., P. H. Gleick, 1988, *Water Resources and Climate Change in Greenhouse Warming: Abatement and Abatement*, Wagonner, et al., Eds., Proceedings of a workshop held in Washington, D.C., June 1988, Resources for the Future, Washington, D.C.

French, R. H., 1989. Effect of the Length of Record on Estimates of Annual Precipitation in NV. *ASCE. J. Hydraulic Eng.*, Vol. 115, No. 4, pp. 493–506.

French, R. H., 1987. *Hydraulic Processes on Alluvial Fans*, Elsevier Scientific Publishing Co. Amsterdam, The Netherlands.

Gleick, P.H., Climate Change, Hydrology, and Water Resources, *Rev. Geophysics*, Vol. 27, No 3, August 1989.

Gleick, P. H., Methods for Evaluating the Regional Hydrologic Impacts of Global Climatic changes, *J. Hydrol.*, 88 (1986), pp. 97–116.

Hains, D. K., H. R. Henry, Issues of Detectability of Climate Change Impacts on the Runoff of a Southeastern River Basin, in *Proceedings of the 1989 National Conference on Hydraulic Engineering*, Michael Ports, ed., New Orleans, LA, 1989.

Intergovernmental Panel on Climate Change, *Climate Change: The IPCC Scientific Assessment*, J.T. Houghton, G.J. Jenkins, J.J. Ephramus, (Ed.), Cambridge University Press, Cambridge, U.K., 1990.

Lettenmaier, D. P., T. Y. Gan, P. R. Dawdy, Interpretation of Hydrologic Effects of Climate Change in the Sacramento-San Joaquin River Basin, California, in *The Potential Effects of Global Climate Change, Appendix A - Water Resources*, J. B. Smith, D. A. Tirpak, (Eds.), EPA Office of Policy, Planning, Evaluation, Washington, D.C., 1989.

MacCracken, M. C., F. M. Luther, *Projecting the Climatic Effects of Increasing Carbon Dioxide*, DOE/ER-0237, December 1985.

Martin, P., N.J. Rosenberg, M.S. McKenney, Sensitivity of Evapotranspiration in a Wheat Field, a Forest, and a Grassland to Changes in Climate and Direct Effects of Carbon Dioxide, *Climatic Change*, Vol. 14, (1989), pp. 117–151.

Mearns, L. O., R. W. Katz, S. H. Schnieder, Changes in the Probabilities of Extreme High Temperature Events With Changes in Global Mean Temperature, *J. Climate App. Meteorol.*, Vol. 23, (1984), pp. 1601–1613.

Michaels, P. J., Observed and Projected Climate Change: Fact and Fiction, Dept. of Environmental Sciences, University of Virginia, 1989.

Miller, B. A., W. G. Brock, Sensitivity of the Tennessee Valley Authority Reservoir System to Global Climate Change, Report No. WR28-1-680-101, TVA Engineering Laboratory, Norris, TN, 1988.

Miller, B. A., V. A. Alavian, M. D. Bender, D. J. Benton, M. C. Shiao, P. Ostrowski, J. A. Parsly, Sensitivity of the TVA Reservoir and Power System to Extreme Meteorology, TVA Engineering Laboratory, 1992.

Nash, Linda L., Peter H. Gleick, 1990, Sensitivity of Streamflow in the Colorado Basin to Climatic Changes, Paper Presented at Workshop: *Implications of Climate Change for the Colorado River Basin*, May 17–18, 1990, Denver, CO, Sponsored by the Pacific Institute for Studies in Development, Environment, and Security and the Institute for Resource Management.

National Academy of Sciences, *Current Issues in Atmospheric Change: Summary and Conclusions of a Workshop, October 30–31, 1986*, National Academy Press, Washington, D.C., 1987).

National Academy of Sciences, National Academy of Engineering, Institute of Health, *Policy Implications of Greenhouse Warming*, National Academy Press, Washington, D.C., 1991.

Nemec, J., J. Schaake, Sensitivity of Water Resource Systems to Climate Variation, *Hydrol. Sc.*, 27(3):327–343, 1982.

Orlob, G. T., G. K. Meyer, J. DeGeorge, C. L. Christianson, 1990, Impacts of Global Warming on Water Quality in River-Reservoir Systems, Proceedings, Conference on Minimizing Risk to the Hydrologic Environment, American Institute of Hydrology, pp. 11–2X

Peterson, Dean F., Andrew A. Keller, 1990, Effects of Climate Change on U. S. Irrigation, *J. Irrigation Drainage*, Vo l. 116, No. 2, March/April.

Revelle, R. R., P. E. Waggoner, Effects of a Carbon Dioxide-Induced Climatic Change on Water Supplies in the Western United States, in *Changing Climate*, National Academy Press, Washington, D.C., 1983, pp. 419–432.

Rind, D., F. Lebedeff, Potential Climatic Impacts of Increasing Atmospheric CO2 with Emphasis on Water Availability and Hydrology in the United States, Prepared for U.S. EPA by the NASA Goddard Institute for Space Studies, April 1984.

Rind, D., R. Goldberg, R. Ruedy, 1989, Change in Climate Variability in the 21st Century, *Climatic Change*, Vol. 14, No. 1.

Rosenberg, N. J., M. S. McKenney, Philippe Martin, Evapotranspiration in a Greenhouse-Warmed World: A Review and Simulation, *Agri. Forest Meteorol.*, Vol. 47, (1989), pp. 303–320 .

Schneider, S. H., The Greenhouse Effect: Science and Policy, *Science*, Vol. 243, (1989), pp. 771–781.

Schnieder, S. H., N. J. Rosenberg, 1988, The Greenhouse Effect: Its Causes, Possible Impacts and Associated Uncertainties, in *Greenhouse Warming: Abatement and Adaptation*, Rosenberg, et al., Eds., Proceedings of a workshop in Washington, D.C., June 1988, Resources for the Future, Washington, D.C.

Sheer, D. P., D. Randall, Methods for Evaluating the Potential Impacts of Global Climate Change, in *The Potential Effects of Global Climate Change, Appendix A - Water Resources*, J. B. Smith, D. A. Tirpak, Eds., EPA Office of Policy, Planning, and Evaluation, Washington, D.C., 1989.

Shiklamanov, J., 1987, Changes in Runoff in Soviet Rivers Due to Climate Change, Paper presented at the International Union of Geodesy and Geophysics Symposium, Vancouver, British Columbia, August 1987.

Smith, J. B., D. A. Tirpak, The Potential Effects of Global Climate Change on the U.S., Draft Report to Congress, Executive Summary, EPA Office of Policy, Planning, and Evaluation, Washington, D.C., October 1988.

Szesztay, K., 1970, The Hydrosphere and the Human Environment: Results of Research on Representative and Experimental Basins, *Proceedings, Wellington Symposium, UNESCO Studies and Reports in Hydrology*, No. 12.

U.S. Geological Survey, 1984, National Water Summary 1983—Hydrologic Events and Issues, Water Supply Paper 2250, Government Printing Office, Washington, D.C.

U.S. Water Resources Council, 1978, *The Nation's Water Resources: 1975–2000, Vol. 2, Second National Water Assessment,* Government Printing Office, Washington, D.C.

Waggoner, P. E., Ed., 1990, Climate Change and U.S. Water Resources, Report of the AAAS Panel on Climatic Variability, *Climate Change and the Planning and Management of U.S. Water Resources,* John Wiley & Sons, New York.

World Meteorological Organization, 1988, Water Resources and Climatic Change: Sensitivity of Water Resource Systems to Climate Change and Variability, *World Climate Applications Program-4,* Geneva, Switzerland.

Chapter 6

SEA LEVEL RISE AND COASTAL HAZARDS: AN ASSESSMENT OF IMPACTS AND COASTAL ENGINEERING RESEARCH NEEDS

Authors: Ashish Mehta
Contributors: Robert Dean, Hans Kunz, Victor Law,
Say-Chong Lee, Zal Tarapore

6.1 SYNOPSIS

Sea level has been rising globally for thousands of years, and there is a high probability that global climate change will cause an increase in the future rate. Relative mean sea level (RMSL), which depends also on the rise or subsidence of the land and other effects, has as well been changing and is site dependent. Increased RMSL will lead to increased shore impacts including, for example, beach erosion, flooding of lowlands, storm hazards and ecological changes. There is substantial uncertainty as to the magnitude and timing of these changes. The "noisiness" of typical tide gauge data and short time-records may make it impossible to state with confidence that an increase in the rate of sea level rise has commenced, before several years have passed. Furthermore, although predictive modeling techniques for forecasting impacts, e.g., shoreline response under assumed sea level rise rate scenarios, have vastly improved in recent years, the confidence level in being able to predict effects over long time scales and short spatial scales of interest to engineering projects continues to be generally low. Also, episodic forcings, such as number, intensity, and paths of tropical cyclones and the El Niño-Southern Oscillation (ENSO) are very important, and these are poorly predicted by existing models. A tempering factor is that the risk of accelerated RMSL rise, although sufficiently established to require consideration in planning and design of coastal facilities, can be responded to by continually evolving engineering planning and design criteria; indeed, present decisions need not be based on any particular RMSL rise scenario. However, when built infrastructure cannot be strengthened and enhanced in the future, planning and design criteria should take into account an appropriate amount for an increased RMSL rise and for altered storm surges and waves. This design aspect must be addressed on a site specific basis, and because of the many uncertainties, solutions must be robust.

To improve confidence with respect to measurement of secular water level changes in the sea, and predict impacts under selected RMSL rise rates and episodic event scenarios, a range of research initiatives will be required. Since these are unlikely to be funded concurrently over, say, the next decade, it will be necessary to rank the research agenda, such as on the prediction of RMSL rise (including subsidence), storms (elevation, duration, number), waves (period, height, direction), the response of the coast on the altered hydrodynamic conditions, the adaptation of the design criteria for established engineering means, etc. While continuing such focused research, we recommend the following three-element plan to enable rational management responses: (1) to use existing data to categorize projected hazard levels based on past and present trends; (2) to collect new data to complement existing information; and (3) to apply existing technology and develop new strategies to eliminate or measurably reduce anthropogenic causes that enhance the effects of RMSL rise, storm surges, etc. It will be necessary to work with policy decision makers and the public to provide them with rational assessments of likely future hazards, their impact on the coast and the engineering responses required to meet the targets fixed by society.

6.2 INTRODUCTION

Some human actions have a significant potential to adversely impact the future global environment. Besides environmental degradation and thus the future quality of life, there is concern that global warming may increase storm frequency and intensity and possibly alter their paths. This scenario, coupled with rapid development along many shores, may be setting the stage for material changes in policies on shore habitation and resource development. The effects of our actions, however, are uncertain. Although prudence might argue for developing future strategies according to the most pessimistic scenarios, engineers pressing this viewpoint risk losing their credibility with the public and alienating decision makers responsible for supporting research and guiding policy (Dean, 1991).

Concerning the most responsible scientific position, the community will, as a rule, express a range of opinions and offer suggestions for technical stewardship. However, the mainstream technical position should be authoritative and justified by data. Following a brief introduction to the sea level rise problem, a few comments on episodic forcings and technical issues related to coastal engineering, this report focuses on recommended tasks and future research needs. In so doing, we have relied significantly upon two reports, one by the Marine Board of the National Research Council (1987a) and the other by the U.S. Department of Energy (Mehta and Cushman, eds., 1989). We have also used our knowledge of events, impacts, responses and effectiveness associated with a number of cases studies. We wish to acknowledge at the outset the leading role played by the Marine Board in publishing several reports that directly or indirectly highlight coastal engineering research needs that bear upon understanding sea level rise concerns (see National Research Council, 1987b; 1989; 1994). We conclude this report by making brief observations on recommended actions concerning technical and other management imperatives.

With RMSL rise the tide, storm surges and waves may be altered. The term "RMSL rise" addresses all of these water level determining components. The term "shore" is used in a general sense and includes sandy beaches of islands and the mainland, estuaries, salt marshes, marshlands, and cliffs vulnerable to erosion. The complexities of shoreline (i.e., land-water boundary) response to sea level rise are contingent upon a wide range of interrelationships between physical/ecological factors. Engineering response must ultimately be based on predictive capability, since we are principally dealing with the question of how shorelines and the shore environment will change with future sea level rise. Prediction in turn requires an understanding of process fundamentals and adequate data. Therefore, much of what follows pertains to these aspects, which typically have to do more with the basics of engineering response to hydrodynamic and meteorologic forcing than to sea level rise. If this response can be explained, then imposing and evaluating the effect of sea level rise becomes a less difficult task.

Organization of basic knowledge of coastal processes is interlinked with the question of resolution of spatial and temporal scales. The desired resolution for the evaluation of a resource is set by criteria that are dependent upon many nontechnical factors. At a built-up shore, a 10 m recession could severely damage a structure, while at a natural shore the concerns will be less stringent. Then again, in low lying areas such as coastal Louisiana, just a few centimeter rise in sea level could prove to be disastrous to water management, and cause extensive ecological changes associated with salinity intrusion (Boesch et al., 1994). A rapidly rising sea level can generate a materially different response than a slow one, an example being the fragile barrier island. Finally, there is the question of absolute sea level rise and the associated shoreline scenarios. By keeping the issues focused on the processes themselves, with a few exceptions we have stayed clear of centering on specific temporal and spatial scales explicitly, even though such considerations are inherent in evaluating the degree of uncertainty in the state-of-the-art knowledge and in future research needs. In any event, because of the uncertainties in predicting changes however, robust solutions are necessary. Examples are coastal dikes and other built infrastructure that can be easily retrofitted or repaired.

6.3 SEA LEVEL CHANGE

6.3.1 Relative Mean Sea Level Change

Engineering response to sea level change in any site-specific case depends on the relative change, i.e., the difference between the eustatic (global) change and any local change in land elevation. There are six long-term causes of RMSL rise, but not all of the processes operate in every locality. The following definitions and associated explanations are helpful to what follows (National Research Council, 1987a):

1. Eustatic rise of world sea level. "Eustatic" means a global change of the oceanic water level. Its most important components presently are regarded to be glacio-eustasy, caused by melting of land-based glacier ice, and the steric expansion of near-surface ocean water due to global ocean warming. "Steric" refers to the specific volume of the medium, which expands when heated or contracts when cooled, at least over the temperature ranges of interest to the problem.

2. Crustal subsidence or uplift of the land surface due to neotectonics, i.e., contemporary, secular, structural downwarping of the earth's crust. Tectonic phenomena occur in five distinctive categories: subsidence of former glacio-isostatic marginal uplift belts (e.g., the Eastern U.S.); cooling crustal belts following rifting (e.g., parts of the Gulf of California); subsidence in regions of long continued sediment loading (e.g., East and Gulf Coasts, especially the Mississippi delta); uplift in regions of active crustal subduction (e.g., Puget Sound); and subsidence due to loading by volcanic eruptions (e.g., Hawaii, Aleutians Islands).

3. Seismic subsidence of the land surface due to sudden and irregular incidence of earthquakes, including mechanisms leading to soil liquefaction.

4. Auto-subsidence due to compaction or consolidation of soft, underlying sediments, especially mud or peat. This process operates along barrier islands as they "roll over" onto lagoonal deposits.

5. Manmade subsidence due to structural loading, and also groundwater, oil and gas extraction. Of the four subsidence processes only this category, namely anthropogenic subsidence, can be reversed or at least partially mitigated by recharge or other management actions.

6. Variations due to climatic fluctuations. These are a consequence of oceanic factors including El Niño-Southern Oscillation (ENSO) effects, and are related to secular changes in the size and mean latitude locations of subtropical high pressure cells. Along mainland coasts (especially east coasts in the Northern Hemisphere), a decreasing current flow associated with warming epochs causes a rise in sea level due to the Coriolis effect, whereas in midoceanic gyre regions there is no mean sea level change. It also appears that tide gauge records contain substantial long-period fluctuations (on the order of 5–100 years), which indicates that accurate extrapolation of small sea level rise values from the data is very difficult. Furthermore, determining changes in rates of rise is even more difficult.

Of these identified causes of sea level rise, only the eustatic rise is a universal, global effect, by definition. For any one area the other causes come into play in various proportions. It should be stressed that no national survey of the local extent of the processes has ever been undertaken, but it is clear that the variations will be highly regional. Various segments of the U.S. coastline experience subsidence or uplift due to factors (2) through (5). Superimposed on this regional subsidence is the global eustatic sea level rise. If the greenhouse effect/glacier melt concept occurs with the magnitudes of some of the predictions, its potential contribution to mean sea level rise could outstrip other causes of relative sea level rise along most of the U.S. coastline by some time in the first half of the next century (National Research Council, 1987a).

6.3.2 Past Sea Level Changes

Sea level fluctuates with a very broad range of frequencies from short-term, but devastating, storm tides up to 6 m in height to variations on a geological time scale exceeding 100 m. These variations affect human patterns of shoreline habitation. Figure 6.1 (adapted from Shepard, 1963) presents the sea level over the past 20,000 years based on carbon-14 dating. Very approximately, the rate of sea level rise was 0.8 m per century from about 20,000 BP to around 6,000 years BP. Subsequently it has dropped to one-tenth of that rate. Figure 6.2 (Gornitz and Lebedoff, 1987) presents an analysis of changes over the past century as determined from tide gauges. This plot illustrates

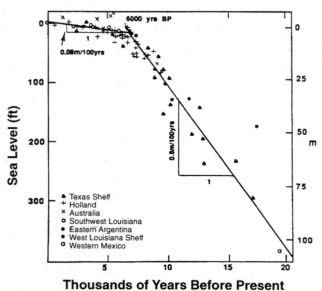

Figure 6.1   Sea level over the past 20,000 years as obtained from carbon-14 dating in relatively stable areas. (Adapted from Shepard, 1963.)

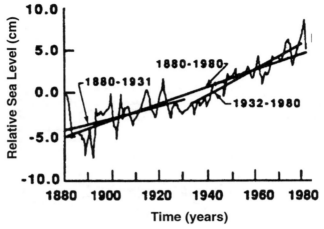

Figure 6.2   Global relative sea level rise from tide gauges. Lines represent least-squares fits to data for time period noted. (From Gornitz, V. and Kebedoff, S., in *Sea Level Changes and Coastal Evolution,* Pilkey, O.W. and Howard, J.D., Eds., SEPM, Tulsa, OK, 1987. With permission.)

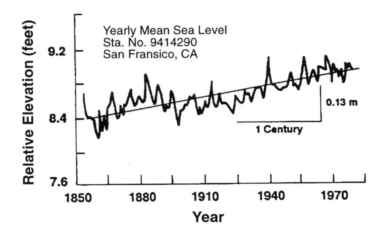

Figure 6.3   Annual means of tide gauge at San Francisco, CA, 1855 to 1980. (From Hicks et al., 1983.)

the sensitivity of mean sea level rise rate calculation to the period chosen for calculation. To illustrate better the character of tide gauge data, Figure 6.3 (Hicks et al., 1983) presents annual means for San Francisco, California, the longest-term U.S. gauge. The "noisiness" of this record is impressive and thus somewhat discouraging for identifying secular trends, such as the determined mean rate of rise of

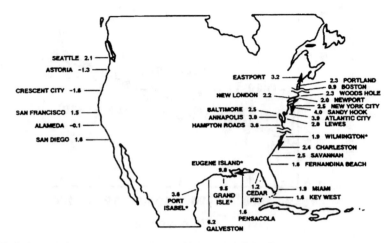

Figure 6.4   A summary of the present best estimates of local relative sea level changes along the U.S. continental coastline in mm/yr. The figures are based on the tide gauge records over different intervals of time during the period 1940–1980. Much regional variablility is evident. (Adapted from Stevenson et al. (1986) as appearing in National Research Council, 1987a. With permission.)

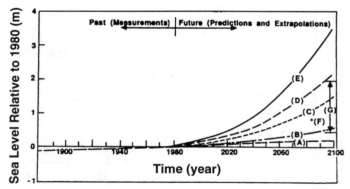

Figure 6.5   Eustatic sea level rise: (A) Rate over last century or so, projected into the future; (B), (C), (D) and (E) are EPA estimates by Hoffman et al. (1983) for conservative, mid-range low, mid-range high and high, respectively; (F) is the single (asterisk) Revelle (1983) estimate; (G) is the National Research Council (1985) estimate augmented with Revelle estimate for thermal expansion. (From National Research Council, 1987a.)

0.13 m per century. Rates of local RMSL changes are quite variable as illustrated by the data along the U.S. continental coastline as shown in Figure 6.4.

Table 6.1   Temporal estimates (in centimeters) of future sea level rise. From Hoffman et al., as appeared in National Research Council, 1987a. With permission.

Source	Qualitative Magnitude	Year 2000	Year 2025	Year 2050	Year 2075	Year 2085
Hoffman et al.	Low	4.8	13	23	38	—
(1983)	Mid-Range Low	8.8	26	53	91	—
	Mid-Range High	13.2	39	79	137	—
	High	17.1	55	177	212	—
Hoffman et al.	Low	3.5	10	20	36	44
(1986)	High	5.5	21	55	191	258

6.3.3 Future Mean Sea Level Rise Scenarios

Various predictions of future sea level rise rates have been carried out based on projected uses of greenhouse-generating gases and on the resulting response of the physical system. Figure 6.5 presents the 1983 U.S. Environmental Protection Agency (EPA) projections (Hoffman et al., 1983), which for the year 2075 range from 38 to 212 cm. Subsequent projections by the EPA (Hoffman et al., 1986) resulted in lower values, as noted in Table 6.1. A Dutch study on responses to rising waters (de Ronde, 1991) is based on estimates that range from 10 to 20 cm (Rijkswaterstaat; Delft Hydraulics, 1990). See also Houghton et al. (1990) for a discussion on the subject by the Intergovernmental Panel on Climatic Changes.

A common characteristic of all predictions is that any increase in sea level rise rate will first occur fairly slowly and then more rapidly in the future. In other words, the curves are concave upward, as in Figure 6.5. This fact and the inherent noisiness in the records, as shown in Figure 6.3, suggest that it may be several decades or more before it will be possible to detect convincingly a noteworthy increase in the rate of sea level rise.

6.4 IMPACTS OF RELATIVE MEAN SEA LEVEL RISE

6.4.1 Processes and Interactions

The interactive nature of coastal processes makes it difficult to isolate engineering impact issues and place them under well-defined "umbrellas" for descriptive purposes. Despite this constraint, we have selected nine areas within which significant research topics have been identified. These include: (1) water level and compaction measurements; (2) tidal range effects; (3) storm surge and waves; (4) natural coastal features and structures; (5) shoreline response modeling; (6) saltwater intrusion in ground water; (7) estuarine saltwater penetration; (8) sedimentary processes in the estuary and wetlands; and (9) coastal ecosystems. While processes and interaction mechanisms in these areas are noted elsewhere (e.g., Mehta and Cushman, eds., 1989), concerning impacts, we wish to elaborate upon the matter of coastal development as a case in point, specifically addressing the questions of: (1) increases in shoreline recession, and (2) increases in hazards associated with living along the shore.

6.4.2 Increased Shoreline Recession of Sandy Beaches

The commonly used method to relate shoreline recession, R, to sea level rise, S, is the Bruun Rule (Bruun, 1962), expressed as

$$R = S \; \frac{W_*}{h_* + B}$$

(Eq. 6.1)

in which W_* and $h_* + B$ are the horizontal and vertical dimensions of the active beach profile, respectively. It is worthwhile to note that the basis for this relationship does not include any net gains to, or losses

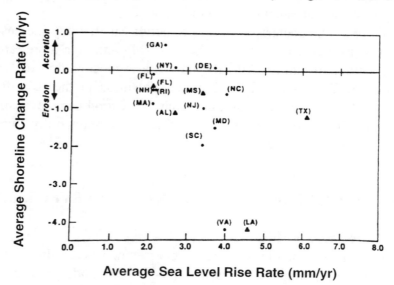

Figure 6.6   Shoreline change rate (From Dolan et al., 1983) vs. sea level rise rate (From Hicks and Hickman, 1988) for U.S. East Coast and Gulf of Mexico states.

from the profile, but only considers sediment shifted seaward to maintain an equilibrium profile. The ratio $W_*/(h_* + B)$ is usually taken to be in the range of 50–100. Based on Eq. 6.1, one would anticipate that, for example, doubling the sea level rise rate would double the recession rate.

The interpretation above presupposes that all beaches are receding and that the primary cause of recession is sea level rise. However, not all beaches are eroding, and there are many causes of shoreline change, including of course several anthropogenic causes. Figure 6.6 presents average shoreline changes plotted versus sea level rise rates for the U.S. East Coast and the Gulf states. Three of the states (Delaware, Georgia, and New York) are accreting on the average; thus, Eq. 6.1 is not directly applicable. Within a state, the range of shoreline changes from the mean is substantial; for example, Figure 6.7 presents a histogram of shoreline changes over approximately the past century for more than 1,600 locations along the east coast of Florida. These long-term changes range from recession rates of 9 m per year to advancement rates of 6 m per year. It is interesting to note that on the average, there has been a shoreline accretion of 0.18 m/yr over a period exceeding 100 years.

To allow prediction of changes in shoreline response to increases in sea level rise rates, it is useful to incorporate effects of deviations as follows (see also Dean, 1990). It is assumed that the present recession

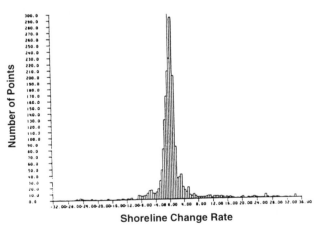

Figure 6.7   Shoreline change rates (feet per year) over the last century for the east coast of Florida. (After Grant, 1992. With permission.)

rate, R_1, is related to the present sea level rise rate, S_1, according to the Bruun Rule plus a site-specific deviation, G_i, due to unknown causes:

$$R_1 = S_1 \frac{W_*}{h_* + B} + G_i$$

(Eq. 6.2)

G_i can be either positive or negative, as can R_1. For a different sea level rise rate, S_2, we assume that G_i will be unchanged. Thus, the associated recession rate becomes

$$R_1 + [S_2 - S_1] \left[\frac{W_*}{h_* + B} \right]$$

(Eq. 6.3)

In a sense the addition of the G_i term can be considered as calibrating the Bruun Rule to conditions at the local site. Thus, we now have an improved basis for evaluating recession impacts due to increased sea level rise rates. Equation 6.3 has been applied to project coastal land loss over the next century at selected sites in the southern U.S. Site selection was based on physical factors that would make these sites significantly vulnerable to inundation at higher stands of mean sea level and associated storm surges (Daniels et al., 1992).

6.4.3 Saltwater Intrusion in Surface and Ground Waters

The effect of an increase in salt water intrusion in estuaries due to sea level rise is somewhat similar to increased penetration due to channel deepening. An example is the deepening of the 60 km navigation channel connecting Lake Maracaibo to the Gulf of Venezuela. Capital dredging for this purpose began in the Thirties, which eventually caused the lake water to become brackish (Mehta and Cushman eds., 1989). This type of impact can be exacerbated by reductions in river runoff, which can increase salt penetration by kilometers. Thus a climate change scenario in which sea level rise is coupled with reduced freshwater outflows can be severe. Salt barriers can be installed to mitigate this problem, e.g., the underwater sill structure in the Mississippi River south of New Orleans (Johnson et al., 1987).

Many coastal cities rely on local groundwater to meet domestic and industrial needs. With increasing demand due to greater population and industrial concentrations along the coast coupled with sea level rise, the potential for saltwater contamination of the aquifer will increase. In The Netherlands increasing seepage of saltwater into inland polders and increasing salt intrusion via the rivers are both matters of great concern. The present natural drainage of Lake Ijssel during low tide may have to be replaced by a pumping system. Storage of drinking water below the dunes may be diminished (de Ronde, 1991).

6.4.4 Estuaries and Wetlands

Besides increased saltwater penetration, the tidal range will be affected by sea level rise. Fuhrboter and Jensen (1985) noted a trend of rising tidal range over the past century at ten stations along the coast in the German Bight area. They concluded that this trend is not due to any long term changes in meteorological conditions, but is possibly due to the morphology of the North Sea, a shallow water body in which the influence of global rise of mean water level is amplified via a standing wave effect. This observation possibly suggests a situation in which the natural frequency of the water body approaches the tidal forcing frequency with increasing water depth and changing boundaries. In the estuary proper these amplifications can be further enhanced by reduced bottom friction in deeper water, coupled with the so-called Green's Law effect due to decreasing estuarine cross section with increasing distance upstream.

Shallow bays surrounded by wetlands will expand rapidly in response to sea level rise both because of the gentle slope and the deterioration of the marshes in response to salinity increases. For example, Barataria Bay in Louisiana has increased its surface area by about 10 to 15% over the last century, in response to about 1 m local relative sea level rise (National Research Council, 1987a). For the same reasons, land loss in this area is predicted to be significant based on future sea level rise scenarios (Daniels et al., 1992). In the event of a rapid rise in water level coupled with reduced sediment supplies, salt marshes, for example, will not be able to maintain their elevations and will be inundated. For instance, inundation of vast areas previously dominated by salt marshes has been occurring along coastal Louisiana, where the relative sea level has been rising at an order of magnitude greater rate than the eustatic rise due to soil compaction (Boesch et al., 1994).

6.4.5 Breakwaters and Seawalls

Many breakwaters and seawalls can be retrofitted (or modified) to perform their functions adequately, in response to RMSL rise and changes in episodic forcings. An example is the breakwater, marina and waterfront in Redondo Beach, California. This coastal area has subsided several feet in the past few decades (Beverly, 1988). In mid-January, 1988, it was subjected to storm waves rarer than a one in a hundred-year event (Shore and Beach, 1989). Remedial work was done on the breakwater (U.S. Army Corps of Engineers, 1989). Also, the buildings at the water's edge within the harbor were repaired. The seawall (actually, a revetment) between a hotel and the water was strengthened with additional rock and a concrete "wave splash wall" was constructed on top of the revetment, and some areas graded and paved. The total cost was about $2.3 million (S. Schoettger, Harbor Director, City of Redondo Beach, CA, personal communication, June, 1991).

6.4.6 Coastal Dikes

Several low-lying coastal areas around the world are protected by dikes. Some of these protection works have been in place for several hundred years. In present practice, essential parts of these protection structures may include storm surge barriers, seawalls, etc., as are found in Germany and The Netherlands. An appropriate planning and design of these features require that we be able to forecast the development of the area located seaward of the protection structures (tidal flats, saltmarshes, foreland, etc.). Forecasting is also needed for the decision that has to be made by society on the development of the coastal zone with respect to targets that need to be matched besides coastal protection (especially preservation of nature, including wildlife, but also recreation, etc.).

The natural rise of the mean high water level as predicted from long time scale tide gauge data has been taken into account for the design of coastal protection works along the North Sea coastline (e.g., 25 to 30 cm for the next 100 years in Germany). This design must be adjusted when an increased rise of the RMSL can be determined (Kunz, 1990). Based on selected scenarios, calculations were made in The Netherlands of the cost for adjusting their coastal protection system against an increased rise of the RMSL of 60 cm/100 yrs. Altogether, a 60 cm rise in sea level would cost about $10 billion (de Ronde, 1991).

6.4.7 Ports, Airports, and Infrastructure

Although the elevations of ports and harbors are established to provide efficient operations within a certain range of sea level fluctuations, the economics of these systems are such that increasing the working surface (land or pile-supported bases) elevations every 50 or so years should not present insurmountable problems within the next century. In fact, the increased water depths resulting from mean sea level rise will provide some benefits in terms of deferred dredging requirements. Airports of many coastal cities are built on landfills in bays. Dikes (levees) are sometimes used to protect the facilities from flooding. A combination of levee retrofitting, landfill and modified drainage systems are available technology. An example of estimated impacts is available in a study of the airport at Honolulu (Hawaii Coastal Zone Management Program, 1985).

The time scales associated with normal replacement or modification of coastal structures and infrastructure are relatively short and many responses would probably be taken care of in the regular sequence of events. Some discussion is available in the National Research Council report (1987a).

6.4.8 Considerations for Developing Countries

A large segment of the population in developing countries resides in low-lying coastal areas, so that the effects of a rise in RMSL on these areas deserve special consideration. Generically speaking, these coastal areas may be divided into two broad categories. The first category includes small islands with low tidal ranges, a virtually nonexistent continental shelf and a low population. Examples are coral islands such as the Maldives, the Marshall Islands, Tokelau, Tuvalu etc., which are barely a meter or two above the present mean sea level. These are more susceptible to inundation than some volcanic islands that rise to tens of meters above the present mean sea level.

The effects of global warming and consequent sea level rise would be quite disastrous for many low relief islands even at the predicted low rise of 36 cm by the year 2075, and would lead to total disappearance at the predicted high of 212 cm. Engineering responses to such eventualities are available today, but techno-economically speaking these responses are perhaps unviable. However, weighting factors other than techno-economic will have to be taken into account, such as the rights of sovereign nations and the anthropological need to preserve ancient cultures/archaeological sites. Volcanic islands that are higher provide the opportunity for managing a rise in RMSL within a longer time frame with various options based on techno-economic considerations.

By far the largest impact, from the population-standpoint, in the developing countries will be in the second category that encompasses the deltas of major rivers such as the Hwangho and Yangtze in China, the Mekong in Vietnam, the Irrawaddy in Myanmar, the Ganges-Brahmaputra in Bangladesh and India, the Indus in Pakistan, the Congo in Africa, and the Amazon in Brazil. These deltas are characterized by relatively high tidal ranges, susceptibility to tropical storms and concomitant storm surges, waves, high upland discharges and a wide continental shelf. The Nile in Egypt is somewhat different as essentially no discharge is permitted now. Being agriculturally fertile, deltaic regions are cultivated, requiring engineering ingenuity for irrigation in periods when the river stage is low, and drainage when the river stage is high. Extensive dike systems protect these low-lying areas against water levels that may occur with return periods of 25 to 50 years. A rise in RMSL would involve raising and strengthening these extensive networks of dikes, the cost of which will go up as the square of the height of the dike, and more if one considers the increase in the frequency and intensity of tropical cyclones that may result from a global warming trend.

Finally, the impacts of a rise in RMSL on barrier beaches, as for example found in southern India, need to be considered. These areas, which are already eroding, would be highly vulnerable to a rise in RMSL and the increased severity of the wave climate resulting from global warming. Since large populations subsist on these vulnerable areas and derive their livelihood from the ocean and the lagoons between the barrier beaches and the mainland, a special consideration needs to be given to protection measures to prevent erosion and destruction of the fragile ecosystem of the lagoon behind the barrier beach (Nair, 1988). The capital and maintenance costs of protecting these beaches would approximately double in the event of a meter rise in RMSL.

6.5 IMPACTS OF CHANGES IN EPISODIC EVENTS

6.5.1 Increased Storm Impacts

It is possible that global warming will bring about an increase in the intensity and frequency of coastal storms and associated storm surge. This effect, should it become significant, will be of critical importance to all issues related to coastal defense. There are two possible reasons for increased storm impacts due to sea level rise. The first is that the elevations of fixed, engineered elements (including roads and buildings), and barrier islands kept at fixed elevations will become lower relative to the rising sea level. The second possible cause is increased storminess, which will not be discussed further since this is believed to be conjecture at this point. However, considering the storm climate to remain the same, it is a relatively straightforward engineering matter to recalculate, for example, the storm tide return periods due to a certain sea level rise rate. For example, Figure 6.8 presents the calculated return period relationship for storm tides relative to the National Geodetic Vertical Datum (NGVD) in Lee County, Florida. An approximately 0.3 m (about 1 ft) increase in mean water level would, if the storm-tide statistics were unchanged, decrease the return period for a 100-year storm tide to 50 years.

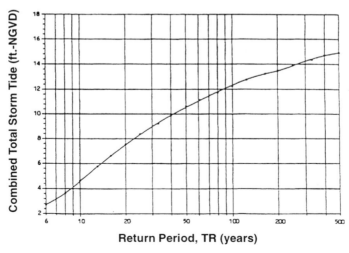

Figure 6.8   Calculated storm-tide (surge plus astronomical tide relative to NGVD) return period for Lee County, Florida. (After Dean et al., 1990. With permission.)

6.5.2 Storm-Resistant Buildings and Insurance

With increased coastal development, major storms can increasingly cause potentially significant damages and financial losses. However, some measures have already been implemented to make the structures more storm-worthy and to place essentially the entire economic burden on those at risk through the Federal Emergency Management Agency (FEMA) insurance programs. There are three primary structural characteristics that are required to make a building storm resistant:

1. The structure must be built on piles so that the main horizontal elements are elevated above the reach of the wave crest during the design storm. FEMA has been requiring this design criterion for more than two decades as a condition on new structures for participation in the FEMA insurance program. This is a robust solution in that the piles can be relatively easily retrofitted if necessary, by increasing the elevation and making them stronger.
2. The piles should be sufficiently deep so that vertical erosion of land during the storm will leave intact the foundation's integrity. Although the FEMA program is not very explicit about this need, some state programs—for example Florida—are specific regarding the need for deep pile embedment. The capability to predict erosion during a storm, and thus pile embedment, is still developing.
3. Structural connections must be specifically designed for high wind loads. Note that this should also be the case for earthquake-induced loads.

As an illustration of the effectiveness of these measures, soon after Hurricanes Elena in 1985, Gilbert in 1988, and Hugo in 1989, fairly extensive field inspections of western Florida, Cancun, Mexico, and South Carolina, were carried out to document erosion and structural damage. With one exception, which was attributed to a possible tornado spawned by a hurricane, no structures built to the modern standards described above were found to have suffered major damage (Dean, 1991). This observation should put to rest the question of whether or not we can design structures to resist a major storm event. We are still left to consider if or how structures should be located on an eroding coast. Certainly our future shores should not have sound structures left intact in the surf zone by a retreating shoreline. Some states address this question partially by requiring that single-family dwellings be set back from the shoreline a distance equal to 30 times the annual erosion; for multi-family or commercial buildings, the factor is increased to 60.

The second way in which increased development along the shore has been addressed is by requiring those incurring the risk to bear the financial burden through participation in the FEMA insurance program. Before Hurricane Hugo, the coastal portion of this program had a $600-million surplus; claims resulting from Hugo reduced this amount to $400 million. However, annual premium income is about $200 million. In the history of this program, the premiums have been increased twice, and it appears that at present the rates are more than sufficient to provide for the average long-term losses related to coastal storms.

6.6 PREDICTIVE CONSTRAINTS

It is apparent that regarding impact prediction, particularly from the perspective of coastal processes, with the possible exception of tidal hydrodynamics and salinity transport, considerable further

Table 6.2   Relationship between temporal and spatial process scales and predictive confidence. (Adapted from Mehta, 1990b.)

	Time Scales	
Process	**Scale**	**Predictive Confidence**[a]
Storm dune erosion	~1 day	6
Local beach erosion	~1 month to ~ 1 year	4
Sea level rise induced erosion	~10 to 100 years	3
	Spatial Scales	
Process	**Scale**	**Predictive Confidence**[a]
Local erosion (e.g. over a lot)	~10 meters	3
Structure induced erosion (e.g., by a groin)	~1 kilometer	6
Barrier island erosion	~10 kilometers	3

[a] Based on a scale of 0 through 10 of increasing order of confidence, 10 being the highest.

research will be required for assessing shoreline and shore environmental response in a confident manner (see, e.g., Mehta and Cushman eds., 1989). Strides made during the past decades have been impressive but, for example, where sediment transport is a key factor, we are significantly limited in long-term predictive capability. This is partly due to the lack of good quality synoptic hydrodynamic/meteorologic data. In turn this problem has an impact on ecological modeling, which is contingent upon a knowledge of flows and sediment movement (National Research Council, 1994).

Concerning sedimentary processes, it is worth commenting on the state-of-the-art in shoreline change prediction capability in the present context, as an indicator of future needs in improving coastal process technology. Consider, for instance, the relationship between the temporal and spatial scales and predictive ability in shoreline prediction as shown in Table 6.2 (Mehta, 1990b). Without elaborating on details that are self-explanatory, it suffices to note that in general confidence levels in prediction are not high, and reflect the fact that a great deal of progress to achieve even these accuracies is actually only about a decade old. Clearly, much research work remains to be done; among factors limiting future progress, the overwhelming one is likely to be the scarcity of research funding; there is no real shortage of needed expertise or innovative concepts.

At the Workshop on Sea Level Rise and Coastal Processes (Mehta and Cushman eds., 1989), considerable attention was focussed on future research needs in nine areas noted earlier, following a review of relevant literature. A significant part of research needs noted in what follows is based on the outcome of that meeting and on subsequent developments.

6.7 RESEARCH NEEDS

6.7.1 Eustatic Mean Sea Level Change/Compaction Effects

Improvements in our understanding of eustatic sea level change can come about through use of the existing data base or development of new data. Extraction of more meaningful results from the existing data base will require either improved analysis procedures or an improved understanding and application of the physics of relative sea level change, including the noise present in the records (Emery and Aubrey, 1991). Enhancement of the existing data base through new measurements will most likely occur through satellite altimetry, once this is proven to centimeter accuracy over the open ocean. Additionally, in some cases much can be learned locally about anthropogenically generated compaction in areas of tide gauges through the installation of simple compaction measuring devices. One feature of new data is the length of time that will be required for such data to "mature" to yield significant meaningful information.

Use of Existing Data: Analysis considering the physics of RMSL change may be the most effective and productive use of existing data. In particular, accounting for the contribution of long period waves, as explored by for example Sturges (1987), would allow interpretation and removal of a major portion of the noise in the RMSL measurements.

A second productive area is a more thorough analysis than presented previously of the contribution of post-glacial adjustment of the earth following the last ice age. Lambeck and Nakiboglu (1984), for example, have inferred from viscous models of the earth that the actual eustatic rise is roughly one half to two thirds the value determined from analysis of records based only on areas of relative stability. Improved estimates of eustatic sea level rise could be based on either more inclusive data sets with or without the use of a viscous earth model. Obviously, more meaningful results could be obtained with the combined approaches simultaneously.

Most approaches of direct analysis attempt to reduce the noise in a record on a station-by-station basis by determining some sort of RMSL estimate through fitting to the data. Unfortunately, the noise in individual records is such that at least 20 to 40 years of data are required for the results to be meaningful. An approach that would make these results meaningful upon their availability is the weighted averaging of many stations along a coastline to establish a more stable value. This averaging length could encompass, for example, the North American or North and South American shoreline(s). Thus, if a wave with length exceeding the expanse of the stations were contributing to the "noise," this process would tend to reduce or, in fortuitous cases, eliminate its contribution. By averaging over long segments of the shoreline, weighting each station by its alongshore influence length, then combining appropriately the results for various such shoreline segments, a much more stable year by year value could be obtained. This procedure would allow effective use of the data available for the east coast of South America, where records from eight of the twelve available gauges are less than 30 years in duration. In general, 30 years is not adequate to obtain a stable estimate from an individual gauge.

Need for New Data: There are two types of new data that would contribute to improved estimates of eustatic sea level rise: those that contribute immediately and those that would require a data base of at least several years before meaningful results could be obtained. It is anticipated that even with the potential benefits of satellite altimetry, at least one decade and possibly two decades will be required before adequate confidence can be placed in the data to yield accepted reliable estimates of eustatic sea level rise. Three research needs in the category of "new data" are described below.

Compaction Gauges: As is well-documented by a number of studies, withdrawal of ground water and hydrocarbons can contribute to substantial subsidence and by that a "relative sea level rise." Note that this is probably the only component that realistically can be controlled by humans. The obvious general but not universal correlation of areas of tide gauge locations and ground fluid extraction near populated areas justifies a concern over this activity. Also, the fact that these are the areas where continued RMSL rise may contribute most to the ultimate response cost (relocation, defense, repair, etc.) makes it important that the significance of anthropogenically induced subsidence be quantified and possibly controlled as early as possible.

Very simple and sensitive compaction meters have been used in quantifying this effect near Osaka and Niigata, Japan among other locations (e.g., Murayama, 1970). Ideally, installations would be made near tide gauges and also remote from cities but, say, inland and in the same geological formations as those near the tide gauges. These gauges would commence yielding valuable data immediately, and it may be possible to supplement the compaction data collected with models using data representing the geological formations and the history of past ground fluids extraction to estimate earlier compaction. Such results would be valuable in providing more reliable estimates of past and future eustatic sea level rise.

Positioning Accuracy/Satellite Altimetry: In many regions, tectonic activity may be equal to or greater than compaction. The only means to resolve the apparent rise in sea level at a tide gauge into its subsidence, tectonic, and eustatic components with a degree of confidence is to relate tide and compaction gauge elevations to an ultra-precise geodetic reference system. Such a program utilizing Very Long Baseline Interferometry (VLBI) and the Global Positioning System (GPS) is described by for example Carter et al. (1986). The GPS or other satellite system should also be used to monitor the deep ocean.

New Tide Gauge Data: The Southern Hemisphere and parts of the Northern Hemisphere are especially deficient in long-term tide gauge data. A number of relatively short-term tide gauge records are available along the east and west coasts of South America; however, there is a need for an international effort to install and maintain additional gauges to provide a representative distribution. In addition to the Southern Hemisphere and parts of the Northern Hemisphere, more insular tide gauges and tide gauges along the open coast are needed, the reason, in part, being that natural and anthropogenic physical phenomena in bays, where most tide gauges are located, typically lead to an underestimation of sea level rise (Mehta, 1990a). A first phase effort could be a survey to identify suitable open coast sites.

6.7.2 Tidal Range Effects

In many cases, it seems, numerical hydrodynamic modeling capabilities have "outstripped" data quality such that inaccuracies in collected data limit the accuracy of mathematical prediction of tides in the nearshore and estuarine areas. Data limitations arise from many causes; it suffices to note two factors. One pertains to a lack of physical understanding, on a microscale, of phenomena that ultimately affect water level prediction. An example is our understanding of bed forms, the manner in which they change with flow, and the precise relationship between their occurrence and the flow resistance they generate. Such forms may range from small ripples to large, migratory sand waves found in estuaries, inlets and in nearshore waters.

The second factor is related to historic tide records. Many records are highly contaminated by effects of land subsidence, poor leveling between gauges, shifting gauge locations, and a general lack of knowledge of the physical surroundings and variations in parameters characterizing these surroundings over the duration of tidal record. Thus, an accurate, quantitative evaluation of superelevation effects would require the deployment of better monitored gauges. For example, superelevation in bays (mean bay level minus mean sea level) vary with natural and anthropogenic factors that may or may not be present along the open coast. Mehta and Philip (1986) noted that our understanding of bay response and its relation to response outside would be considerably enhanced by: (1) establishment of additional primary stations along the open coast; (2) collection of long-term records at several presently designated secondary stations in bays; (3) accurate geodetic leveling connecting additional outside and inside stations; and (4) publication of relevant present and historic data in a user-oriented format. The National Ocean Service's marine boundary programs and tidal datum survey programs appear to be directed toward this type of effort, particularly with respect to the first three items.

6.7.3 Storm Surge and Waves

Besides an increase in the base level on which storm surge occurs, long-term sea level rise is likely to have a measurable effect on storm surge and wind wave generation in locations where the continental shelf is shallow and its length fixed by a naturally hard shoreline, or one that has been stabilized with structures (Mehta and Cushman, eds., 1989). Therefore, the aspects of storm surge and wind-wave generation that require research have less relation with a long-term sea level rise than with the basic phenomena themselves. There have been intensive theoretical and numerical studies of storm surge over the past three decades, and several sophisticated numerical models for storm surge prediction exist. However, there is a lack of field measurements of hurricane and extratropical storm surge with which to calibrate and verify these models. Required are concurrent time series from devices placed along the coast at intervals small enough to resolve the behavior of the surge as a storm moves out of the open ocean and makes landfall. The ability to model and predict storm surge cannot improve significantly without such data. Also, several phenomena associated with storms such as the superelevation of water level before arrival of the storm (often called a forerunner) are still not understood.

Research on wind-wave generation in deep and shallow water has progressed well. However, there is a lack of detailed, high quality wind and wave data with which to verify these models. The basic process of damping of wind-waves as they cross the continental shelf due to bottom effects (friction and others) and breaking induced by wave-wave interaction are other areas in which field data are needed.

It is also necessary to stress the spectral approach to investigating wave damping, as most methods available to date are limited to the assumption of monochromatic waves. Because waves refract in response to variations in water depth, basic research on the directionality of wave spectra, in both deep and shallow water, is also necessary before a better understanding of the effects of sea level rise on ocean waves can be assessed accurately.

Finally, if sea level is rising in response to changes in climate, these changes may alter the frequency of occurrence, severity and behavior of tropical cyclones and other storms. Research is needed not only to investigate these possibilities and improve predictive techniques, but the statistics and behavior including landfall locations from the historical record should continue to be examined so that a reliable baseline can be established with which future records can be compared. Some work of this nature has already been done in conjunction with the FEMA-Federal Insurance Administration studies for coastal counties.

6.7.4 Natural Coastal Features and Structures

Research needs in the area of modifying shoreline response to sea level rise and its effects on the design of protective works lie in the realm of ongoing basic studies of natural shore processes, and have little requirement for specific treatment of sea level rise. If, for example, engineers had an accurate surf zone sediment transport model capable of reproducing and predicting beach response to storms and structures, including the effects of sea level rise, it would be a relatively simple matter to increase the mean water depth in the model. However, until we develop a better understanding of the basic processes and better models, there is little reason to expect accurate prediction of the response of beaches to sea level rise.

There are six major areas requiring research in basic coastal physical processes: (1) wave refraction/diffraction; (2) wave breaking; (3) undertow and longshore currents (nearshore circulation); (4) sediment motion under shoaling and breaking waves; (5) erosion/sedimentation of fine materials (tidal basins, tidal flat areas, salt marshes, marshforeland); and (6) forces on and responses of coastal protection structures. The knowledge gained from research in these areas would then be used as input to improved beach profile and planform response models.

Once a reasonable expertise in shoreline modeling has been developed, the greatest research need is for the analysis and quantification of the performance and costs of the available alternatives for dealing with sea level rise, and the determination of their cost-effectiveness. Studies should be implemented that are specifically devoted to dikes and artificial dunes, offshore breakwaters, and beach fill design. These measures appear to be the most promising for confronting sea level rise.

Research needs in structural design also require studies of a basic nature to increase performance and decrease costs. Most importantly, the most likely sea level rise scenarios must be refined and agreed upon before engineers will be inclined to include them in their designs. Until then, research on economical methods to retrofit existing structures must be pursued. The modes and methods of failure of coastal and estuarine structures, as well as self-induced subsidence must also be studied. Better mathematical descriptions of the processes by which, for example, earth mound structures including levees are overtopped and eroded will be needed if the methods of setting protection criteria are changed.

6.7.5 Shoreline Response Modeling

The development of capability to predict reliably shoreline response to future sea level rise rates requires the consideration of cross-shore sediment transport fundamentals and applications, and a quantitative understanding of the transport components. As noted previously, the Bruun Rule is deficient in not allowing for the onshore transport of sand that is clearly occurring at some locations and undoubtedly occurring at many less evident locations. The three types of research needs identified fall in the categories of analysis of existing data, new data, and new technology.

6.7.6 Analysis of Existing Data

Isolation of Anthropogenic Effects: The substantial effects that navigational structures and sand management practices at entrances have on shoreline stability are well documented, e.g., at Folly Beach, SC, Tybee Island, GA, Santa Barbara, CA, and Assateague Island, MD (Mehta and Cushman, eds., 1989). Besides the effects at entrances, the effects of groins, seawalls, etc. should be considered. A straightforward methodology could be applied; however, it is believed that development of new, more effective methodology would be worthwhile.

Regional Correlation of Shoreline and Sea Level Change Rates: Previous estimates of long-term shoreline change have been developed (U.S. Army Corps of Engineers, 1973: Dolan et al., 1983). These estimates are available on a statewide basis and regionally (such as the entire Atlantic coast). It would be a useful and instructive first broad-brush effort to correlate these estimates with local estimates of sea level rise over the past 50 years or so.

Shoreline Monument System: The state of Florida, for example, maintains a monumented baseline along 1,000 km of its sandy shoreline. Since the early to mid 1970s, comprehensive surveys have been conducted on approximately a decadal basis and post-storm studies carried out when appropriate. This type of system provides the only basis for obtaining high quality data of shoreline change. It would be worthwhile, in anticipation of the increasing concern over shoreline response to sea level rise, to encourage other coastal states to install, monitor and maintain a monumented system similar to that in Florida.

6.7.7 New Data

Quantification of Cross-Shore Sediment Budget Components: The complete methodology for this purpose for the modified Bruun Rule application has not been developed, but would probably consist of long-term observations of offshore stakes to determine total vertical change, studies of biogenic production and attrition and deposition rate by suspended sediment traps. It would be useful to conduct this element in conjunction with the experimental element described below.

Evidence from the Continental Shelf: The seafloor of the continental shelf contains information relating to past shoreline response to sea level rise and potential future response. For instance, Swift (1975) has shown that along much of the Mid-Atlantic Region, there is a "lagoonal carpet" of mud that could not contribute significantly to the sediment budget of the active shoreface. Additionally, the shape of the offshore profiles, together with the availability of sand-sized material contains information (although as yet not completely understood) pertaining to whether the offshore profile will serve as a source or sink of sand.

In addition to the above, it would be worthwhile to conduct measurements of long-term sediment movement on the continental shelf. These measurements would be conducted along a representative profile; they would document the forcing function (waves, currents, tides, stresses, etc.) and sediment transport (response function). Sediment transport would best be documented through passive means, such as to the use of sediment tracing techniques.

6.7.8 New Technology

An example is laser-based profiling. A laser profiling system has been developed in which an airborne laser oscillates in a conical pattern sweeping out a swath of one half the airplane height, with the ground level pattern being nearly circular overlapping trajectories. The laser return establishes the dry beach elevation, the water surface and the below water profile to approximately two "secchi" disk depths. The potential of this technology to conduct regional and tidally controlled surveys of shoreline change, systematically and periodically, is extremely encouraging.

6.7.9 Saltwater Intrusion in Ground Water

Given the ambient flow conditions in a coastal aquifer, the transmissive properties and porosity of the aquifer and various scenarios of extraction "demand" on the aquifer, it appears that the characteristics of salinity intrusion, including any time dependencies can be predicted reasonably reliably by state-of-the-art numerical modeling. However, at present the effects of sea level rise on saltwater intrusion have been examined only for special cases. Any decisions by those responsible for planning related to the need to relocate well fields, modify usage, implement remedial measures, etc. must be based on realistic estimates of the effects of sea level rise and other causes such as increases in extraction rates.

A research program is recommended to develop and exercise simple models with the objective of illustrating the relative effects of sea level rise. The remaining efforts would be more comprehensive and concentrate on evaluating the reliability of numerical models for predicting salinity intrusion and on case studies of areas where the potential for salinity intrusion is high. These studies would focus on the applications of models, parameterized and calibrated for the local aquifer characteristics and using various sea level rise scenarios to predict effects on extracted water and on the need for and effectiveness of various types of remedial measures.

Long-term persistence of Pleistocene-age water-quality impacts, as for example in parts of the Floridan Aquifer, indicates that parts of the aquifer system may respond more slowly than expected to changes in head and water quality due to sea level rise, at least in areas where the aquifer is confined and at a great depth. Thus, the transient response of the saltwater-freshwater interface in deep confined aquifers needs to be understood better to determine whether the predicted sea level rise will have any significant impact on such systems. Also, many of the analytical and numerical models in the literature are limited to describing the location of the saltwater-freshwater interface under equilibrium conditions, or they do not describe in detail the response to mitigation measures. An evaluation of the analytical and numerical models in the literature is needed to determine which models can be used to investigate transient conditions and various protective measures. In addition, the large number of case histories and descriptions of protective measures could be analyzed systematically to evaluate the effectiveness of these measures and to provide guidelines for site-specific applications in areas where saltwater intrusion is expected to be a significant problem.

6.7.10 Estuarine Saltwater Penetration

Scientific work in understanding the basic processes of mixing between salt and fresh waters is likely to continue well into the future, in order to improve steadily the predictive methodologies, which have at present attained a reasonable degree of sophistication. However, further improvements are desirable. An area that deserves further consideration is mixing under wave action, or combined wave and tide action. At present the effect of waves in this respect is only poorly understood. Much of the analysis carried out appears to have been directed toward situations dominated by tides alone. It is noteworthy that with deeper water associated with sea level rise, waves would be admitted more freely, thereby decreasing fluid stratification by virtue of greater vertical mixing, particularly in estuary mouths.

Another important research area is related to the development and motion of fronts, zones of surface convergence, due to salinity, temperature and turbidity. Understanding the behavior of these fronts is vitally important to a range of water quality and ecological issues.

Most of the work done to date appears to have dealt with such effects related to channel deepening or changes in upstream river hydrology, rather than sea level rise. As noted, sea level rise can also increase the coastal tide and wave action, particularly if the coast is rocky, and does not recede, while water depth increases. A similar situation can also arise if the shore is erodible, but sea level rise is so rapid as to prevent depths at the mouth from achieving quasi-equilibrium with the hydrodynamic forcing. This would lead to a situation in which the estuarine mouth would be in deeper water, and where the tide (and waves) would arrive with lesser hindrance due to reduced bottom friction and lesser chance of wave breaking in the offshore waters. In most work carried out so far, the coastal tide (and waves) is typically assumed to practically remain unchanged.

6.7.11 Sedimentary Processes in Estuaries and Wetlands

Present day capability in predicting the evolution of the morphology of the estuary mouth and adjacent shores including tidal flat areas and salt marshes due to sea level rise, is limited. So is our ability to evaluate quantitatively wetland response to sea level rise (see, e.g., Mehta, 1990b). Where fine sediments are involved, the evolution of mud flats and their seasonal variability is difficult to model, and in situations involving vegetative canopies for instance, we know very little about hydrodynamic forcing and its relationship to sediment movement. Marsh sediments often contain organic as well as inorganic matter derived from tidal creeks. The vegetation in the marsh serves as an autochthonous source of sediments, as most marsh sediments contain 10–90% organic matter. It is not clear how the productivity of these plants may respond to sea level increases, or the degree to which transport of organic material from the marsh to the open estuary might be modified by such increases. With respect to this question, human-induced input of nutrients seems to have a major impact (Boesch et al., 1994).

Predicting capabilities for sedimentation within the estuary proper are better, although in areas where high suspension concentrations occur, the physics is poorly understood. Furthermore, the precise nature of chemical and biological variability is not clearly known. As a result, we are unable to establish the quantitative significance of physical versus chemical versus biological control in estuarine sedimentation processes.

As with most other areas, the basic research issues are not particularly related to sea level change, but with physics, chemistry and biology, as they interact and influence estuarine processes. Advances in knowledge have been diluted by too much site-specific and empirical work. A part of the problem appears to lie with the fact that a great deal of effort has been directed to address specific problems unrelated to basic questions posed by scientists. In recent years, several scientific reports have identified relevant research areas that must be tackled for ultimately improving our predictive capabilities. Greater attention needs to be paid to research requirements outlined in these reports. In a report of the National Research Council (1987b), research areas in fine sediment transport have been identified.

6.7.12 Coastal Ecosystems

Prediction of the effects of sea level rise on the future extent of various coastal ecosystems would be greatly enhanced by the development of models that incorporate the determinants of the vertical and horizontal growth components of submerged, intertidal, and supratidal coastal ecosystems. To understand the human consequences of changes in extent of each ecosystem, a quantitative understanding of the value of these systems to humans is essential. Energy analysis and natural resource economics techniques (e.g., Odum et al., 1987) are promising in this regard if specific fish, wildlife, ecological diversity, and aesthetic values can be included.

Synthesis of existing information on growth determinants and quantitative ecosystem values is the first step. The responses of some of the more prevalent coastal plants and animals to some environmental changes are known. As predictions improve the effects of sea level rise on ecologically important physical variables, better assessments of the nature and timing of ecological changes can be made. For better assessments, new ecological and physiological information may be required. The most important information needs can be identified with the aid of a literature synthesis and subsequent simulation modeling of ecological responses to predicted environmental changes, e.g., rising water level, encroaching salinity, increased tidal range and wave energy. Estimates of functions and parameter values can be based on the best available information, but some guesswork is anticipated. Uncertainty in ecological generalizations can be explored by a sensitivity analysis of the model(s). Needed information can be ranked according to a combination of the uncertainty involved in an estimate and the sensitivity of the models to changes in the estimate. Thus, not only can the most important information needs be identified, but also the consequences of a lack of this knowledge can be demonstrated with the model (Montague et al., 1982).

Although literature synthesis and exploratory models will ease the identification of specific research needs, the most relevant research will undoubtedly include several general areas. The time required for full development of subtidal, intertidal, and very nearshore supratidal ecosystems should be established with greater certainty. Under the most rapid sea level rise scenarios, conditions may not remain constant long enough for full development of an ecosystem. If so, the production of fish and shellfish and the stability of shore may decline.

Knowledge of nearshore topography and predictions of tidal range are essential to predictions of aerial extent, as is an understanding of the level of suspended sediments to be expected and the trapping rate of sediment by subtidal and intertidal plants and microbes (Montague, 1986). Knowledge of the major regulators of the production of principal animals and plants is essential for coupling predictions of ecological changes to predictions of physical changes. Factors that determine the type and productivities of organisms in the coastal zone include: light (turbidity), temperature, nutrients (including carbon dioxide), salinity, water level, and biochemical oxygen demand. All of these will be influenced by sea level rise, global warming, and increased levels of atmospheric carbon dioxide.

Physical uprooting and erosion of present ecosystems should be a major agent of ecological change. Predictions are needed both for shores and for tidal flat areas, tidal creeks, and estuaries. Knowledge of the resistance to erosion of these systems is also required. Although human values of coastal ecosystems may be compared using various energy and economic analyses, the variation in value within a general category must also be evaluated. Intertidal marshes and mangroves, for example, have been highly touted as good habitat for the growth of juvenile fish and shellfish of commercial and recreational importance. The Wadden Sea barrier islands, tidal flats and salt marshes extending along the North Sea coast from The Netherlands to Denmark are protected as a unique ecosystem. In addition, exchange of materials between the marsh and the estuary is believed to control supplies of nutrients in adjacent estuarine waters. Not all marshes are equivalent in their habitat value, and not all exchange significant quantities of materials with surrounding waters (Montague et al., 1987). Perhaps the most important factor in the accessibility of marshes to organisms, and in the exchange of materials, is the density of tidal creeks (Zale et al., 1987). The density of tidal creeks can be defined as the ratio of length of the edge of tidal creeks to surface area of marsh. Knowledge of the influence of creek density on habitat use and material exchange may be essential to understanding the relative value of marshes that develop in response to sea level rise. Comparative studies of the effects of creek density have never been reported, however.

Valuable predictions of coastal changes may be obtained from empirical models, if sufficient data can be collected. Four areas of study are needed. First, paleoecological analysis of cores from various coastal ecosystems can assess responses to past sea level rise (Kurz and Wagner, 1957). Second, analysis of ecological zonation along gradients of salinity and elevation should reflect the kinds of ecosystems to be expected as salinity encroaches and water becomes deeper. Third, analysis of the effects of "experiments of opportunity," in which human or natural events have altered local sea level or caused salinity intrusions, may simulate future effects of sea level rise. Fourth, greater knowledge of the environmental variation under which each major type of system can now exist is needed. Detailed physiometric studies of coastal ecosystems are limited to a few areas, usually near marine research laboratories. Results are often extrapolated to other sites. A given set of predicted environmental conditions, however, may not match those of these few study sites. Each type of coastal system may exist in a much broader range of environments than is now documented, and gradual changes probably occur between system types. Greater regional knowledge of the variety of ecosystem types, and of the variety of environments that support the same ecosystem, will enhance the resolution of empirical ecological predictions.

6.8 ELEMENTS OF MANAGEMENT RESPONSE

6.8.1 Should We Respond Now?

Coastal impacts from sea level rise can be far reaching in terms of the number and type of resources that would be affected, and the magnitude of impact in each case. Since a typical engineering project is designed for a life span of 25–100 years, and since growth management is an issue in most cases, there is a seeming need to incorporate sea level effects at the design stage. Nevertheless, the Marine Board Committee on Engineering Implications of Changes in Relative Mean Sea Level (National Research Council, 1987a) had these cautious comments:

> "The prognosis for sea level rise should not be a cause for alarm or complacency. Present decisions should not be based on a particular sea level rise scenario. Rather, those charged with planning or design responsibilities in the coastal zone should be aware of and sensitized to the probabilities of quantitative uncertainties related to future sea level rise. Options should be kept open to enable the most appropriate response to future changes in the rate of sea level rise. Long-term planning and policy development should explicitly consider the high probability of future increased rates of sea level rise."

What should be recommended in terms of elements of management response? Clearly, considering the diversity of geographical, social and political settings, and particularly the wide range of engineering issues that are sure to arise in the event of a relative sea level rise, we cannot possibly develop response criteria in a "nutshell." We do wish to point out, however, that perhaps no country has initiated a more exhaustive effort toward arriving at technical and other management responses than The Netherlands (see, e.g., de Ronde, 1991), even though in other countries such as Bangladesh and small island nations, the impacts can be more devastating than say in The Netherlands, where engineering has been directed historically toward the water level problem.

6.8.2 Shore Development

Concerning recommendations for technical and other response elements, we will focus on an important issue in the U.S., namely the management of shorelines having sandy beaches, particularly those along barrier islands (Dean, 1991).

1. *Understand the Natural System:* Although the case has not been made in this report, it should be clear that the appropriate management decisions on an eroding coastline depend greatly on the ambient background erosion rate. Those shorelines with high background erosion rates should be identified and restricted to uses that do not allow permanent structures. These uses could include public parks or, if in the private domain, readily moveable single-family structures with a contractual understanding that if erosion persists and results in the shoreline receding up to the structure, the structure will be moved. Thus, coastal engineers and geologists should document authoritatively the long-term erosion rates and other coastal response characteristics to sea level rise.

2. *Eliminate or Reduce Anthropogenic Erosion:* Inlets improved for navigation have been shown to be responsible for much of the erosion occurring to downdrift shorelines (Dean, 1988; Dean and Work, 1993). As technical stewards of these resources, we should strive to develop effective and economical sand management technology capabilities and then persist in seeing that they are implemented on a continuing basis. We should continue to document the effects of navigational entrances and the benefits of improved sand management technologies.

   A second cause of anthropogenic erosion, which can be locally significant, is ground subsidence induced by extraction of hydrocarbons or ground water. This effect has been well documented in many areas, with the most extreme example in the U.S. being more than 6 m subsidence at Terminal Island, California due to hydrocarbon extraction from the Wilmington Oil Field. A subsidence of, say, 30 cm will cause beach recession equivalent to a local sea level rise by the same amount.

3. *Quantify Shore Hazard Zones and Inform the Public:* All shores of the U.S. should be assigned a hazard factor that characterizes the long-term, persistent erosion component and the short-term storm component based primarily on long-term erosion rates but secondarily on land elevations, storm frequencies, etc. This information could be presented in various levels of sophistication with the more advanced including annual probabilities of damage that, on an eroding shore, would increase in future years.

This hazard level should serve both as a basis for regulation with restrictions against permitting large, permanent structures on an eroding shoreline and for informing the public with special emphasis when a purchase of property is being considered or consummated.

4. *State-of-the-Art Engineering Solutions to Beach Erosion:* Changes and responses will be site specific, with beach nourishment, sea walls or revetments, detached breakwaters, groins, or selective retreat being options. An estimate has been made of the cost of nourishing with sand the approximately 675 km long east coast of Florida, in response to a 0.60 m rise in RMSL during 100 years. It is based on a mean equilibrium beach profile representative of Florida's beaches and a profile change closure depth of 8.2 m. It is estimated that 500 m^3 of sand would be needed per meter of coast. If sand placement costs were about $8 per cubic meter, the total cost would be approximately $2.7 billion. For a discussion of the costs of defense in more sheltered areas, see for example Smith and Tirpak (1989).

5. *Continue Focused, Prioritized Research:* Given the wide scope of research needs presented earlier, it is evident that research prioritization will be a major need as far as technical improvements through scientific and engineering research is concerned. Here, we wish to make a plea for focussed research with obviously high benefits over the next decade. In general, with respect to improving our response to climate-induced hazards along our coastline, many uncertainties of future effects exist and all modern means should be employed to explore them. Examples include:

   (1) early detection of an increased rate of sea level rise,

   (2) early detection of the effects of sea level rise, perhaps in terms of beach recession averaged over large sections of shoreline,

   (3) additional hazards of sea level rise due to, for example, maintaining barrier islands static rather than allowing them to move landward and upward as they would in their natural state, and

   (4) the development of rational strategies for coping with these effects.

The latter include the development of a decision-making procedure in which the public would play a leading role. Questions include the designation of areas where, if a storm destroys structures, those structures would not be allowed to be rebuilt. Also, the question of compensation should be addressed.

Funding to carry out research should be a result of factual communications between the technical community and the authorities responsible for allocating research funds for worthy projects. It is essential that our primary concentrations be on the need for developing a scientific understanding, on developing an early framework for responsible action, and improving that framework with additional knowledge gained by further research. We should not attempt to overstate dramatically either the urgency or the degree of the hazard. While this may serve scientific effort in the short term, policy decision makers have a corporate memory as to scientists and wish to be provided with justifiable and reliable needs and plans for their resolution.

6.9 GOAL AND RECOMMENDED TASKS

6.9.1 Overall Goal

The overall goal can be stated as: evaluation and improvement of engineering technology for the assessment and mitigation of, and/or adaptation to, the impacts of RMSL rise and possible increase in the number, intensity and paths of episodic forcings, and the development of a range of robust options for mitigation and adaptation.

6.9.2 Tasks

To meet this goal, the following tasks are recommended:

1. *Update Objective Evaluation of Worldwide Tide Gauge and Satellite Water Level Data:* An engineering response is techno-economic in nature. The highly nonlinear relationships between coastal protection costs and mean sea level rise, coupled with the variation in the predicted value for mean sea level rise, e.g., by the year 2075, which ranges from a low of 36 cm to a high of 191 cm, highlight the need for a periodic review, say every 5–10 years, to provide an objective evaluation of the rise in mean sea level due to global climatic changes. There is a need

to add tide gauges in appropriate locations along the open coast and where there are presently few, such as in the Southern Hemisphere. Data from all available tide gauges should be made freely available to a committee of experts with international representation, for analyses and periodic updating to enable long-term engineering responses to the problems associated with the rise of mean sea level.

2. *Evaluate Possible Changes in Episodic Forcing:* An evaluation should be made of possible changes in episodic forcing, such as the number, intensity and paths of tropical cyclones (hurricanes, typhoons), storm weather fronts, and of ENSO conditions. For instance, wave conditions responsible for erosion and structural damage in California, Oregon, and Washington are strongly dependent on the jet stream strength and configuration and associated storms. It appears that these causes are poorly predicted, if at all, by existing models. Scenarios should be developed for a reasonable range of forcings.

3. *Develop and Implement Technology for Remotely and Periodically Monitoring Shorelines:* In order to establish the magnitudes of the effects of RMSL rise, it is necessary first to develop a data base of current shoreline rates of change. Such data, correlated with sea level and weather systems, would provide the means for improved understanding and modeling of the dynamics of the nearshore system, including the effects of engineering projects. Yearly, remotely sensed data would first provide useful results on a worldwide basis, and as additional flights are available, will allow valid assessments on progressively smaller regions. At present, photography from U2 aircraft reportedly can provide resolution of at least 0.5 m, while that of satellite imagery is considerably poorer. Thus, at present the U2 is the preferred platform; however, this may change with advancing technology.

4. *Conduct Worldwide Studies of RMSL Rise (Subsidence and/or Mean Sea Level Rise), Resulting Responses and Engineering Solutions:* Studies should be made of areas in which there has been a recent RMSL rise (mostly due to land subsidence of about the amount predicted for mean sea level rise by global climate change models), and associated effects, together with engineering responses that have been used to mitigate or adapt to the

effects. Evaluation of the effectiveness (including costs and benefits) of the actions, and estimation of their usefulness for future changes are necessary components of the studies. Modifications should be recommended where appropriate.

5. *Develop a Database and Evaluate Likely Problems of Coastal Ports, Airports, Power Plants, and Infrastructure:* A database should be developed of the elevations and exposure of components of coastal ports, airports, power plants, other structures and infrastructure. A technical and economic evaluation should be made of mitigation and adaptation options necessary for continued operation of these facilities.

6. *Develop Improved Predictive Capability for Modeling:* Areas of interest include but are not limited to: (1) outercoast shoreline response; (2) saltwater penetration in estuarine and ground water in the coastal zone; (3) interior shoreline response including shallow water margins; and (4) coastal ecosystem modeling.

7. *Develop and Apply Procedures for Evaluating Sediment Excess/Deficit Conditions of Dynamic Beach Profiles:* Beach profiles can be considered to contain an excess or a deficit of sediment compared with the equilibrium condition, with the long-term potential of shoreline advancement or recession, respectively. If beach profiles could be so classified and the response scales of disequilibrium established for various water depths, the impacts of various sea level rise scenarios on changes in long-term shoreline rates of change could be established. It is recommended that field studies of equilibrium profiles and the associated sediment bottom be conducted on a worldwide basis and compared with known shoreline changes to improve and evaluate the existing technology on equilibrium beach profiles. Once developed, the methodology can be applied to quantify the degree of change for various sea level rise scenarios.

8. *Establish the Effects of Sea Level Rise on Inlet Response:* The projected rise in mean sea level and its possible effects on destabilizing coastal inlets, deserves special attention. The migration of inlets due to anthropogenic causes has led to negative effects on heavily inhabited beaches. A study of these cases could lead to a valuable understanding of the behavior of stable inlets in response to a rise in mean sea level, particularly the growth of

the ebb tidal delta, the channel geometry, salinity distribution and beach erosion. A comprehension of this phenomenon would be a first step in determining the engineering response to such changes.

9. *Understand Sediment Motion over Tidal Flats, Intertidal Zones and Salt Marshes:* In these areas the material in motion is often fine-grained and cohesive. The development of engineering technology for the enhancement and management of coastal wetlands under different water level and sediment supply scenarios requires a better understanding of the transport properties of these sediments under various levels of hydrodynamic, chemical and biological controls. Process models developed from this understanding can then be used to improve the performance of predictive mathematical models useful for protecting wetlands.

10. *Develop Technology for the Enhancement and Management of Coastal Wetlands:* This approach involves: (1) The application of coastal process principles to understand and model the physical dynamics of estuarine and open coast habitats such as salt marshes; and (2) The development of hydraulic and structural means to enhance the extent and viability of such habitats, and to improve their management.

11. *Improve the Design of Lowland Drainage:* Drainage of lowlands is possible through sluices during low tides, and by pumping. RMSL rise makes drainage more difficult and more expensive, as both low and high tide levels are likely to be rising also. Information on existing systems, such as those along the North Sea coast of Germany and The Netherlands, should be obtained and evaluated accordingly.

12. *Develop and Implement Effective Sediment Management and Stabilization/Accretion Technology:* Dredging is the predominant method for the transport of sand/sediment from one location to another. Often, when dredging is performed for navigational purposes, the dredged material is disposed of in locations that are not optimal with respect to beach or shoreline stabilization or enhancement. It is recommended that dredging permits include plans for the most effective placement of the dredged materials.

    Beach/shoreline stabilization and enhancement technologies currently include sea walls, groins, bulkheads, etc. While these can be locally effective they can also have detrimental effects, such as increased down-current erosion. Improved designs of these existing devices would be an important goal.

The engineering community should encourage new and innovative methods of sediment transport/bypassing and beach/shoreline stabilization. Some examples of these, whose effectiveness is yet to be proven, are "beach cones" and the electrodeposition of minerals, such as calcium carbonate, from sea water onto wire mesh substrates placed in locations of very low energy, where accretion is required to thwart the effects of mean sea level rise. Another example is the use of a "perched beach" for which prototype trials and evaluations should be made in several different types of regions.

13. *Improve Cost-Effective Sediment Transfer and Delivery Systems:* The need for considering beach protection measures is related to beach use and the population residing in these valuable areas. Beach nourishment is perhaps one of the most desirable forms of protection in that it is the least interfering with respect to beach use. One of the inhibiting factors, particularly in developing countries, is the cost, ranging from US$2 to US$10 per cubic meter of sand placed. The need for developing low-cost sediment transfer and delivery systems gains importance in the light of the additional erosion of the beach associated with expected rise in mean sea level due to global climate change.

14. *Emphasize Robust Engineering Solutions in Response to Global Climate Change:* Because of the many uncertainties in the quantitative assessment of the forcings, including mean sea level rise and possible changes in the numbers, intensities and paths of major episodic forcings such as tropical cyclones (hurricanes and typhoons), storm weather fronts, and ENSO conditions, emphasis should be given to robust engineering solutions. By robust is meant, for example, beach nourishment or structures that can be effective for a range of changed conditions, or modified and/or repaired rather easily.

15. *Improve the Design and Maintenance of Navigation Channels in Harbors and Entrances:* Under rising waters channels will face increased salt water penetration and sedimentation. Increased sedimentation in turn will increase the problem of

sediment disposal. Design of channels and forecast of the required dredging frequency and means to control sedimentation can be improved by better understanding the mechanism of sediment transport by tides and especially by episodic action. Experiments on the collection of long-term signatures of forcing and response parameters constitute the first essential step toward the ultimate objective of developing predictive capability for channel stability and infilling rates.

16. *Improve the Design of Dikes and Evaluate Effects of Mean Sea Level Rise and Extreme Events on Dikes:* Lowlands bordering the sea and estuaries in some areas are protected against inundation due to storm floods by dikes, walls and storm surge barriers. Site-specific design criteria must take into account the sea level during a significant storm flood, the magnitude of the wave runup, and the magnitude of the future rise in mean high water level (50 to 100 years). The design return period must achieve the protection goals depending on the number of people and built infrastructure to be protected.

17. *Consider Effect of Sea Level Change on Coastal Structures and Construction Methodology (Including Port Facilities):* The effect of sea level change should be considered together with other factors over the design life of coastal and port structures. When not constrained by resources, legal issues and decisions, construction of structures for coastal defense can be carried out over comparatively very short periods. Therefore, there appears to be no rational basis for emergency measures now to mitigate the effects of anticipated future increases in the RMSL.

18. *Promote Equity Considerations in Assessing Responsibility for Mitigating Effects of Sea Level Rise:* While data on eustatic sea level rise appear to show an increase, its magnitude is the subject of much controversy. Perhaps more important is the need to assess the magnitude of relative sea level rise, which is caused by a number of constituent processes such as natural subsidence, oil/gas/water extraction, etc. A scientific/engineering assessment of the relative magnitudes of these contributory effects associated with relative sea level rise would be useful for equity considerations.

19. *Develop a Database on Valuation of Shore Properties and Estimation of Response Costs:* The valuation of shoreline

properties should include not only real property values, but also ecological, cultural and historical values. In addition to obtaining monetary values from tax records, the emphasis should be to develop rational bases for establishing environmental, cultural and historical values.

20. *Develop Rational Decision-making Methodology in Environmentally Sensitive Areas:* The maintenance of stable shorelines at their present locations may only be possible by engineering means, including beach nourishment, accretion of fine material by "active protection," armoring the forelands in front of sea dikes and strengthening them by revetments. These means may affect or be viewed as affecting the environment in a negative manner through modification of the natural development of wetlands and marshes, loss of wildlife habitats by erosion, and development toward more shoreline armoring. For such cases, engineering solutions should be sought that meet the protection goals as well as the environmental quality desired. Acknowledging that this approach will not be possible in every case, definitive and well-monitored experiments must be carried out to address environmental concerns. They must lead to prognoses with respect to the development of the total affected system. Public education on coastal protection as well as environmental aspects are necessary to develop an adequate answer from the affected society whether to retreat or to maintain the existing protection system. The problems addressed arise more in highly developed areas and where national parks or wildlife preserves are established in the coastal zone such as in Germany and The Netherlands. In addition, utilizing areas with large RMSL rise, such as the coast of Louisiana, as a laboratory for studying mitigation options is also a worthwhile endeavor.

21. *Improve Public Perception of Engineering Potential for Dealing with Future Coastal Hazards:* Three elements of this task include: (1) development and participation in public education forums; (2) cooperation with decision makers to enable them to make management decisions relative to response to future hazards based on sound data bases and rational principles; and (3) promotion of interdisciplinary academic programs and research.

6.10 CONCLUDING COMMENTS

There is a high probability that global climate change will cause an increase in the presently experienced rate of sea level rise. This increase will in general lead to increased loss of shoreline, storm hazards, salt penetration and alteration of coastal ecology. However, there is substantial uncertainty as to the magnitude and timing of these changes. Given these uncertainties yet the potentiality of increased coastal hazards, a three-element plan is proposed. The first element would be to use existing data to categorize projected hazard levels based on past and present trends. The second, concurrent element would be to apply existing technology and develop strategies that would not only eliminate or at least measurably reduce anthropogenic causes that enhance the effects of sea level rise, e.g., beach erosion, but also ensure that such anthropogenic practices are not pursued in future. The final element would be to form a working relationship between engineers-scientists and policy decision makers to provide them with authoritative assessments of related needs and associated hazards, to convince them of the importance of continuing focussed research to allow early and accurate detection of sea level changes and their effects, and to provide rational bases for coastal zone management.

Participation by Professor Robert Wiegel of the University of California at Berkeley in the committee work is sincerely acknowledged.

REFERENCES

Beverly, H. W., 1988. Discrepancies between bench mark elevations at Redondo Beach, California and the mean sea level of the Pacific Ocean, Los Angeles, CA: U.S. Army Corps of Engineers, Los Angeles District, 5pp.

Boesch D. F., Josselyn M. N., Mehta A. J., Morris J. T., Nuttle W. K., Simenstad C. A., and Swift D. J. P., 1994. Scientific assessment of 333 coastal wetland loss, restoration and management in Louisiana. *J. Coastal Res.,* Special Issue No. 20, 103pp.

Bruun, P., 1962. Sea level rise as a cause of shore erosion. *J. Waterways Harbors Div.,* 88(WWI): 117–130.

Carter, W.E., Robertson D. S., Pyle T. E., and Diamante J., 1986. The application of geodetic radio interferometric surveying to the monitoring of sea-level. *Geophys. J. R. Astronom. Soc.,* 87:3–13.

Daniels R. C., V. M. Gornitz, A. J. Mehta, S. C. Lee and R. M. Cushman, 1992. *Adapting to sea-level rise in the U.S. southeast: the influence of built infrastructure and geophysical factors on the inundation of coastal areas.* Report ORNL/CDIAC-54, Environmental Sciences Division Publication No. 3915, Oak Ridge, TN: Oak Ridge National Laboratory.

Dean, R. G., 1988. Sediment interaction at modified coastal inlets: processes and policies. In: *Hydrodynamics and Sediment Dynamics of Tidal Inlets,* D. G. Aubrey and L. Weishar, eds., New York: Springer-Verlag: 412–439.

Dean, R. G., 1990. Beach response to sea level change. In: The Sea, Volume 9, Ocean Engineering Science, Ch. 25, B. LeMehaute and D. M. Hanes, eds., New York, NY: Wiley: 869–887.

Dean, R. G., 1991. Impacts of global change: Engineering solutions. In: *Our Changing Planet: Joining Forces for a Better Environment.* Proceedings of Symposium in Commemoration of the 20th Anniversary of the Graduate College of Marine Studies, Newark, DE: University of Delaware: 13–17.

Dean R. G., Chiu T. Y., S. Y. Wang, 1990. Combined total storm tide frequency analysis for Lee County, Florida. Report submitted to the Florida Department of Natural Resources, Tallahassee, FL: Beaches and Shores Center, Florida State University, 69pp.

Dean R. G., P. A. Work, 1993. Interaction of navigational entrances with adjacent shorelines. *J. Coastal Res.,* Special Issue No. 18, 91–110.

de Ronde, J. G., 1991. Rising waters: impacts of the greenhouse effect for The Netherlands. *Report gwao 90.026,* The Hague, The Netherlands: Rijkswaterstaat, Tidal Waters Division, 40pp.

Dolan, R., B. Hayden, S. May, 1983. Erosion of the United States shorelines, in: *Handbook of Coastal Processes and Erosion,* Ch. 14, P. D. Komar, ed., Boca Raton, FL: CRC Press: 151–166.

Emery, K. O., D. G. Aubrey, 1991. *Sea Levels, Land Levels, and Tide Gauges.* New York, NY: Springer-Verlag, 251pp.

Fuhrboter, A., J. Jensen, 1985. Longshore changes of tidal regime in the German Bight (North Sea), *Proceedings of the 4th Symposium on Coastal and Ocean Management (Coastal Zone '85)*, Vol. 2, New York, NY: American Society of Civil Engineers: 1991–2013.

Gornitz, V., S. Lebedoff, 1987. Global sea level changes during the past century. In: *Sea Level Changes and Coastal Evolution*, O. W. Pilkey and J. D. Howard, eds., Special Publication No. 41, Tulsa, OK: Society of Sedimentology and Geology: 1–14.

Grant J. R. H., 1992. Historical shoreline response to inlet modifications and sea level rise. M.S. thesis, Gainesville, FL: University of Florida, 152pp.

Hawaii Coastal Zone Management Program, 1985. Effects on Hawaii of a world-wide rise in sea level induced by the greenhouse effect, Paper prepared in response to Senate Resolution 137, 1984, Honolulu: Department of Planning and Economic Development, 10pp.

Hicks, S. D., A. Debaugh, and L. E. Hickman, 1983. *Sea level variations for the United States, 1855–1980*. Rockville, MD: National Oceanic and Atmospheric Administration, 170pp.

Hicks, S. D., L. E. Hickman, 1988. United States sea level variation through 1986. *Shore and Beach*, 56(3): 3–7.

Hoffman, J. S., D. Keyes, J. G. Titus, 1983. *Projecting Future Sea Level Rise: Methodologies, Estimates to the Year 2100, and Research Needs*. Washington, DC: U.S. Environmental Protection Agency, 121pp.

Hoffman J.S., Wells J.B., Titus J.G., 1986. Future global warming and sea level rise, *Proceedings of Iceland Coastal and River Symposium '85*, G. Sigbjarnarson, ed., Reykjavick: National Energy Authority.

Houghton, J. T., G. J. Jenkins, J. J. Ephraum, eds., 1990. *Climatic Change: The IPCC Scientific Assessment*. Cambridge, England: Cambridge University Press, 200pp.

Johnson, B. H., M. B. Boyd, G. H. Keulegan, 1987. A mathematical study of the impact on salinity intrusion of deepening the Lower Mississippi River navigation channel. *Report TR HL-87-1*, Vicksburg, MS: U.S. Army Engineer Waterways Experiment Station, 76pp.

Kunz, H., 1990. The impact of increased sea level rise on the German Wadden sea and how the global climate change may affect the coastal zone management for this region. *Proceedings of Littoral 1990 Symposium*, Marseille, France: EURO-COAST Association: 319–324.

Kurz, H., K. Wagner, 1957. *Tidal marshes of the Gulf and Atlantic coasts of northern Florida and Charleston, South Carolina*. Florida State University Studies, No. 24, Tallahassee, FL: Florida State University, 168pp.

Lambeck, K., S. M. Nakiboglu, 1984. Recent global changes in sea level. *Geophys. Res. Lett.*, 11: 959–961.

Mehta, A. J., 1990a. Significance of bay superelevation in measurement of sea level change. *J. Coastal Res.*, 6(4): 801–813.

Mehta, A. J., 1990b. Role of coastal sedimentary processes in marine habitat protection and enhancement. Paper presented at the meeting of Marine Board Committee on the Role of Technology in Marine Habitat Protection and Enhancement, Washington, DC: National Research Council, 30pp., unpublished.

Mehta, A. J., R.M. Cushman, eds., 1989. Workshop on Sea Level Rise and Coastal Processes. *Report DOE/NBB-0086,* Washington, DC: U.S. Department of Energy, 308pp.

Mehta, A. J., R. Philip, 1986. Bay superelevation: causes and significance in coastal water level response, Report *UFL/COEL-TR/061,* Coastal and Oceanographic Engineering Department, Gainesville, FL: University of Florida, 65pp.

Montague, C. L., 1986. Influence of biota on erodibility of sediments. In: *Estuarine Cohesive Sediment Dynamics,* A. J. Mehta, ed., New York, NY: Springer-Verlag: 251–269.

Montague, C. L., W. R. Fey, D. M. Gillespie, 1982. A causal hypothesis explaining predator-prey dynamics in Great Salt Lake, Utah. *Ecolog. Modelling,* 17: 243–270.

Montague, C. L., A. V. Zale, H. F. Percival, 1987. Ecological effects of coastal marsh impoundments: a review. *Environ. Manage.,* 11: 743–756.

Murayama, S., 1970. Land subsidence in Osaka, In: *Land Subsidence: Proceedings of the Tokyo Symposium,* Vol. I, Tokyo, Japan: IAHS/UNESCO: 105–129.

Nair A. S. K., 1988. Mudbanks (chakara) of Kerala—a marine environment to be protected. Proceedings of the National Seminar on Environmental Issues, Golden Jubilee Seminar, Trivandrum, Kerala, India: University of Kerala: 76–93.

National Research Council, 1985. *Glaciers, Ice Sheets, and Sea Level.* Polar Research Board, Washington, DC: National Academy Press, 330pp.

National Research Council, 1987a. *Responding to Changes in Sea Level: Engineering Implications.* Marine Board, Washington, DC: National Academy Press, 158pp.

National Research Council, 1987b. *Sedimentation Control to Reduce Maintenance Dredging of Navigational Facilities in Estuaries.* Marine Board, Washington, DC: National Academy Press, 352pp.

National Research Council, 1989. *Measuring and Understanding Coastal Processes for Engineering Purposes.* Marine Board, Washington, DC: National Academy Press, 119pp.

National Research Council, 1994. *Restoring and Protecting Marine Habitat: The Role of Engineering and Technology.* Marine Board, Washington, DC: National Academy Press, 205pp.

Odum, H. T., E. C. Odum, M. T. Brown, D. LaHart, C. Bersok, J. Sendzimir, 1987. Environmental Systems and Public Policy. Ecological Economics Program, Phelps Laboratory, Gainesville, FL: University of Florida, 237pp.

Revelle, R., 1983. Probable future changes in sea level resulting from increasing carbon dioxide. In: *Changing Climate, Report of the Carbon Dioxide Assessment Committee,* Washington, DC: National Academy Press: 433–448.

Rijkswaterstaat; Delft Hydraulics, 1990. Impact of sea level rise on society; a case study for the Netherlands, *6WAO Document 90.016 Tidal Waters Division; Report H 750,* The Hague, The Netherlands: Delft Hydraulics.

Shepard, F. P., 1963. *Submarine Geology.* New York, NY: Harper and Brothers, p. 575.

Shore and Beach, 1989. An issue of the journal dedicated to the storm of 17–18 January 1988 off Southern California, 57(4), 84pp.

Smith, J. B., D. A. Tirpak, eds., 1989. *The Potential Effects of Global Climate Change on the United States.* New York, NY: Hemisphere Publishing, 648pp.

Stevenson J. C., L. G. Ward, M. S. Kearney, 1986. Vertical accretion in marshes with varying rates of sea level rise. In: *Estuarine Variability,* D. Wolf, ed., New York, NY: Academic Press: 241–260.

Sturges, W., 1987. Large-scale coherence of sea level at very low frequencies. *J. Phys. Oceanogr.,* 17(11): 2084–2094.

Swift, D. J. P., 1975. Barrier island genesis: evidence from the central Atlantic shelf, eastern U.S.A. *Sediment. Geol.,* 14: 1–43.

U.S. Army Corps of Engineers, 1973. *National Shoreline Study.* Vols. I-V, Washington, DC: Corps of Engineers.

U.S. Army Corps of Engineers, 1989. Storm damage reduction for King Harbor (Redondo Beach), Redondo Beach, California. *General Design Memorandum No. 3,* Los Angeles, CA: Los Angeles District (Draft).

Zale, A. V., C. L. Montague, H. F. Percival, 1987. A synthesis of potential effects of coastal impoundments on the production of estuarine fish and shellfish and some management options. In: *Waterfowl and Wetlands Symposium: Proceedings of a Symposium on Waterfowl and Wetlands Management in the Coastal Zone of the Atlantic Flyway,* W.R. Whitman, W.H. Meredith, eds., Delaware Coastal Management Program, Dover, DE: Delaware Department of Natural Resources and Environmental Control: 424–436a.

Chapter 7

AGRICULTURAL AND BIOLOGICAL SYSTEMS

Authors: Norman R. Scott
Contributors: John N. Walker, Gerald F. Arkin,
James A. DeShazer, Gary R. Evans, Glenn J. Hoffman,
James W. Jones

7.1 INTRODUCTION

Much has been written about global climate change during the past several years with varying projections about future impacts of observed and predicted changes in climatic parameters (EPA, 1991b). Some models predict significant global warming, particularly in the upper latitudes with significant melting of the polar ice caps, while others predict little change or even global cooling. The predictions are based upon general circulation models (GCMs) that simulate the dynamics of atmospheric processes on a global basis. The disparity in the predictions and the inability of the models to reproduce past recorded events raises questions about the validity of the models or the understanding of phenomena which have major climatic impacts, as discussed in Chapter 1.

Though the impacts of the climatic change are subject to dispute, there is little debate about the fact that at least some atmospheric parameters that affect climate have changed in the past century (Committee on Science, Engineering, and Public Policy, 1991). Concentrations of carbon dioxide, methane, and other greenhouse gases have clearly risen, and ozone in the upper atmosphere has declined. The major source of these gases is from energy use and production, as shown in Figure 7.1, and agricultural activities contribute about 14% of the total rate of increase in the atmospheric greenhouse gases on a global basis (Committee on Alternative Energy Research and Development Strategies, 1990). Historical evidence has shown that the atmospheric concentration of carbon dioxide has changed during past major climatic changes.

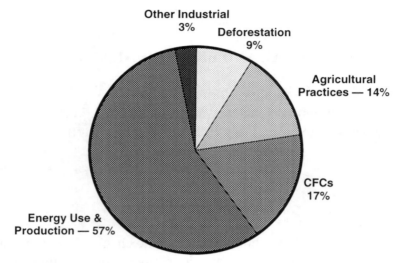

Figure 7.1   Contribution of agriculture and other activities to greenhouse gases. (From Committee on Alternative Research and Development Strategies, 1990.)

It is not the purpose of this chapter to debate the nature of global climate change, the magnitude or direction of change, or the consequences of the change of one climatic parameter on other parameters. Rather, an attempt is made to identify the effects on the agricultural sector if climate changes, the role of agricultural activity on greenhouse gas emissions, engineering issues and approaches that might be pursued to reduce emissions, and ways to adapt to anticipated changes in climate. The focus is on the intensely managed systems for food, fiber, and forest products. Ecosystems such as wetlands, natural woodlands, and other specialized natural areas that receive only small amounts of management are not discussed. Where the knowledge base as it relates to managed systems is inadequate to define corrective strategies, research is suggested.

7.2 CLIMATIC FACTORS OF IMPORTANCE TO AGRICULTURE

7.2.1 Carbon Dioxide

Atmospheric carbon dioxide as measured worldwide is clearly increasing (Committee on Science, Engineering, and Public Policy, 1991) as discussed extensively in Chapter 2. This increase is attributed to the increasing dependence on fossil fuels and the combustion of wood and other biomass. Carbon dioxide (CO_2) is an essential component in the photosynthesis process of plants. However, not all plants respond equally to CO_2-enriched environments. Most tree species, grasses, and many food plants display a significant response. These plants are known as C_3 plants. Assuming environmental parameters such as temperature, light, soil water potential and soil nutrients are adequate, a doubling of the atmospheric CO_2 concentration will lead to a doubling of the photosynthetic rate (Kimball, 1983; Allen et al., 1987; Parry, 1990; Idso, 1993) and an increase in yield of up to 30 to 40% in some species (Department of Energy, 1990).

Such levels of increase, demonstrated in controlled environmental studies, are not common to all plants. Some plants which are known as C_4 plants, such as corn, sorghum, sugarcane and millet, use a different photosynthetic pathway. In these plants, the CO_2 level within the leaves, where photosynthesis takes place, is less affected by the atmospheric CO_2 concentration than by cellular leaf structure of the plant. For these plants, increased CO_2 in the atmosphere will have a less positive impact on yield (10% or less) (Kimball, 1983; Intergovernmental Panel on Climate Change, 1990). Since corn is one of the major crops and is one of the primary crops used to make up food deficits, the lack of significant response to CO_2 enrichment may result in a shift from corn to other crops, or it may result in the corn producing regions of the world not benefiting from an expanded production potential on the available land (Smit et al., 1988; Wilks, 1988).

In studies on carbon dioxide enrichment other factors known to affect plant growth have normally been held at levels where they would not limit growth. Growth or yield increases observed in these studies cannot simply be projected to large (field) plant communities. In natural or field settings, temperature, soil water potential or radiation will often be less than optimum and may be a controlling climatic factor. Unfortunately, the difficulty in maintaining the ambient CO_2 levels above normal atmospheric levels has resulted in few definitive studies on the aggregate response which might be obtained in natural settings.

CO_2 concentration also affects plant water use. When all other factors are the same, plant leaves exposed to elevated CO_2 concentrations will have lower transpiration rates because of increased resistance to water vapor diffusion through stomata (Rosenberg, 1981; Allen et al., 1985). Under rainfed conditions, a decreased rate of water use by crops will result in less water stress and increased yield. Thus, water use efficiency, defined as biomass produced per unit of water consumed, will increase in crops under elevated CO_2 conditions because of high photosynthesis rates and lower water use. In C_3 plants, the main impact of CO_2 is on water use efficiency and photosynthesis, whereas in C_4 plants, water use efficiency is affected more by decreased transpiration rates. Although the magnitude of stomatal response measured on individual leaves under doubled CO_2 may result in a 20–30% reduction in transpiration, savings in water use by crop canopies integrated over the season may be only 5 to 10% because of more rapid crop growth and other factors (Department of Energy, 1990). Stomatal response to increased CO_2 will also result in slightly higher plant temperatures because of a decrease in evaporative cooling of the leaves with resultant decline in transpiration. Although these general responses are understood, the combined effects of these responses for increased atmospheric temperatures is unclear. Little research has been done to quantify the combined effects of CO_2 and high temperatures on water use, growth, and yield of most agricultural crops.

Although plants utilize carbon dioxide and other elements to form carbohydrates by photosynthesis, plants also release carbon dioxide back to the atmosphere through respiration (Public Service Research and Dissemination Program, 1989). Respiration is the process of using carbohydrates for plant growth. It is temperature-sensitive, and increases as temperature increases. Ideally, when light during the day is limiting and carbohydrate production is suppressed, the resulting cooler temperatures reduce respiration and the consumption of carbohydrates. Similarly, if high levels of carbohydrates exist within the plant tissue, higher temperatures would increase photosynthesis.

Carbon dioxide is also released during the aerobic decay or digestion of plant material. This natural recycling of carbon back to the atmosphere is the prevalent carbon pathway. However, the carbon dioxide recycled back into the atmosphere on an annual basis, through the biological processes, is less than the carbon dioxide utilized in photosynthesis. Carbon stored in trees and other perennial plants is not recycled annually and represents a net withdrawal of carbon dioxide from the atmosphere. In addition, not all decay or digestion of plant material is aerobic. Anaerobic decay or digestion results in the conversion of the plant carbon to methane rather than carbon dioxide (ICF Incorporated, 1989). Anaerobic decay occurs in marshes and paddy cropping systems such as rice where water limits adequate oxygen. In addition, methane is released from anaerobic digestion of plant material that occurs in ruminants and termites. Anaerobic decay and digestion are reported to account for about one half of the methane released to the atmosphere annually (ICF Incorporated, 1989).

The decay of organic material; such as roots, buried plant material, and manure, below the soil surface is normally an aerobic process. The carbon dioxide produced within the soil profile may diffuse to the surface and escape to the atmosphere; it may be picked up by the roots of growing plants; it may be absorbed in water moving through the soil; or it may be utilized by soil microorganisms as a source of carbon. The relative magnitude of the different pathways varies greatly, but generally the majority of the carbon dioxide released within the soil profile eventually finds its way into the atmosphere as carbon dioxide.

Within both fresh and salt water bodies, plant growth utilizes CO_2 in the photosynthesis process. Photosynthesis below the water surface involves the utilization of carbon dioxide dissolved in the water for the formation of carbohydrates. Much of the photosynthesis in water bodies occurs within small, cellular size plants, such as phytoplankton. Though the size of the photosynthetically active plant is small, the large number of plants in a unit volume of water is such that the total activity is significant. Though sunlight can only penetrate a few feet below the water surface, the vertical movement of water due to thermal, wind, or gravity forces results in active plants of a much greater volume than the thin upper layer of water where sunlight can activate photosynthesis. Kelp and other large aquatic plants also utilize carbon dioxide and can contribute to the removal of carbon from the atmosphere.

Since the equilibrium level of carbon dioxide in water is a function of atmospheric level of carbon dioxide, an increase in atmospheric carbon dioxide would be expected to increase the carbon dioxide level in water bodies and consequently to increase aquatic plant growth.

7.2.2 Temperature

Changes in the atmospheric and surface temperature as a result of increased greenhouse gas concentrations has been the subject of much debate and considerable disagreement. Many GCMs suggest that a significant increase in the annual mean temperature is already unavoidable because of greenhouse gases already introduced into the atmosphere (International Council of Scientific Unions, 1990). Other models, using different assumptions suggest smaller change.

Regardless of the answer, the impact of temperature changes upon agriculture can be severe, particularly if the changes result in an increase in the magnitude and frequency of temperature extremes. Both plants and animals suffer from extremely high and extremely low temperatures — with death occurring if the temperatures reach a level beyond which the plant or animal can maintain biological functions. Plants, and to a lesser extent animals, also are affected by changes in the mean temperature. A plant or animal adapted to one temperature regime can often be stressed in another temperature regime that is not greatly different.

The effect of mean temperature on plants is most evident on the development cycle. There is a threshold level beyond which the biological processes will not function. Above the threshold level, rate of development will occur linearly up to some temperature and then development will begin to decline as temperatures increase further (Muchow and Bellamy, 1991). This can clearly be seen in seed germination and initial plant development. Assuming adequate soil water potential, seeds placed in the soil will not germinate until the soil temperature rises to a level specific for plant species. After germination, root and shoot development are strongly dependent upon the soil temperature. Instances of successful germination and then the subsequent rotting of the seedling due to a protracted cool spell are not uncommon.

Soil temperature is a factor in root development. Cool temperatures restrict development and hence moisture uptake. If warm, dry conditions follow a protracted cool period during which soil temperatures decline, plants are more susceptible to moisture stress. Conifer and evergreen landscape plants also frequently endure moisture stress during winter periods when soil temperatures are low. Such plants, particularly when young, are subject to moisture stress when the ground is frozen and water availability to the plants is restricted.

Excessively warm soil temperatures can also be a significant problem with young transplants or newly emerging seedlings. Warm soil temperatures accelerate evaporation and the rapid drying of the upper layers of soil. Soil water stress for plants at this stage of development can be devastating. As soil drying occurs, intense sunlight and hot ambient temperatures can result in temperatures which exceed the upper limits, resulting in death of the plants.

After emergence, a seedling's continued development is dependent upon the temperature degree-days. The concept of a discrete number of temperature degree-days (the sum of the differences between the mean daily temperature and a reference temperature) for emergence, flowering, and production of seeds is documented for a number of plants. If temperatures are warmer than normal, the plants mature earlier and seed yield may be suppressed. Similarly, if temperatures are cooler than normal, growth may be retarded and plants may enter the reproductive stage too late in the year to develop fully before a killing frost. The mean daily temperature is the key determinant in establishing the geographical limits and optimum cropping zones for crop species.

The process of photosynthesis, though temperature dependent, is less affected than development or respiration. As discussed earlier, from a viewpoint of respiration, cool temperatures are favored when plant carbohydrate levels are low due to low light levels. Conversely, warmer temperatures are desired when the inverse is true. Though both respiration and photosynthesis are affected by temperature, these are not normally the environmental factors that control plant development in economically important food crops. The effect of temperature on the initiation of flowering and seed or tuber formation is much more important. Not only does the accumulated degree-days affect the initiation of flowering, but also the pollination process and hence the subsequent seed development can be affected by temperature. Both above and below normal temperatures can have an adverse effect. Warm temperatures can also cause problems with leafy crops. If temperatures rise above an optimum level, lettuce will enter the reproductive stage and will produce seeds rather than continue to produce leaves.

The ambient temperature also significantly affects overall growth by virtue of its role in evapotranspiration (IPCC, 1990). As temperatures rise, evapotranspiration increases, thus increasing the water demand. For equal water availabilities, higher temperatures will result in a water stress condition occurring earlier. This may be alleviated if water is abundant and irrigation can be practiced to maintain high water availability. If high temperatures are combined with very low humidities, water stress may occur even when soil water potential is adequate.

Temperature extremes can be particularly catastrophic to plants. Once plant growth is initiated in the spring, freezing temperatures can be devastating. Similarly, an early fall frost can stop plant growth before the plants have fully matured and optimal growth has been achieved. High temperatures can also be harmful. A number of plants will cease development at high temperatures (IPCC, 1990). Severe water stress leading to plant death can also result from high temperatures. There is evidence that the optimal temperature for photosynthesis is higher under elevated CO_2 level (IPCC, 1990).

Animals are also affected by temperatures. For mature animals, high temperatures have a greater impact than cool temperatures. Though young animals may respond adversely to cold temperatures, animal housing at a reasonable cost can be provided to protect young animals. Air conditioning can be used to reduce the impact of high temperature but it is a much more expensive option. It is, however, used particularly with breeding males. Under stress, animals reduce feed intake and increase their evaporative heat loss in an effort to maintain thermal balance. This results in reduced rates of growth. An animal in milk production will reduce milk output and chickens will reduce egg production. Reproductive efficiency is also lowered with reduced conception rates and lower survival live birth weights under heat stress.

Animal performance is also adversely affected if the supply and quality of feed is insufficient. This is particularly true for ruminants where forage quality can deteriorate under high temperatures. Coarse, stemmy forage produced under high temperature conditions is less nutritious than leafy, succulent forage produced in cooler conditions.

The adverse effects of cool temperatures can be reduced or mitigated by animal housing. Though this is true, the cost of providing housing and the increased cost of managing and feeding animals in confinement increases the cost compared to the open range system utilized in warmer more temperate regions. This is particularly true of beef and dairy systems, and to some extent poultry operations.

High temperatures are more likely to be serious than are cold temperatures. Short periods of high temperature have resulted in the death of a large number of poultry. Sprinkling with water can alleviate or at least reduce heat stress with swine and dairy; however, this has not been practiced with poultry, although misting has proven effective. When high temperatures are combined with low humidities, evaporative cooling methods have been effective with all animal species, but they are less effective when the humidity is high.

Not only do aerial temperatures affect the growth and development of plants and animals, they also affect food and feed products in storage. Potatoes, apples, other fruits, and vegetables continue to respire after harvest. The respiration rate is temperature sensitive; with respiration increasing with increased temperature. Product quality is directly related to post-harvest respiration. Storage-life therefore depends upon the storage temperature. Most fruits and many vegetables are currently stored in refrigerated storage after harvest. For those crops, an increased ambient temperature would not have an adverse impact providing the products were moved rapidly from the field after harvest into the refrigerated storage. For potatoes, cabbage, and other products not kept in refrigerated storage, an increase in the ambient temperature results in reduced product quality. High temperatures during storage can also result in decreased seed quality. This would be particularly important in developing countries where on-farm seed storage is practiced.

Warm temperatures in temperate climates also intensify problems with insects in grain and food storage. Currently, grain can be stored on-the-farm throughout the winter with little deterioration if aeration is provided to prevent temperature stratification. Insects are not a major problem until warm spring and summer conditions occur. Fumigation can be practiced to control insects and vermin.

In addition to affecting germination and root development, soil temperature affects microorganism development. This includes organisms involved in plant development as well as organisms involved in the breakdown of organic matter in the soil and the subsequent release of nutrients. Higher temperatures favor the biological breakdown. The breakdown, however, can be too rapid, with a complete breakdown of all organic matter and a loss of soil tilth and nutrients.

7.2.3 Water

The effects of an increase in greenhouse gases on local water resources is highly uncertain. The various GCMs predict quite different magnitudes and directions of change in water resources and wide variances with both time and space of events. Regardless of the impact of global climate change, water is essential for the normal growth and development of plants and animals. It is so critical that even short delays in providing the required amount of water can result in death of the plant or animal. For rainfed agriculture, reductions in rainfall amount or changing the timing and frequencies of rainfall events can have devastating effects on production and could necessitate changes in crops, practices or even shifts to other industries. In areas where irrigation is practiced, agriculture is such a large user of water that restrictions in availability are always suggested when water shortages occur. Politicians and the public are very reluctant to restrict water from home use or industry, particularly if little or no cuts are proposed for agriculture. The recent drought in California is just one example where agriculture is pitted against urban and industrial water needs. Unfortunately, when rainfall is small agriculture's demand is the greatest. If water is not available to agriculture for even a few days, crop production for the entire year can be lost and years would be required to replace a livestock herd that was forced to be sold due to water shortage.

Agriculture has instituted conservation practices to improve water use efficiency in both rainfed and irrigated crops by reducing evapotranspiration, using more efficient irrigation systems such as downward spraying sprinkler nozzles, and installing animal-activated no-drip waterers. Beyond such conservation practices water must be available in adequate quantities when needed, or agriculture cannot be practiced.

Soil water, in addition to providing water required for plant growth, is important to enhancing activity of microorganisms within the soil profile. Soil water is also affected by the drainage potential of the soil. If an impervious soil layer exists the upper soil profile can become saturated during a heavy rain. If the soil remains saturated, aeration within the rooting zone can be restricted resulting in reduced plant growth and plant death if saturated conditions persist. High soil water levels also affect soil trafficability. Not only can vehicles cause deep ruts moving across saturated soils, but also soil compaction of the soil profile is intensified. If high soil water occurs at harvest time, reduced machine trafficability can delay or prevent harvest. This is particularly true for root crops.

Water runoff increases as rainfall intensity increases above the infiltration rate of the soil and soil water reaches saturation levels. Increased runoff leads to a greater loss of topsoil by the erosion process. Erosion is increased even more if the frequency and magnitude of severe, intense rainfall events increase. Soil compaction, because it reduces drainage, also intensifies runoff and therefore increases erosion. Runoff can be reduced and infiltration increased by land forming and practices that enhance infiltration.

Run-off can be trapped and stored in surface impoundment's where it can be utilized for crop or livestock watering. In this case, crop yield or livestock vitality may not be adversely affected. Pumping of water from impoundments for irrigation of crops is an added production cost. However, if it prevents severe crop water stress, the cost is normally justified. Increased construction of water impoundments, therefore, represents an option for mitigating the impact of cycles of excessive rainfall and periods of inadequate rainfall.

Impoundments of water also permit the practice of aquaculture. Fish farming is increasing dramatically and the production of protein per unit of land area can be as high with fish farming as it is with most efficient agricultural enterprises. The use of an impoundment for fish farming, however, prevents the full use of the impoundments for irrigation since a major drawdown of the water level would limit the fish carrying capacity of the impoundment.

Atmospheric moisture content (humidity) is also an important environmental parameter affecting agriculture. First, as atmospheric humidity decreases, evapotranspiration increases, thereby increasing the water demand if optimum plant growth is to be maintained. Atmospheric humidity is also critical in the drying of mature crops during the harvest period. The drying rate of forage crops such as alfalfa and grasses are dependent upon atmospheric humidity as well as the atmospheric temperature and wind speed. The same is true of grain crops.

Increased atmospheric humidity also decreases the ability of animals to lose internally produced heat and hence during warm or hot periods can result in a level of heat stress that causes reduced animal production. High humidities also reduce the effectiveness of evaporative cooling systems. Thus, in the humid Midwest, mid-Atlantic, and Southeast, such systems are not as effective. High humidities also favor the spread and development of many plant diseases.

7.2.4 Atmospheric Gases Other Than CO_2 and H_2O

Nitrous oxides are both a product of the natural decay of organic matter and a product of an industrialized society. The natural production is actually larger than the man-made contribution; the man-made output, however, is produced in areas where mankind is concentrated and where high usage of fossil fuels occur. Nitrous oxides at higher than normal concentrations can be damaging to lungs of humans and animals. Nitrous oxides are a key component in photochemical smog and acid precipitation.

Sulfur oxides are also a product of industrialized societies, particularly where fossil fuels containing sulfur are utilized as a primary energy source. Though much is being done to reduce sulfur oxides from electrical generation and primary steam production facilities by the increased use of low sulfur fuels and smokestack cleaning equipment, sulfur dioxides along with nitrogen oxides are considered to be major components in acid rain. Whereas procedures exist for the continued reduction of sulfur dioxide discharges, no equally effective solutions exist for the reduction of nitrogen oxides. The impact of atmospheric acidity on lakes and on soil without innate natural buffering is well documented. The disappearance of fish and other aquatic life in water bodies where the acidity has risen significantly due to acid rainfall is common throughout the northeastern U.S. and Canada. In these same regions, tree death due to increased soil acidity is widely reported.

Another atmospheric gas that can impact plant growth adversely is ozone. Ozone is produced by photochemical reaction of hydrocarbons and nitrogen oxides released into the atmosphere. The problem with atmospheric ozone is most severe near cities or where large numbers of vehicles are concentrated. Though catalytic converters reduce the discharge of hydrocarbons from passenger cars, large amounts of nitrogen oxides and hydrocarbons are still emitted into the atmosphere. Paint compounds, refiners, cleaning solvents, and asphaltic materials, in addition to cars and trucks all contribute to the problem. Ozone damage to agricultural production has been reported in areas distant from industrial and urban areas. The ozone damage to tobacco in North Carolina, for example, has been estimated to exceed $50 million annually.

Methane is another atmospheric gas considered to be a key contributor to the greenhouse effect. Methane at current or projected levels of atmospheric concentration does not adversely affect agriculture, but agriculture is a major producer of atmospheric methane (IPCC, 1990). Methane is produced in the anaerobic decay of organic matter such as occurs in wetlands and in rice paddies. Substantial amounts of methane are also produced by ruminants animals during the digestion of forages within the rumen and gut. Methane is also produced by termites during digestion of wood cellulose. The amount of methane released to the atmosphere from termites, wetlands, and rice paddies is not well determined.

7.2.5 Solar Radiation

Solar radiation is the primary driving force for all plant growth and can be the limiting factor, particularly in the early morning and late in the afternoon. Any significant reduction in solar intensity during these periods, whether due to increased atmospheric particulate matter or increased cloud cover, results in reduced plant growth. Reduced sunlight has some benefit in that evapotranspiration would be reduced, thereby reducing water requirements; however, photosynthesis and thus growth would also be reduced.

The spectral quality of light is also important. The most widely discussed change in spectral quality due to possible climate changes is the increase in the intensity of ultraviolet radiation due to a reduction of the upper atmospheric concentration of ozone. Increased ultraviolet radiation can adversely affect both animals and plants. For animals, ultraviolet radiation is related to the incidence of skin cancers and for plants, ultraviolet radiation can affect reproductive development. Ultraviolet radiation may affect photochemical reactions of chemical compounds used in agriculture for cultural and management practices. These reactions are only recently being investigated.

7.2.6 Combined Effects

An understanding of the effects of climate change on agricultural production is complicated by the likelihood that several climate variables will change and that there may be variations with time. Most experiments on seasonal crop responses have included only one (e.g., carbon dioxide) or at most two variables (e.g., carbon dioxide and temperature) (Acock and Allen, 1985; Kimball, 1983; Cure and Acock, 1986). Because of this, crop simulation models have been used to estimate the possible changes in production and water use by crops growing in various regions under future climate scenarios (Curry et al., 1990; Rosenzweig, 1990). These crop simulation models integrate the effects of temperature, solar radiation, carbon dioxide and water availability on photosynthesis, development, dry matter growth and partitioning, and evapotranspiration to estimate seasonal yield and water use by crops (Reynolds and Acock, 1985; Jones and Ritchie, 1990). Results from these models suggest that significant changes in U.S. crop production will occur if climate changes as suggested by the GCM's and that regional increases in irrigation needs and shifts in production will occur (Adams et al., 1990). Although more experiments are needed to test and further develop the crop models, they are serving as tools to provide first-order estimates of climate change effects on agriculture.

7.3 POTENTIAL ADAPTATIONS BY AGRICULTURE

7.3.1 Crop Plant Production

Many of the climatic factors discussed can either enhance or degrade agricultural productivity, particularly plant growth. Changes that would increase production would generally be deemed advantageous and no changes would be required. Changes that improve quality would also be advantageous. In contrast, any change that would reduce the quantity or quality of crop production must be dealt with if agricultural productivity is to be sustained.

The selection, breeding, or genetic manipulation of new varieties adapted to the changed climatic condition would be one alternative. If the cropping season is shortened or lengthened, varieties that are adapted to the changed season could be grown. The same is true of a change in the mean daily temperature. Some varieties could be shifted from other parts of the world to match the new temperature regime. This alternative may require additional breeding for disease, pests and water stress tolerance and photoperiod response (IPCC, 1990). There is a limit, however, to the level of change that can be overcome by genetic manipulation. Timeliness is critical because standard plant breeding techniques require many years. Changes in climate could also provide favorable conditions for crops that previously could not be profitably grown.

From an engineering point of view, modifying the level of the atmospheric variable(s) that limits plant growth can be another alternative. For instance, if reduced sunlight retards plant growth, sunlight can be supplemented with artificial lighting. This practice is currently used in off-season production of high-valued plants in greenhouses. Supplemental lighting may be used to increase overall light intensity above the plant threshold level so that photosynthesis exceeds respiration. Lighting can also be used to extend the day length. Low levels of light for short durations can actually cause large changes in plant development due to the photoperiod effect. This can be advantageous where warm temperatures cause rapid plant development.

Carbon dioxide supplementation has also been used effectively to enhance growth on selected crops within plant growth structures. Carbon dioxide supplementation is only effective when light levels do not limit growth. Sources of carbon dioxide used successfully include bottled CO_2, discharge of exhaust gas from methane burners, and biological production from the decay of organic matter.

Day and night temperatures can be managed in greenhouses to enhance the relationship between photosynthesis and respiration. Night temperatures can be increased following a bright sunny day when photosynthesis was enhanced or reduced following days when photosynthesis was low. Similarly, day temperatures can be adjusted to maximize the desired plant response.

When rainfall is inadequate or when intense evapotranspiration results in low soil moisture, irrigation can be used if a water source is available. Similarly, if rainfall is excessive, drainage can be used to reduce soil moisture and minimize the duration of soil saturation conditions and the lack of adequate soil aeration. Soil water can also be affected by limiting evapotranspiration. This can be done by applying soil surface coverings such as mulches, including previous crop residues, plastic film or fouler chemicals that reduce transpiration. If such chemicals reduce transpiration by restricting the stomatal opening, plant growth may also be reduced because of the restriction of carbon dioxide movement into the leaves.

As indicated earlier, plant growth is affected dramatically by climatic extremes, particularly freezing. A genetically developed bacterium has been developed which reduces freeze injury on strawberries and other crops, but this material has not been approved for general use. A wide variety of freeze injury techniques have been developed and are used to protect crops in such states as Florida, California and Georgia. Irrigation, foams, heaters, air inversion fan systems, helicopters, and plastic row covers are some of the present protection techniques. These are generally only economical for highly valued crops due to the expense of installation and operation of such systems.

For highly valued crops, shading is sometimes employed to reduce heat stress. Mixed-or inter-cropping is another method where one crop shades another. Irrigation can also provide cooling if water is available.

Hail damage to plants is also a serious consequence of extreme weather conditions. Severe hail can defoliate a plant completely. Little can be done to protect field crops from hail. Greenhouse crops will not be damaged unless the hail is large enough to break or puncture glass or plastic coverings.

7.3.2 Animal Production

As with plants, different animal species can be raised that are better adapted to climate changes, particularly elevated mean temperature, temperature extremes, and high atmospheric humidity. A change in species may also be desirable if climate changes cause a change in the available forage composition within the area.

Animals are sometimes housed to reduce cold weather stress. Such housing can be simple and animal heat production is usually adequate to raise the temperature within the structure to levels that reduce cold stress. Housing or shading can also be helpful in hot periods to reduce heat stress. Air movement is also used to prevent heat stress. Ventilation can be provided either naturally or by the use of fans. The protection of the animals from direct sunlight by shade can be very beneficial. In many regions of the U.S., but particularly in the dry, low humidity areas of Southwestern U.S., the addition of evaporative coolers or misting systems is used to reduce heat stress.

The most stressful climate for animals is high temperatures combined with high humidity. Mechanical air conditioning, which cools and dehumidifies air, can be very effective, but frequently is not economically feasible.

Heat stress in animals can also be reduced by changing the feed ration. A high energy, low roughage feed will provide the necessary energy for digestion while minimizing heat production. Production practices of reducing animal density, decreasing feed intake, and moving animals to market at lower weights are other methods of decreasing heat stress problems. Producers can also avoid transporting animals to market during high temperature periods or they can use new transportation methods that minimize heat stress. Refrigeration during transportation of eggs, milk, and meat preserves quality. Earthen cooling techniques to protect the quality of animal products is a potentially inexpensive cooling strategy that needs development.

7.3.3 Quality of Stored Products

The quality of stored agricultural products is influenced by temperature, humidity, and gaseous composition. Increased temperatures in combination with high humidities can be particularly harmful to grain products stored in on-farm storages or grain elevators. Under these conditions, formation of molds, including aflotoxin, will be enhanced. Even if mold growths are prevented, high temperatures increase respiration, which leads to deterioration in grain quality.

Problems with vermin and insects also increase with increased temperature and humidity. Problems with these two pests are greatly reduced by cold temperatures.

Fluctuating temperatures can also create problems in grain storages. Cool fronts generated within the stored commodity as temperatures fall generate condensation fronts within the grain mass. Wetting due to condensation in turn increases spoilage. Aeration of the stored grain to eliminate or minimize thermal gradients can eliminate the danger of condensation.

Storage of forages is also affected by environmental conditions. If harvest occurs during rainy or high humidity periods, forages may require an extended drying period before storage. This prolonged drying time in the field increases the danger of rainfall damage and increased leaf loss. Storage of forages at high moisture content is one method used to minimize the nutrient harvest loss. High moisture storage can also be utilized with grain crops. In both instances the process involves the exclusion of oxygen from the stored material until the product is fed. High moisture storage is seldom used when forage or grain is intended for sale and utilized off the farm. Chemical additives are sometimes used to retard the development of molds within forages.

Drying systems that use some form of heated air to reduce the moisture content of harvested grain or forage are common in many areas of the U.S., particularly the more humid regions. These systems are effective, but result in a higher product cost. The cost per unit of nutrient stored may, however, be lower with heated air drying, systems than with natural drying due to the potentially high nutrient loss during field drying.

7.4 ENGINEERING ISSUES RELATED TO THE IMPACTS OF CLIMATE CHANGES ON AGRICULTURE

7.4.1 Plant Systems

In the discussion of the influence of climate on plants, the opportunity for plant geneticists to develop varieties or cultivars which can favorably respond to the changed climate was presented. Engineering can play a role in the application and development of plant genetic manipulation procedures. Engineers are involved in basic gene transfer methods of microinjection and electroporation. A novel "shotgun" approach to genetic engineering was developed by an engineer and biologist to fire millions of tiny metal (tungsten) pellets coated with DNA into plant cells at 1000 miles per hour. The pellets are so small they do not injure plant tissue as they move through the cell walls. This technique has been used to fire bacterial and viral genes into onion, corn, and eggplant cells. Future applications include rice and animal cells. Electroporation is a technique for gene transfer where cells are mixed with DNA in solution and subjected to a brief pulse of electric current. Both the gene gun and electroporation represent biological techniques which engineers and scientists can use to insert individual genes for specific traits for adaptation to climatic change. Engineers can play an increasing role in the effective use of the tools and resources used by the plant geneticists. This includes sophisticated instrumentation for expanding the number of new varieties and for evaluating genetic material.

Engineers can also play a major role in developing simulation models for predicting plant response to projected climate changes. Such models, if reliable, can indicate the magnitude of changes in plant growth and yield as climatic variables change and thereby indicate opportunities for mitigating the negative impact of changes or the opportunity for enhanced plant growth. Because all models depend upon mathematical algorithms to describe accurately the response of plants, and because engineers are specifically educated to use physical models and mathematical procedures to describe phenomena, engineers can contribute substantially to the development of agricultural and biological system simulation models. These models can help define agricultural and biological strategies on a farm and/or a regional basis that provide for the efficient use of energy, nutrients, and water and that do not contribute to environmental degradation.

Engineers have also been principal scientists involved in the conception, design, and installation of improved water management systems. Though irrigation, drainage, impoundment, and water management techniques are highly developed, continued opportunities exist for improved efficiency of such systems. Competition for water is intensifying, and improved efficiency will be a key if agricultural productivity is to be sustained should reduced precipitation or increased evapotranspiration be a result of climate change. Increased storage and reduced water loss from storage are methods that will need additional attention. Also storage within the soil profile resulting from reducing runoff and increasing infiltration, on-farm reservoirs and major impoundments will need to be considered.

Improved water use efficiency is another area needing special consideration. This includes the use of techniques that reduce evaporation, minimize deep percolation below the rooting zone, and enable irrigation at the proper time and place in the appropriate amount to maximize water use efficiency. Evaporation from the soil surface can be reduced by using ground covers, by modifying the soil surface, and by increasing the boundary layer resistance to water vapor transfer at the soil surface. Transpiration from plant surfaces can be reduced by coating plant surfaces, and by techniques that reduce stomatal opening. An increase in atmospheric carbon dioxide levels would allow this technique to be more effectively utilized because a carbon dioxide enriched atmosphere promotes a higher diffusion rate into leaves through smaller stomatal openings while restricting water vapor movement from the leaves.

Optimal timing of irrigation will depend upon well-formulated simulation models which in addition to assessing plant response to various levels of soil moisture include economic criteria that permit projection of economic cost and benefit of watering alternatives. Such models would optimally be used with daily or hourly input of local climatic data and would control irrigation equipment. Another technique to control precise water application is the development of a microminiature sensor to sense plant water needs.

Deep percolation is restricted naturally by impervious soil layers. Procedures for restricting excessive water loss while allowing for the leaching of undesirable salts need additional study. The restriction of deep percolation should not, however, be instituted at the expense of impacting adversely either the quantity or quality of water. Deep percolation of water can also be reduced in irrigated soils by better irrigation management and by use of irrigation systems that permit application of water uniformly across the entire field as contrasted to the highly variable application that occurs with some irrigation systems.

Engineers have played a key role in the design and development of freeze protection systems. Such systems can be very energy intensive, particularly when air currents cause a rapid transport of heated air away from the crop being protected. Though many systems are available, and procedures for design and operation are understood, improved designs are required and totally new methods should be studied. Use of infrared heaters, new and innovative covering materials, new methods of applying and removing covers on a daily basis, methods of applying foams or freeze protecting bacteria, and other innovations need investigation.

Though carbon dioxide (CO_2) supplementation has been practiced in plant growth structures, the cost of CO_2 has restricted its use to a few selected crops. Wider adaptation will depend upon cheaper sources of CO_2. If carbon dioxide enrichment is to be practiced with field crops, techniques must be developed to prevent the dissipation of the CO_2 from the crop canopy into the atmosphere. Control of air currents by using covers or plant foliage may have a role. Other innovations need to be investigated.

Lighting strategies for both photoperiod control and increased growth deserve more study. Simulation modeling of plant response with economic considerations should precede any intensive work in this field, but a continuous evaluation of lighting alternatives needs to be undertaken as new advances occur in lighting technology.

Effective use of day/night temperature control to optimize plant growth depends upon the accurate simulation of carbohydrate production and the determination of respiration. Temperature control to optimize both photosynthesis and respiration processes requires engineering design and management. Though temperature control is straightforward in plant structures, it is seldom attempted in field settings. Yet some of the same processes used for freeze control might be effective for field temperature modification.

To maintain desirable soil tilth, maintenance of organic matter in soils is required. One effective means of doing this is through periodic (at least annual) incorporation of organic matter into the upper soil horizon. High soil temperatures combined with normal soil moisture will, however, result in rapid decomposition, of organic matter. The rate of decomposition is however, affected by the type of organic matter involved. The optimal level of organic matter breakdown is needed for developing models useful for identifying practices that sequester carbon in soil and optimize crop production.

Finally, soil structure and depth affect plant growth. Though required plant nutrients can be added to soils lacking natural fertility, the cost of producing organic material in such soils may be high. A loss of topsoil represents a decrease in plant productivity and an economic loss. Control of erosion from wind and water has, therefore, been pursued since the earliest history of farming. Much is known about the processes of erosion of agricultural soils and on the design of structures and cultural practices that can assist in controlling erosion. Full use of such structures and practices has not, however, taken place. Because design and best management practices are site specific, engineers will be needed if, due to climatic changes which intensify the erosion threat, more aggressive steps need to be implemented.

7.4.2 Animal Systems

As is true with plant systems, the engineer is a valued member of any team devoted to the development of production systems for animal species that will operate better under modified climatic conditions. Engineers can design housing in which animals are essentially disease and stress free. To this end, they can develop instrumentation which monitors health, reproductive, behavioral, and growth status. They can design laboratory tools and apparatus for rapid, accurate evaluation of animal tissue and serum.

The development of new animal species may not be required. Animals are currently grown in a wide geographical range with the nature of the animal food supply often being the determining factor in the selection of the animal species being grown. The impact climatic change has upon plants grown within a geographic region may well, therefore, dictate the nature of animal agriculture more than the desires of the farmer, markets for the animal produced, or the location of processing facilities.

As with plant systems, a major effort is likely to be devoted to development of modified animal systems to maintain or improve productivity if the climate changes. Improved simulation models which can be used to assess animal performance under varying climate conditions must be developed and validated. Much work has been done in this field, but existing models cannot describe the animal production cycle from birth to death, while considering all factors that influence production. Current models project rates of gain and feed efficiency in response to major climatic variables such as temperature and humidity. These models can also handle a wide variety of feedstocks, but they cannot handle all possible feeds, especially new, underutilized feeds or feeds modified by some new processes. For instance, there is a recurring interest in feeding waste products of ruminants to nongastric animals and vice versa. This system cannot be simulated at present.

All animals are adversely affected by stress. This can be climatic, disease, social, noise, etc. Housing is one method of minimizing animal stress. Housing, however, requires an expenditure of energy for environmental control, feed and watering systems, waste management, and a method for handling animals. Each of these inputs represents a cost, which can only be justified if the return in increased animal productivity is sufficient to pay for the added cost. Developing a procedure for estimating the true costs and benefits of housing or environmental modification will require engineering input. Once these procedures have been developed, continued efforts to develop improved housing systems are needed.

In addition to looking at animal facilities from the viewpoint of the impact upon animal productivity, the engineer needs to play a major role in evaluating the use of buildings to reduce the negative perceptions of the public relative to animal agriculture. The number of people with on-farm experience continues to decline. Individuals are less willing to accept odors, dust, insects, and pesticides emanating from animal production intruding upon their personal environment.

Engineers can design housing and facilities systems that fully control each of these potential sources.

The reproductive cycle of animals is influenced by environmental factors. It is known, for instance, that the egg laying of chickens is influenced by light and that the virility of male sheep is influenced by high temperatures as well as the photoperiod. The fertility of a cow may be influenced by the quality of the roughage available. The engineer works with the animal scientist to design housing systems or field units that enhance the reproductive response. In addition, sensors can prove useful to detect estrus. Environmental modification of the total structure is not necessary in all cases. The male sheep can be kept cool in a small air conditioned shelter. Milk cows can be cooled by being sprayed or misted with water under shade. Swine can be provided wallows. Each of these alternatives can reduce heat stress and lead to better animal performance. To design such systems, the physiological and behavioral response of the animal to environmental stress must be understood.

The engineer also plays a significant role in exploring methods for utilizing non-traditional feedstocks. Fermentation, thermal, or acid degradation of coarse unpalatable feeds may make such feeds useful. Such modifications will be especially necessary if the consumption of red meat continues to decline for health and consumer preference reasons. Currently, the direct use of forages by chickens and swine is very limited, yet there are vast tracts of land that are not adapted to row crop agriculture but are suitable for forage production. Forages currently are processed into human food through ruminants animals. If the number of ruminants were to decline significantly, some means of using this biological material would need to be found. Pigs and chickens are inherently more efficient in converting feed into weight gain. A process of converting forages for suitable monogastric animal or human food may be a means of significantly expanding the worldwide food supply. Another alternative is diversion of forage land to biomass energy production.

The engineer also plays a major role in the handling and use of agricultural waste. If climatic changes occur that alter rainfall and thereby soil moisture conditions, or if frozen conditions prevail for longer periods of time, the impact upon waste pollution problems will likely intensify. Heavy rainfall after spreading wastes on the soil surface or upon feedlots can result in runoff containing fecal waste. If deep percolation occurs, particularly in the Karst areas of the country, the waste laden water may enter underground aquifers, seriously contaminating water supplies. If waste ends up frozen on the soil surface it will remain in an exposed state and be subject to transport by rain or melting snow into streams or impoundments. Livestock waste can add desired organic matter to soil when properly incorporated. The manner in which it is introduced is critical if waste is not going to be transported through the soil profile by infiltrating water. These problems can be severe, but the processes involved are understood by appropriately trained engineers and procedures are available for negating undesired events.

Engineers can develop totally controlled and contained waste management systems where none of the biologically active waste is misapplied. Such systems are currently expensive and the benefits of the waste as a soil amendment are low. Less expensive systems where the products produced during the waste digestion process are utilized need to be developed. Work is also needed to enhance the production of these products which have the highest potential for use and which present the least environmental impact. For instance, the production of methane in an anaerobic waste system should not be emitted to atmosphere. The trapping of nitrogen components in the waste in some form such as ammonia or urea should also be pursued so these products could be collected and used for fertilizer.

7.4.3 Aquaculture Systems

Engineers have not played an extensive role in the development of aquaculture systems. They have assisted in the design and construction of ponds or channels but have contributed less to the development of aeration, feeding, and pond management equipment. It is a testament to the skill of farm operators and local craftsman that the industry has shown rapid growth during the past several decades. Engineers should contribute to the design of equipment for the measurement and control of water quality determinants such as oxygen level, temperature, pollutants, and waste products.

Engineers are working on new designs of modularized, contained systems. These systems include recycling water to develop controlled environmental conditions. Mathematical capabilities and an understanding of physical principles make engineers key contributors to development of simulation models of aquacultural processes. They are able to guide physiologists and marine scientists in defining variables that need to be quantified to model and design aquaculture systems.

7.4.4 Product Storage

The storage of farm products both on-and off-farm is crucial due to the seasonal nature of agriculture and the cyclic occurrences of devastating climatic variations. Droughts, periods above or below normal temperatures, early fall or late spring freezes, or major floods are common occurrences. The storage of produce is essential to provide food and animal feed during periods when supply directly from the field is interrupted.

Product protection must continue from the point of product maturity, through harvest and transport to storage, within the storage and to the consumer. Adverse climate in terms of excessive moisture, high humidities, or very high temperatures can significantly reduce or endanger product quality. The protection of products during harvest and transport to the storage facility can be difficult for the farmer and better systems are needed. Once the material is in storage, engineered systems can permit aeration, fumigation, drying, or cooling to maintain product quality. Though systems are available for these purposes, changes in climate may expand the need for storage and make additional facilities and capabilities necessary. For instance, forced air drying of hay with unheated air may be practical when rainfall is limited and ambient humidities are low. However, if rainfall becomes more extensive and ambient humidities increase, heated air drying may be required.

Similarly, potatoes can be held in unheated storage in northern states due to cool nights leading to acceptable storage temperatures. An increase in daily mean temperatures particularly in the fall could necessitate the use of storage with a greater control of temperature within the storage.

7.5 REDUCTION OF AGRICULTURE'S CONTRIBUTION TO ATMOSPHERIC CHANGE

7.5.1 Carbon Dioxide

Agriculture is both a primary producer and a primary sink of atmospheric carbon dioxide. Plants utilize atmospheric carbon dioxide in the photosynthesis process to produce carbohydrates. Subsequently, carbon dioxide is released as plants utilize carbohydrates in the formation of plant tissue. During growth more carbon dioxide is captured than is released. The net carbon stored in plant tissue can be released subsequently by burning or other processes which result in an oxidation of plant carbon and a release of carbon dioxide. Long-term reduction in the carbon dioxide level in the atmosphere can be achieved if plant tissue carbon is not recycled back into the atmosphere. Trees, where the volume of timber per unit of land area increases, represents one system where the net carbon stored on an annual basis exceeds that released into the atmosphere. Even when wood is used as a fuel source and the sequestered carbon is released, a benefit is derived because the biomass is renewable with a subsequent recapturing of an equivalent amount of carbon from the atmosphere. This also is true for biofuels such as ethyl alcohol and vegetable oils. When an equal amount of energy is obtained from fossil fuel, the released carbon is not sequestered.

Agriculture therefore plays a positive role in reducing atmospheric carbon dioxide if the cultivation of crops and plant systems where long-term storage of biological material occurs is expanded. The maximum impact will be provided by the growth of rapidly growing softwoods where the largest amount of annual growth occurs. Similarly, increasing fertility, optimizing water availability, controlling pests, and implementation of cropping practices which enhance growth will all contribute positively. Once a stand reaches maturity with annual recycling equaling the annual growth, contribution of the stand in reducing atmospheric carbon dioxide will cease.

Considerable carbon is also stored in soil, principally as humus. Practices are needed to increase the storage of soil organic carbon, not only to help reduce atmospheric levels of carbon dioxide, but also to increase soil fertility.

7.5.2 Methane

With many plant systems aerobic degradation of refuse plant material is not a recycling process. Rather, material is broken down through an anaerobic process. In anaerobic processes, carbon is released as methane gas rather than as carbon dioxide. Agriculture's contribution to the emission of methane to the atmosphere is shown in Figure 7.2 and discussed in Chapter 2.

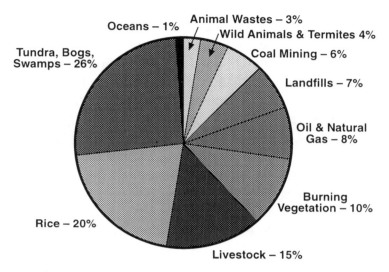

Figure 7.2   Estimated shares of atmospheric methane.

Anaerobic breakdown of plant tissue occurs in the gut of ruminants animals, in the stomach of termites, and in the decay of plant material submersed in water. Decay of biological material in wetlands and in rice paddies results in significant quantities of methane being produced. Though the amounts of methane produced can be substantial, not all methane produced reaches the atmosphere. Some of the methane is utilized by microorganisms in the soil, and there are a number of cultural practices that can be used to reduce methane production (EPA, 1991b).

Anaerobic digestion of plant material and the subsequent release of methane can be reduced by modifying the type and amount of forage fed to animals. It must be remembered, however, that the dominant method of converting forages into human food is through ruminants. If the same level of forages continue to be produced and is not fed to ruminants, it must either be converted to feed for non-gastric animals, used as green manure for crops, or anaerobically digested for methane; otherwise it will be aerobically digested with the subsequent production of carbon dioxide. If carbon dioxide production is favored over methane production, this may be a viable alternative. A reduction in methane production by termites would require either a control over the termites or a reduction in the amount of wood available to termites. The latter could be accomplished by the clean-up of fallen timber or timber clippings produced during logging operations.

7.5.3 Water

The atmosphere holds a substantial amount of water vapor. Agriculture has impacted the amount of water vapor in the atmosphere in large areas of the U.S. and the world through irrigation. In dry areas where intensive irrigation has been practiced, increased rainfall has occurred downwind from these areas and mean atmospheric humidity has increased. Restriction of evapotranspiration would result in a reduction in water vapor released into the atmosphere. Ground covers, foliate sprays, and irrigation systems which apply water efficiently and reduce temperatures are all methods that can be used to reduce evapotranspiration.

7.5.4 Dust and Particulates

Agriculture is a major producer of airborne particulates. Soil left bare is subject to wind erosion with resulting dust storms. The process is aggravated during tillage and planting operations when the soil surface is dry, and hence, more readily transported by wind. Ground covers and management to keep the soil surface moist when vegetation is absent can be practiced. A method currently practiced is conservation tillage where reduced or no tillage and crop residue management minimize the amount of exposed soil, and thereby, minimize soil erosion.

Substantial dust can also be generated during harvest operations, particularly for forages and grain. Though the dust intensity may be large at the point of machinery operation, the total discharge would be small when compared to that which can be generated through wind erosion over a large area of exposed, bare soil. Because the impact in terms of climate change is small little change in harvest practice would appear necessary.

The burning of wood, grain, or forage residue can create a large discharge of particulates into the atmosphere. Forest fires have resulted in serious reductions in sunlight many miles away. These discharges can most easily be controlled by the elimination of grass or straw burning and the effective control of forest fires which do not contribute to improved forest stands.

7.6 RESEARCH GOALS AND OBJECTIVES

The agricultural sector has as its first responsibility the continued production of food, fiber, and forest products for the sustenance of humankind. Regardless of the magnitude of change in climate, food, the most basic need of humankind, must continue to be made available in sufficient quantity and quality to meet the nutritional needs of the global community.

Beyond supplying food, fiber, biofuels and forest products, the agricultural sector has a responsibility to reduce its contribution to atmospheric degradation and contribute to environmental improvement.

Agriculture's contribution to greenhouse gas emission, and its capacity to affect those emissions, is discussed in the preceding sections. In addition to reducing undesirable emissions by adopting appropriate practices, agriculture can also increase the amount of carbon sequestered in the soil profile or in biological materials thereby reducing the level of atmospheric carbon dioxide.

Finally, agriculture must conduct its activities while acting as a good steward of the natural resources. The concept of sustainability of the world's resource base, which assures a viable ecosystem for future generations, is a goal that all of agriculture must embrace.

The specific goals and objectives adopted at this workshop which reflect these concerns and societal responsibilities are:

7.6.1 Goal 1

To provide food, fiber, forest products, and energy for the global community (projected to be 10 billion people in the next century) through ecologically sustainable systems while enhancing the environment.

1. Design and demonstrate ecologically sustainable rural communities (20,000 to 50,000 people) and transfer this technology to the global community.

Biomass for energy production; food production via plant and animal systems; recycled nutrients, water, residuals, and carbon; manufacturing and processing food and non-food products; land use planning; forests; retention of wetlands and natural ecosystems; landscape ecology; local governments; economic infrastructure; social infrastructure (schools, parks, recreation, etc.); business groups; high technology industries (biotech, electronic, etc.); health services; transportation services; service industries; communication systems; energy conservation practices; etc. are proposed to be integrated into a sustainable community based on carbon sequestering and recycling. The objective is to design, develop and demonstrate ecologically sustainable communities (approximately 20,000 to 50,000 people/community) and transfer this knowledge to the global community. Some examples of anticipated subsystems to be applied are:

- wood and cellulose to ethanol
- agricultural, municipal, and other residuals to bio-fuels for manufacturing and processing industries, transportation etc.
- recycling of water, nutrients, residuals, and carbon
- co-generation systems for electrical and thermal energy conversions
- compatible and interdependent business enterprises
- revitalization of rural communities

2. Build a national Biotron facility wherein atmospheric and soil parameters can be controlled to develop quantitative relationships for describing plant, animal, and other biological and organismal responses to global climate change.

A next generation controlled environment facility is needed to quantify relationships that describe plant, animal, and organismal responses to global change. Today's phytotron facilities do not allow for simultaneous control of key atmosphere and rhizosphere variables. Rhizosphere studies will be used to develop and validate needed biophysical models and test sensors and control devices. Of primary importance will be the ability to simulate infrequent, but extreme environmental conditions. Present-day phytotrons are not designed to reproduce the conditions which frequently lead to catastrophic losses because of an extreme, short-term event or a series of less extreme conditions.

Scale and flexibility will be among the important design features enabling the Biotron facility to accommodate diverse and innovative plant and animal system studies.

3. Develop and validate process models for predicting agricultural and biological responses to climate change and for assessment of climate change, decision making, and feedback information for climate modeling efforts.

Process models will be developed and improved for predicting the effects of climate, soil, and management practices on the performance of agricultural and biological systems. These models will perform the carbon, water, and nutrient balance calculations necessary to describe crop and animal productivities under existing and anticipated conditions and to provide more precise earth surface feedback responses to climate modeling efforts. Development and testing of these models will rely on biotron experiments that span the anticipated ranges and combinations of environmental conditions and on field research and monitoring efforts.

Decision support systems will be developed in which the agricultural and biological models will be integrated with other models (water resources, economics) and with spatial data bases of soil, weather, land use, and other land features. The decision support systems will assist planners and policy makers in assessing the impacts of agricultural practices on regional production, net carbon gains, net energy use/production and on the types, amounts, and distribution of residuals. The decision support systems will also be used to design farming systems that recycle carbon, nutrients, and water and to assist in the design of ecologically sustainable rural communities.

The models will also be used as decision aids that assist managers of agricultural and biological systems in improving the efficiency of resource use for food, fiber, and energy production.

4. Coordinate the development of international and national monitoring networks for standardized climate and biological variables.

A coordinated land and satellite-based global monitoring system is needed to provide a data base of climatic and biological variables from which assessments and predictions of global climate change can be made. This knowledge is essential for developing strategies to sustain the biosphere. In general, existing climate monitoring networks are not designed nor intended to monitor variables paramount to interpreting biological systems response to climate change. Equally, if not more important, is the total lack of any monitoring network to observe biological responses. Such a monitoring undertaking is essential. Unfortunately, although many attempts are now in progress, they are uncoordinated. Also, it is uncertain to what extent climatic variables to be measured are of biological significance. Conversely, the impact of biological systems on climate change can be ascertained.

5. Integrate farming systems for efficient utilization of resources through recycling nutrients, water, and carbon, and use of new and mixed species production systems.

Natural resources need to be used more efficiently for food and fiber production in response to global climate change and to enhance the environment. The integration of farming systems includes double cropping of various plant species to supply nitrogen through biological nitrogen fixation and provide mulch and plant shading to modify temperature and conserve water for the crops. Management systems will also be developed to increase soil carbon and to sequester carbon in woody plants. As the climate changes, new gene pools or plant and animal species for food, including meat, eggs, and milk production, from warmer climates will be adapted to higher latitudes through the development of new facility designs and management systems. Wastes resulting from fish farms can be used as nutrients for crop production. Wastes from animal production, e.g., poultry, can be used as a feedstock for cattle. These systems require the engineering of materials handling and processing systems.

6. Develop systems for energy production from energy crops and on- and off-farm wastes.

Energy produced from crops, agricultural and forest product residues, and municipal wastes can play an important role in reducing demand for fossil fuels. Additionally, emissions from biofuels have little sulfur dioxide and reduced particulates. On-farm technologies will be developed and/or improved to convert biological materials to energy for supplying local energy needs with minimal transportation costs. In addition, systems for large scale production of energy crops and for the conversion of these energy crops and/or farm and municipal wastes into energy sources (methane, vegetable oils, alcohol, etc.) will be developed and demonstrated as a prototype system. This will include the planning of land use, management practices for feeding the conversion facility, transportation, and distribution of the energy.

7. Develop on-farm storage, processing and packaging technologies adapted to anticipated changes in climate.

The potential increase in atmospheric temperatures will dictate better environmental control strategies and methods such as the use of low energy drying of grain and processing and packaging of horticultural crops. The processing and packaging of food on the farm will allow the waste from the processing to be recycled on the farm to provide energy or soil amendments to increase soil organic matter. This technology will decrease the transportation costs and it will also improve soil nutrients, thereby reducing the amount of inorganic fertilizer required to produce crops, improve soil water holding capacity, and reduce soil erodibility. The processing systems need to be evaluated closely to assure that the energy used for low quantity processing plus the lower energy requirements for transportation results in a net energy savings when compared to larger processing plants.

7.6.2 Goal 2

To develop bioremediation technologies to limit or sequester atmospheric gases that exceed desirable levels.

1. Improve forest and halophyte management systems for sequestering carbon dioxide.

The objective of these management systems is to increase the sequestration of carbon dioxide through use of forest and arid lands. Rapidly growing trees are capable of taking up a significant quantity of carbon and sequestering it in wood. Salt tolerant and upland shrubs (halophytes) also have the capacity to sequester carbon in high density woody stems. Such material has a long life and slow decomposition rate. Halophyte species can provide forage, fuel wood and other wood products of value to those regions where they occur.

Research is needed to improve planting equipment, especially for seedlings, that reduces hand labor. Research to improve the design and function of containers for seedlings to enable automated or robotic feeder mechanisms for shoot and tree seedlings is also required.

Some engineering design and research is required to develop land forming technologies to assure tree and shoot survival in arid and semi-arid regions. Technologies for improving the efficiency of cultivation (mechanical and spray) suitable for tree and shrub plantings, until seedlings have closed canopies and reduced competition are also needed.

2. Enhance beneficial microbiological systems for soil environments and waste recycling to increase net carbon content in soils.

The objective is to develop microbiological organisms that have the capacity to decompose organic material and organic waste to provide a net increase in soil organic carbon. It has been estimated that management systems alone may increase soil organic carbon by approximately 0.5%. The goal is to biologically engineer microorganisms to be compatible with the soil system that can increase soil organic carbon content whether through crop biomass or by adding waste materials from other sources.

Research is needed to develop a broad spectrum of micro-organisms that can be introduced into organic wastes. Management systems will include soil surface forming to relocate organic material until it is incorporated into the soil. This will prevent loss by wind and water erosion. Application to the soil must be shallow for aerobic microbial decomposition or deep for aerobic decomposition. The management systems will reflect the processes that result in the highest net soil carbon retention.

3. Develop management systems which reduce methane emissions from rice paddies, ruminant livestock, and lagoon systems.

Agriculture is a major producer of atmospheric methane. It is produced in the anaerobic decay of organic matter such as that which occurs in wetlands, rice paddies, and lagoons. Modification of the microorganism nutrient availability, microbial population and other such techniques need to be studied and/or developed which can reduce the methane produced or which will utilize the methane so that it is no longer released as an atmospheric gas. Substantial methane is also produced by ruminants in the digestion of forages. Research needs to be conducted on the modification of animal diets and ruminant microbial populations which reduce methane.

4. Develop new structural and other engineering components utilizing biological materials to retain sequestered carbon.

The sequestering of atmospheric carbon in biological material will only be of benefit for the period of time that the cellulose structure is maintained. The use of lumber in structures is one example. Other structural and engineered components which consist of biological materials need to be developed to sequester carbon. Paving materials, wood chip structural panels, filler materials, insulation, decorative panels, and flotation materials are just some of the long-term uses which may have potential to provide a useful product.

7.6.3 Goal 3

To develop strategies and technologies for sustainable agricultural and biological systems that optimize the use of water and soil resources in response to global climate change.

1. To use water resources efficiently in anticipation of changes in climate and/or water use policy.

Sensors are required to detect in real time the onset of plant water stress. These sensors would be incorporated into computer-controlled irrigation systems. To schedule irrigations on demand, distribution systems are needed that supply pressurized water continuously. The distribution systems should be of closed-pipe construction to prevent evaporation losses. The irrigation system, such as a buried trickle system, should also minimize soil evaporation losses. The system should apply the water, soluble nutrients, and chemicals for pest control based upon the specific needs of each parcel of land. The irrigation system must also accommodate degraded waters and provide the appropriate degree of leaching. In both irrigated and humid areas, soil surface modifications and crop residue management are needed to efficiently capture and store precipitation (both rainfall and snow) for crop water use to minimize irrigation, reduce water stress, and minimize soil erosion.

2. Develop strategies and technologies to conserve soil resources; improve organic, physical, and chemical properties; and enhance the rhizosphere for crop roots and beneficial microorganisms to adapt to climate change.

Extremes of weather related to global climate change present the greatest challenge to developing strategies and technologies for managing soil resources. Technological adaptations will likely, in most instances, be dependent on a better understanding of the relationships between biological and organic ecosystem components and environmental variables. To capitalize on soil organic systems to enhance the rhizosphere, sequester carbon and increased humus, a more basic understanding of below-ground biota is needed. With this understanding it should then be expected that improved tillage, crop residue and nutrient management equipment and practices can be designed and manufactured.

Current methods developed to minimize soil erosion, such as conservation tillage and terracing, will have to be adapted to ameliorate the effects of climate change. If higher temperatures and reduced rainfall occur, technologies to enhance the rhizosphere will be required to store and conserve water and to minimize the impact of increased soil surface temperatures by mulches and tillage.

3. Modify the plant and/or its microenvironment to minimize drought, temperature, and air quality stresses.

A number of strategies could be employed to minimize stress from various types of climatic change. If water stress is to be minimized, one might:

- withhold water at the least critical growth stage
- spray the crop canopy with various substances to reduce transpiration
- cover or spray the soil surface to minimize evaporation
- remove leaves or a portion of the canopy
- mist the canopy rather than apply copious quantities of water
- breed plants for a larger root to shoot ratio

To minimize temperature stress, possibilities include:
- delay or advance the seeding time
- mist the canopy to induce evaporative cooling
- shade the crop by physical or biological means
- breed plants for heat or cold tolerance
- pollinate mechanically

For protection from degraded air, one could:
- breed plants that close stomata to avoid polluted air
- breed plants that are tolerant to polluted air

REFERENCES

Acock, B., L. H. Allen, Jr. 1985. Crop Responses to Elevated Carbon Dioxide Concentrations. In *Direct Effects of Carbon Dioxide on Vegetation*, eds., B. R. Strain, J. D. Cure, pp. 53–97. DoE/ER-0238. U.S. Department of Energy, Carbon Dioxide Research Division, Washington, D.C.

Adams, R.M., C. Rosenzweig, R.M. Peart, J.T. Ritchie, B.A. McCarl, J.D. Glyer, R.B. Curry, J.W. Jones, K.J. Boote, L.H Allen, Jr. 1990. Global Climate Change and U.S. Agriculture. *Nature*, 345:219–224.

Allen, L. H., Jr., K. J. Boote, J. W. Jones, P. H. Jones, R. R. Valle, B. Acock, H. H. Rogers, R. C. Dahlman. 1987. Response of Vegetation to Rising Carbon Dioxide: Photosynthesis, Biomass, and Seed Yield of Soybean. *Global Biogeochem. Cycles*, 1:1–14.

Allen, L. H., Jr., P. Jones, J. W. Jones. 1985. Rising Atmospheric CO2 and Evapotranspiration. In *Advances in Evapotranspiration*, Pub. 14–85, pp. 13–27, American Society of Agricultural Engineers, St. Joseph, MI.

Committee on Alternative Energy Research and Development Strategies. 1990. *Confronting Climate Change: Strategies for Research and Development.* National Research Council. National Academy Press. Washington, D.C.

Committee on Earth and Environmental Sciences. 1990. *Our Changing Planet: The FY 1991 Research Plan.* A Supplement to the U.S. President's Fiscal Year 1992 Budget. Office of Science and Technology Policy. Washington, D.C.

Committee on Production Technologies for Liquid Transportation Fuels. 1990. *Fuels to Drive Our Future.* National Research Council. National Academy Press. Washington, D.C.

Committee on Science, Engineering, and Public Policy. 1991. *Policy Implications of Greenhouse Warming.* National Research Council. National Academy Press. Washington, D.C.

Cure, J. D., B. Acock. 1986. Crop Responses to Carbon Dioxide Doubling: A Literature Survey. *Agric. Forest Meterol.* 38:127–145.

Curry, R. B., R. M. Peart, J. W. Jones, K. J. Boote, L. H. Allen, Jr. 1990. Simulation as a Tool for Analyzing Crop Response to Climate Change. *Trans. ASAE,* 33:981–990.

Department of Energy. 1990. Effects of Air Temperature on Atmospheric CO2 Plant Growth Relationships. Report TRO48. Office of Energy Research, Office of Health and Environmental Research, Washington, D.C.

Energy for Planet Earth. 1990. Special Issue, *Sc. Am.*, September 263 (3).

Environmental Protection Agency. 1991a. Global Comparisons of Selected GCM Control Runs and Observed Climate Data. Office of Policy, Planning, and Evaluation, Washington, D.C.

Environmental Protection Agency. 1991b. Sustainable Rice Productivity and Methane Reduction Research Plan. Office of Radiation, Washington D.C.

Environmental Protection Agency. 1990. Workshop on Greenhouse Gas Emissions from Agricultural Systems, PCC-RSWG, AFOS Subgroup. Draft Summary Report, Washington, D.C.

Environmental Protection Agency. 1989. The Potential Effects of Global Climate Change on the United States, Report to Congress. Office of Policy, Planning and Evaluation, Washington, D.C.

Environmental Protection Agency. 1989. The Potential Effects of Global Climate Change on the United States, Appendix C Agriculture Volume 2. Office of Policy, Planning and Evaluation, Washington, D.C.

Environmental Protection Agency. 1989. Policy Options for Establishing Global Climate, Executive Summary. Office of Policy, Planning and Evaluation, Washington, D.C.

Flavin, C. 1989. *Slowing Global Warming: A Worldwide Strategy.* Worldwatch Institute. Washington, D.C.

Geophysics Study Committee. 1990. Commission on Physical Sciences, Mathematics, and Resources. National Research Council. *Sea-Level Change— Studies in Geophysics*. National Academy Press. Washington, D.C.

Idso, S. B., B. A. Kimball, 1993. Tree Growth in Carbon Dioxide Enriched Air and Its Implications for Global Carbon Cycling and Maximum Levels of Atmospheric CO2, Paper #93GB01164, *Global Biogeochemical Cycles*.

Intergovernmental Panel on Climate Change, 1990. Potential Impacts of Climate Change, Report from Working Group II to IPCC.

ICF Incorporated, Dec. 1989 Workshop on Greenhouse Gas Emissions from Agricultural Systems IPCC-RSWG Summary Report, Sub Group on Agriculture, Forestry and other Human Activities. Environmental Protection Agency, Washington, D.C.

International Council of Scientific Unions. 1990. The Land-Atmosphere Interface, The International Geosphere Biosphere Programme: A Study of Global Change (IGBP), Report No. 10. Stockholm, Sweden.

International Council of Scientific Unions. 1990. Proceedings of the Workshops of the Co-ordinating Panel on Effects of Global Change on Terrestrial Ecosystems. The International Geosphere-Biosphere Programme: A Study of Global Change (IGBP), Report No. 11. Stockholm, Sweden.

International Council of Scientific Unions. 1990. Coastal Ocean Fluxes and Resources. The International Geosphere-Biosphere Programme: A Study of Global Change (IGBP), Report No. 14. Stockholm, Sweden.

International Council of Scientific Unions. 1990. Change System for Analysis, Research and Training, (START). The International Geosphere-Biosphere Programme: A Study of Global Change (IGBP), Report No. 15. Stockholm, Sweden.

Jones, J. W., J. T. Ritchie. 1990. Crop Growth Models. In *Management of Farm Irrigation Systems*. G. J. Hoffman, T. A. Howell, K. H. Soloman, eds. ASAE Monograph. American Society of Agricultural Engineers. St. Joseph, MI.

Kimball, B. A. 1983. Carbon Dioxide and Agricultural Yield: An Assemblage and Analysis of 430 Prior Observations. *Agron. J.*, 75:779–778.

Lynd, L. L., H. Cushman, R. J. Nichols, C. E. Wyman. 1991. Fuel Ethanol from Cellulosic Biomass. *Science* 25:318–1323.

Managing Planet Earth. 1989. Special Issue, *Sc. Am.*, September 261 (3).

Muchow, R. C., J.A. Bellamy. 1991. Climate Risk in Crop Production: Models and Management for the Semiarid Tropics and Subtropics, CAB International. Wallingford, U.K.

Parry, Martin. 1990. *Climate Change and World Agriculture*. Earthscan Publications Limited. London, U.K.

Public Service Research and Dissemination Program, July 10–12, 1989. Research Needs and Recommendations: Global Climate Change and Its Effects on California, University of California, Davis, CA.

Reynolds, J. F., B. Acock. 1985. Predicting the Response of Plants to Increasing Carbon Dioxide: A Critique of Plant Growth Models. *Ecol. Modeling* 29:107–129.

Rosenberg, N. J. 1981. The Increasing CO2 Concentration in the Atmosphere and its Implication on Agricultural Productivity, I, Effects on Photosynthesis, Transpiration, and Water Use Efficiency. *Climate Change*, 3:265–279.

Rosenzweig, C. 1990. Crop Response to Climate Change in the Southern Great Plains: A Simulation Study, *Prof. Geog.*, 42:20–37.

Smit, B., L. Ludlow, M. Brklacich. 1988. Implications of a Global Climate Warming for Agriculture: A Review and Appraisal. *J. Environ. Quality* 17:519–527.

USDA. 1990. U.S. Department of Agriculture. Global Change Strategic Plan. Washington, D.C.

Utah Agricultural Experiment Station. 1986. Limiting the Effects of Stress on Cattle. Western Regional Research Publication #009, Utah State University, Logan, UT.

Wilks, Daniel S. 1988. Estimating the Consequences of CO_2-Induced Climate Change on North American Grain Agriculture Using General Circulation Model Information. *Climate Change* 13:19–42.

Yousef, M. K. 1985. *Stress Physiology in Livestock. Principles*, Vol. I, CRC Press, Inc. Boca Raton, FL.

Yousef, M. K. 1985. *Stress Physiology in Livestock. Ungulates*, Vol. II, CRC Press, Inc. Boca Raton, FL.

Yousef, M. K. 1985. *Stress Physiology in Livestock. Poultry*, Vol. III, CRC Press, Inc. Boca Raton, FL.

Chapter 8

GEOENGINEERING CLIMATE

Authors: Brian P. Flannery, Haroon Kheshgi,
Gregg Marland, Michael C. MacCraken
Contributors: Hioshi Komiyama, Wallace Broecker,
Hisashi Ishatani, Norman Rosenberg, Meyer Steinberg,
Tom Wigley, Michael Morantine

8.1 INTRODUCTION

One possible response to concerns of climate change from the buildup of greenhouse gases in the atmosphere is to evaluate the potential to offset climate change, or its impacts, through intentional control of climate or of the concentrations of atmospheric greenhouse gases. These approaches are often referred to as "geoengineering." That humans may be capable of altering climate through intentional intervention may seem far-fetched, but technically feasible options exist that appear (in principle and in models) to be capable of reducing or eliminating many of the impacts of greenhouse gas emissions (NAS 1992). However, with today's limited knowledge of the climate system, we urge caution and a focus on research rather than implementation. Geoengineering may well have unintended and unforeseen consequences.

The intent of this chapter is to discuss geoengineering options that may be capable of offsetting the negative impacts of climate change, and to propose research to assess their potential and to improve their chances of being effective. Our strongest recommendations are for continuing efforts to understand better the myriad interacting processes in the climate system and to improve our ability to predict climate change. These are essential prerequisites to evaluation, much less implementation, of any strategy for direct intervention. Increasing confidence in model predictions of climate change, an important criteria for decision making on any response to the greenhouse issue,

379

will lead to increased confidence in the ability to consider geoengineering approaches.

Just as today's climate models offer limited insight into the actual magnitude, timing and regional distribution of any future climate change, they are poor guides to the response of the climate system to intentional manipulation. Nonetheless, the same models that suggest risk from climate change can be used to estimate the sensitivity of the climate system to intentional manipulation. In model simulations it is straightforward to examine the extent to which strategies for intervention offer the potential to offset changes in the model's climate produced by the buildup of greenhouse gases.

Similarly, our understanding and models of the global carbon cycle provide limited guidance on the flows of carbon through the oceans and biosphere. We cannot now predict with accuracy the consequences of intentionally altering major carbon fluxes. At the level of system interaction, we would like to be able to predict how changes in climate would affect carbon flows and how changes in carbon flows would affect climate (e.g., massive reforestation could change surface albedo and changes in temperature could change rates of soil respiration).

Although it would be imprudent to attempt to implement any of these approaches at this stage, serious evaluation of the possibility for intervention offers multiple benefits:

- It seems likely that climate will change to some degree as a result of human actions that have already occurred or to which we are committed. Geoengineering could offer some relief from negative aspects of any change, or could strengthen positive aspects.
- Society may be unable or unwilling to prevent the long-term buildup of greenhouse gases, or climate change may be even more serious than current models predict. If so, it would be advisable to have additional response options available, as insurance, to curtail climate change.
- Studies of the impacts of climate change indicate the vulnerability of society to natural climate variability as well as man-made change. Intentional methods to alter climate may provide relief from future climate variability from natural causes for which control of emissions may offer little protection. Such methods may also present us with options to deal with future cooling as well as future warming.

- Without investigation it will be impossible to assess the merits and demerits of intentional climate modification. Intervention in climate may be more feasible, effective, desirable, and/or less costly than proposals that would alter the world's existing economic structure, which carry their own risk for potentially adverse consequences for society and the environment.
- Pursuit of research into means to alter climate may offer new scientific insight into the operation of key climate processes. Because these processes may be important in determining the course of climate change it is prudent to understand them as thoroughly as possible.
- Finally, decision makers may resist making commitments to reduce the threat of climate change until all feasible alternatives have been analyzed.

In any case, before geoengineering can be considered seriously, a great deal of fundamental research needs to be conducted to assess carefully any scientific opportunities, technical capabilities, and the overall effects of intentional manipulation of climate and concentrations of atmospheric greenhouse gases. For these reasons, this chapter should be regarded as exploratory.

We discuss five general categories of intervention:

1. collecting and disposing of CO_2 from flue gas streams— the disposal aspect involves geoengineering concerns;
2. increasing net uptake of CO_2 into the terrestrial biosphere;
3. increasing net uptake of CO_2 into the oceans;
4. changing the Earth's energy balance by altering the albedo (the fraction of incident solar energy scattered or reflected back to space without being absorbed); and
5. altering internal processes in the climate system.

Society is not without experience with modifications of this sort. Potential climate change from an enhanced greenhouse effect is an example of geoengineering driven largely by human intervention in the geochemical cycle of carbon. Of course, the contributing emissions are a byproduct of essential human activities in energy use, agriculture, industry, etc., rather than a direct intervention aimed at climate modification for its own sake. Also, naturally occurring events such as volcanoes and El Niño perturb climate along lines that may provide heuristic information on certain schemes for climate alteration. There is evidence that anthropogenic aerosols, associated with sulfur dioxide emissions and biomass burning, have influenced climate. By increasing or decreasing sulfur emissions, humans intervene in the climate system. The National Academy of Sciences (1992) suggests that one initial test of climate modification strategies could be that they stay within the range of natural phenomena, e.g., aerosols injectioned into the troposphere would not exceed limits observed during volcanic eruptions.

8.2 THE POTENTIAL FOR CLIMATE MODIFICATION

Estimates for greenhouse gas concentrations, their annual fluxes, and related factors as reported in various sources differ slightly (e.g., for atmospheric concentrations of CO_2 in the mid-1800s) or substantially (e.g., for annual emissions of CO_2 from deforestation). For the most part, at the level of discussion appropriate to this overview of geoengineering, such differences are largely immaterial. We will not seek to resolve inconsistencies among different sources; rather, throughout this chapter different estimates from different sources will be cited, as appropriate to the material under discussion.

For reference (Watson et al., 1990), the atmosphere today contains approximately 750 GtC as CO_2 at a concentration of 353 ppmv (1 GtC = 1 x 10^{15}g carbon, 1 ppmv = 2.12 GtC). In the period 1980–1989 the observed annual increase in the concentration of atmospheric CO_2 was about 1.6 ppmv (or 3.4 GtC), compared with estimated anthropogenic inputs of 1.6 = 1.0 $GtCyr^{-1}$ from deforestation and 5.4 = 0.5 $GtCyr^{-1}$ from fossil fuel use. Between 1850 and 1986, CO_2 concentrations rose from (about) 288 ppmv to 348 ppmv (+130 GtC). Over the same period inputs from fossil fuel use and deforestation amounted to 195 ± 20 GtC and 117 ± 35 GtC, respectively. We do not yet fully understand why the annual rate of concentration increase (in the 1980s) is only 3.4 ± 0.2 $GtCyr^{-1}$ when the estimated human input is 7.0 ± 1.5 $GtCyr^{-1}$, nor do we know in quantitative detail the disposition of the excess carbon, but it is distributed between the oceans and the terrestrial biosphere (Tans et al., 1990; Post et al., 1991).

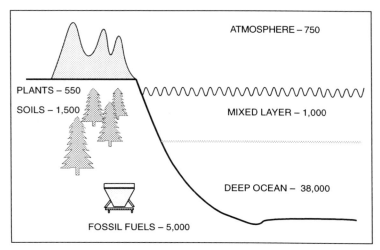

Figure 8.1   Carbon reservoirs capable of significant exchange on time scales of a century or less. (From IPCC, 1990.)

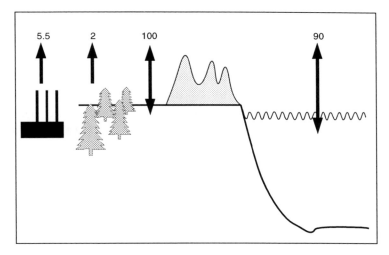

Figure 8.2   Annual exchange rates between carbon reserviors. (From IPCC, 1990.)

Attempts to understand the natural processes that control the atmospheric concentration of CO_2 indicate that levels are maintained by a delicate balance of large exchanges of carbon between the air, marine, and terrestrial reservoirs (Figures 8.1, 8.2). Although the absolute flux levels cannot be measured directly, the annual exchange between the atmosphere and terrestrial biosphere is estimated to be approximately 100 $GtCyr^{-1}$. The annual exchange between the atmosphere and surface waters of the oceans is estimated to be comparable,

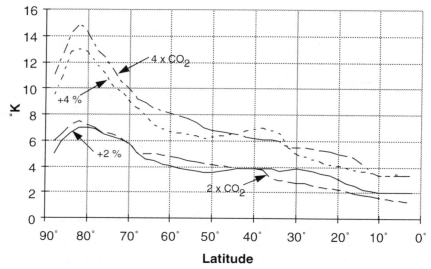

Figure 8.3   Comparison of zonal average temperature changes produced in a simplified general circulation model by increases in the concentration of atmospheric CO_2 and by increases in insolation. (From Manabe, S. and Wetherald, R.T., *J. Atmos. Sci.*, 37, 99, 1980. With permission.)

90 $GtCyr^{-1}$. The underlying mechanisms that control these fluxes are not well quantified, but there are a number of options for planned intervention. If, for example, planning could alter land use to return global forest coverage to levels that existed, e.g., in 1800, then some 100 GtC could be removed from the atmosphere, an amount equivalent to 15 years of human intervention.

A balance between incoming shortwave solar radiation and outgoing longwave infrared (IR) radiation maintains the global heat balance, which energizes the climate system. Atmospheric radiative transfer models estimate that a doubling of atmospheric CO_2 would increase the global average radiation flux at the top of the troposphere by 4.4 Wm^{-2} (Houghton et al., 1990: IPCC). For comparison, a 2% decrease in sunlight would decrease the (absorbed) heat input to the surface-troposphere system by 4.8 Wm^{-2}. (On average Earth absorbs about 240 Wm^{-2} of the 340 Wm^{-2} solar energy flux.) Climate models suggest that equivalent changes in IR and solar radiation induce similar patterns of temperature change (Manabe and Wetherald, 1980, Figure 8.3). As a first approximation, then, anything we could do to produce a 2% decrease in absorbed solar radiation would approximately counteract a doubling of atmospheric CO_2. This suggests that intentional manipulation of Earth's albedo could contribute to limiting global warming. The goal would be to increase albedo at reasonable cost and without introducing undesirable side effects on climate or on atmospheric chemistry.

Attempts to reconcile the historical record of change in global average temperature with models of climate change induced by greenhouse gases show clearly that other factors contribute to the observed variations. For instance, the eruption of large volcanoes, such as Mount Pinatubo, induce cooling by of order 0.5°C, for periods of a few years, through injection of materials to the upper atmosphere which increase the Earth's albedo. Recent evidence, discussed below, suggests that human emissions of SO_2 (Charlson et al., 1991) and biomass burning (Penner et al., 1992) produce aerosols which cool the Earth, by an enhancement of albedo. In both cases estimates suggest that solar heating might be reduced by as much as 2 Wm^{-2}.

8.3 GOALS AND CRITERIA

When exploring ways to engineer the global climate, it is important to consider the ultimate objective. The general perception is that both natural and human systems are adapted to the Earth's current distribution of climates, and that any large or sudden changes would be disruptive and, therefore, undesirable. Acknowledging that climate is continually variable on many time scales and for a variety of geophysical reasons, we would like to focus on proposals that stabilize climate at current conditions, with which we have experience, and to which society and ecosystems are more or less adapted.

When we contemplate purposeful climate manipulation, we need to consider a variety of scientific, technical, economic, and social issues. At this time it is not possible to evaluate how various options rate in a broad range of criteria, but research should improve society's capability to conduct such an assessment. Since geoengineering options could have global, as well as regional effects, criteria need to be developed with a global perspective. Three criteria most often discussed are: effectiveness, cost, and side effects. While these are dominant considerations, we suggest that an additional set of criteria should also be addressed: time to implement; time to achieve potential; technical maturity of approach; control (e.g., is the option reversible, adjustable, and incremental); ongoing commitment to continue the approach (e.g., in perpetuity); owner (who implements, pays, and is responsible for direct and indirect effects); equity (temporal and spatial); winners and losers; ethics (of intentional and unintentional climate change, and of doing nothing). These criteria are particularly relevant to geoengineering options, but they also are relevant to other policies that have been proposed to limit the risk from global change. Our position is that options exist which might prove beneficial to society, and that they require research to improve analyses of their feasibility and desirability. Once these are more clearly defined, such options also require research, education, and discussion to determine the technical and political issues associated with implementation.

8.4 OPTIONS FOR SEQUESTERING CARBON

Here we review mitigation technologies with a view to reducing the appearance in the atmosphere of CO_2 from the combustion of fossil fuel. A number of approaches exist for supplying energy end-use services with lower levels of net CO_2 emissions than at present; many of them are discussed elsewhere in this volume. The question is whether we can continue to use fossil fuels but ensure that carbon byproducts end up other than in the atmosphere as CO_2. The three possibilities addressed in this section are: the recovery and disposal of CO_2 from large combustion facilities, pre-treatment of fossil fuels to produce a carbon-rich fraction and a hydrogen-rich fraction with sequestering of the carbon-rich fraction, and removal and sequestering of CO_2 from the ambient atmosphere by photosynthetic plants. This last section is treated in three parts depending on whether carbon would be stored in terrestrial plants, soils, or oceans. Our emphasis is on the carbon disposal aspects of these schemes.

8.4.1 Collecting and Disposing of CO_2

Because CO_2 can be a useful product in some contexts, and because it is routinely removed from hydrocarbon gas streams and from submarine and spacecraft cabins, there are many developed processes for its separation. Some of these processes are applicable to flue gas streams and there is some experience in extracting CO_2 from flue gases for use in enhanced oil recovery (e.g., Golomb et al., 1989).

There have been a number of studies on various methods for removal of CO_2 from central power station flue gases (e.g., Marchetti, 1975; Baes et al., 1980; Steinberg et al., 1984; Golomb et al., 1989; Hendriks et al., 1990; Herzog et al., 1991) by both physical and chemical means. These have included absorption/stripping with alkanolamines, absorption/stripping in activated charcoal or molecular sieves, membrane separation, cryogenic separation, and air separation/flue gas recycle systems. All of these can be achieved at some cost but the ultimate problem involves disposal of the CO_2 once collected. As well, note that these approaches apply only to large combustion facilities, which account for approximately 30% of global CO_2 emissions today. Relatively few sites have been identified that could handle the massive amounts of CO_2 to be sequestered. The realistic possibilities suggested to date include depleted oil and gas wells, porous geologic substrates, mined salt domes, and the deep ocean. For any source and disposal site the issue of transport also affects cost and efficiency, but appears to present no new technical challenge: CO_2 is already being transported long distances by pipeline as part of enhanced oil recovery projects (e.g., Decker, 1986). The deep ocean seems to be the only reservoir with sufficient capacity to accommodate the very large quantities that might be released from fossil fuels over time. The success of any such disposal scheme depends, of course, on the cost and efficiency of CO_2 collection, but the geoengineering aspects of the problem are primarily in the disposal.

The most recent analyses of processes appropriate for collecting CO_2 from large flue gas streams have been those of Golomb et al. (1989) and Hendriks et al. (1990). Golomb et al. re-examined the earlier literature and summarized five feasible process schemes. Table 8.1 summarizes the efficiency of CO_2 separation, the net CO_2 emissions, and the energy balance for CO_2 collection (with compression to 150 bars pressure) from a coal-fired power plant. Energy demand for CO_2 collection and compression is very large and hence the effect on net plant thermal efficiency is very large. In the extreme case, where CO_2 collection is less than 100% efficient and the power demands are very high, net plant CO_2 emissions per kwh can be higher than for the base case. The most attractive system described by Golomb et al. involves air separation and subsequent coal combustion in a stream of oxygen and recycled flue gas. The process produces a flue gas stream that is 90% CO_2 when dried. Generally, the lower ranges of energy

Table 8.1   Energy requirement comparison.

Process	Energy Requirement		Coal Requirement		Net CO_2 Emissions		Recovered CO_2	
	% of Combustion Energy of Coal	Thermal Efficiency	kg Coal per kWh	Relative to Base Case	kg CO_2 per kWh	% of Base Case	% CO_2 Recovery	% CO_2 Purity
Base Case (No CO_2 Removal)	0	35	0.35	1	0.88	100	0	–
Air Separation/FG recycling	26–31	24–26	0.48–0.51	1.35–1.45	0	0	100	90
Amine Scrubbing	47–79	7–19	0.66–1.68	1.89–4.76	0.17–0.42	19–48	90	99+
Cryogenic Fractionation	55–95	2–16	0.78–7.04	2.22–20	0.20–1.76	22–200	90	97
Membrane Separation	50–75	9–18	0.71–1.41	2.00–4	0.35–0.70	40–80	80	90

requirements in Table 8.1 result from increased efficiency of the processes and involve larger capital investments in equipment. Wolsky et al. (1991) describe additional analyses involving CO_2 collection after coal combustion in oxygen with flue gas recycle. Interestingly, the cost of CO_2 recovery would be greater for oil and gas fired boilers than a coal-fired boiler because the cost is dominated by the air-separation plant and oxygen is required for combustion of both the hydrogen and the carbon in the fuel. Golomb et al. estimate that use of the air separation/flue gas recycle system would add 80% to the cost of electricity and this does not include any cost for CO_2 transport and disposal. The system would result in simultaneous control of both SO_2 and NO_x.

Another process has been suggested to minimize CO_2 collection costs by producing a higher concentration CO_2 stream for separation (40% CO_2 vs. the 10–13% in a typical flue gas). This would involve coal gasification and CO_2 collection prior to combustion (Hendriks et al., 1990). Steam and oxygen gasification of coal produces a CO and H_2O mixture which could be shifted to CO_2 and H_2. The CO_2 would be recovered from the intermediate synthesis gas and the H_2 sent to a combustion turbine. Net plant efficiency of the integrated gasifier, combined cycle power plant would drop from 43.6 to 38.1% (based on fuel lower heating values) as a result of CO_2 collection and compression to 60 bars and the cost of electricity would increase by about 25% (Hendriks et al., 1990). Overall CO_2 recovery would be at least 88% and there would be almost 100% sulfur recovery as well.

Taken together, these studies demonstrate the potential to collect flue-gas CO_2 without adding over 1/3 to the cost of electricity. The frequently envisioned disposal site is the deep ocean. In fact, Marchetti pointed out in 1975 that the CO_2 problem is basically a kinetic problem, that the ocean has a large capacity to dissolve additional CO_2 but mixes and equilibrates very slowly.

According to Baes et al. (1980), the addition of dissolved CO_2 increases the density of seawater (at constant temperature), and a stream of seawater with 1 mol/kg excess CO_2 would sink if injected below about 300 m in the ocean. At 1 mol/kg of excess CO_2, a 1000 MW(e) coal-fired power plant would need to inject about 5200 kg/sec of enriched seawater from either an offshore power station or a long pipe. Dilution of CO_2 along the centerline of the plume varies directly as the 1/3 power of the density excess and inversely as the 2/3 power of the discharge velocity (Baes et al., 1980). Golomb et al. (1989) maintain that seawater is the only practical non-recyclable solvent for CO_2 recovery but that the capital and pumping costs would be very high and the total cost therefore greater than for use of a regenerable solvent and disposal as liquid CO_2.

Liquid CO_2 is more compressible than seawater and has a higher thermal coefficient of expansion. At comparable temperatures, liquid CO_2 should be more dense and hence negatively buoyant if injected below about 3000 m. Whereas liquid CO_2 injected below 3000 m should sink in the ocean, Golomb et al. (1989) show that if a jet is injected at 750 m in a manner that produces drops less than 0.5 cm in diameter, the drops will be buoyant but should fully dissolve in the seawater before rising to the 500 m depth level. Both solid CO_2 hydrate and dry ice are more dense than seawater (1.12 and 1.5 gm/cm^3, respectively) and should sink if dropped into the ocean. For the solid phases, bubbling would cease by 500 m and the decomposition product would then be a CO_2-rich solution. Baes et al. estimated that the energy cost to compress CO_2 to 150 bars and transport it 40 km, to liquefy it, or to convert it to dry ice, would be about 3, 4, and 13%, respectively, of the energy value of the coal. Waters 1000 m deep are available within 100 km of shore for less than 10% of the U.S. east coast, but for other countries, like Japan, potential disposal sites would be more readily available.

There are a number of issues that have to be resolved before we inject large quantities of CO_2 into the deep ocean. For example: for gaseous or liquid CO_2 injected below the thermocline, how long will it take to be re-equilibrated with the atmosphere by natural circulation processes; for solid or liquid CO_2 deposited on the ocean floor, what processes might cause it to be catastrophically released to the atmosphere; and, what are the ecological and ocean circulation consequences of injecting large quantities of CO_2 into the deep ocean?

Possible disposal sites other than the deep ocean include depleted gas fields, depleted oil fields, porous geologic reservoirs and mined salt domes (see, for example, Horn and Steinberg, 1982). These are possibilities that might have locally significant potential. Baes et al. estimated in 1980 that "if carbon dioxide miscible flooding were used for half of present U.S. oil production, the carbon dioxide from U.S. coal-fired utility plants could be consumed." Hendriks et al. (1990) have looked more closely at the opportunities for The Netherlands to

Table 8.2   Technical issues to be resolved for CO_2 collection and disposal.

System Element	Technical Issues
CO_2 recovery	Improve efficiency
Liquefaction	Improve efficiency
CO_2 transport and storage	Location, security, safety, reliability
Deep-sea pipeline	Construction, maintenance
Pipe outlet	Behavioral analysis, optimum depth
Plume behavior	Behavioral analysis, density, diffusion, miscibility,environmental impact

inject CO_2 into natural gas reservoirs. The Dutch have large gas reservoir capacity, including an estimated 2500 x $10^9 m^3$ (at STP) in the Groningen Field, and anticipate the first field exhaustion to occur by the year 2000. Use of the extensive offshore fields would not necessarily become possible at the time of field exhaustion because facilities for gas collection and transport are shared with fields that would still be in production. Hendriks et al. envision piping CO_2 back to the fields as a liquid under conditions similar to those at which the natural gas is now transported (60–80 bars and about 15° C). The Hendriks et al. estimate is that the system could work and that the total capacity for CO_2 storage in Dutch natural gas reservoirs is 40 times the current national emissions rate. Adding disposal in gas wells to the cost of collection would raise the cost of CO_2 handling to 30% of the current cost of electricity.

Table 8.2 summarizes the principal system components for CO_2 disposal and the technical issues that need resolution before the system could be implemented.

Recognizing the difficulty of disposing of gaseous CO_2, Steinberg et al. (1989) have suggested process schemes that would produce from coal a hydrogen-rich and a carbon-rich fraction (carbon black) with subsequent oxidation of only the hydrogen-rich fraction (methanol). The carbon black could be stored in depleted mines and ultimately recovered for use if the enhanced greenhouse effect turned out to be less serious than anticipated. By this scheme only 19% of the energy in coal is available with no net CO_2 emissions but 56% of the energy in natural gas is similarly available. Using such schemes in combination with biomass fuels, Steinberg (1991) envisions the possibility of deriving usable methane or methanol fuel while creating a net withdrawal of CO_2 from the atmosphere. By combining biomass farming and co-processing of biomass and fossil fuels to produce methanol and solid carbon black, CO_2 emissions could be reduced significantly, with some combinations even resulting in net storage of CO_2. The motivation for such processes is to allow society to utilize abundant reserves of coal, even if greenhouse concerns force limits on CO_2 emissions. Of course, these schemes also reduce overall utilization of the energy available in coal reserves and increase costs. Yet another possibility is to remove CO_2 directly from the ambient atmosphere via green plants and to sequester the carbon in either living biomass or wood products. This results in a very different set of possibilities for carbon storage and is discussed further below. We do not consider the direct removal of CO_2 from the ambient atmosphere by chemical processing because Albanese and Steinberg (1980) have shown clearly that the energy costs alone for such a scheme are prohibitive.

8.4.2 Protection and Expansion of Forests

A possibility first suggested by Dyson (1977) and Dyson and Marland (1979) is that we could augment storage of carbon in organic matter by calling on solar energy and the photosynthetic process to extract additional carbon from the atmosphere. It has been estimated that the global biomass currently stores some 1500 ± 200 GtC and that another 1500 ± 200 GtC are stored as organic matter in soils. Annual fossil-fuel emissions of 6 GtC could thus be offset if we could engineer an increase of 0.3% annually in the amount of reduced carbon stored in the biosphere and soils. The most attractive choice appears to be carbon uptake and storage in trees. We know how to grow trees, we know that woody biomass stores a large quantity of carbon with minimal amounts of other nutrient elements, and we can envision a larger mass of global forest. Protecting and expanding forests also promote other social objectives: protecting biodiversity, improving watershed quality, providing other forestry products and recreational opportunities, enhancing aesthetic values, etc.

With current concern over the ongoing destruction of tropical forests, this carbon storage objective is well-served by either establishing forest where forest does not now exist or by arresting the loss of biomass in existing forests. We might also consider means to enhance the ability of existing forests to absorb and store carbon, through improved management practices.

Recent studies estimate that forest destruction, mostly now in developing tropical nations, is a net annual source of 0.4 to 2.6 GtCyr^{-1} to the atmosphere (Dale et al., 1991; see also Watson et al., 1990, who quote 0.6 to 2.6 GtCyr^{-1}). Clearing of tropical forest is driven by forces which differ dramatically among countries and regions, but which include the expansion of crop and pasture land, harvesting of lumber and fuel wood, and the expansion of human infrastructure. Growing populations, large national indebtedness, and aspirations for economic development are problems that will have to be confronted in any effort to limit forest clearing.

In thinking of the role of forestry to offset CO_2 growth, an important issue is the overall target to be achieved. Without clear understanding of the impact of global change one might design targets for specific national goals, e.g., to stabilize net emissions, or to stabilize emissions from a specific sector, e.g., coal fired electrical utilities. This has the advantage that projects can be defined in specific settings and possibly manageable scale.

On a global scale, a (conceptually) simple target is the goal of completely offsetting total human emissions. Several analysts have speculated on the magnitude of new forest area that would be required to offset current rates of fossil-fuel CO_2 emissions. Their conclusion is that $(0.5–1.0) \times 10^9$ hectares of new, fast-growing forest would be required (OTA, 1991). Although this is a very large number, it is not out of scale with the area of existing closed forest (2.8×10^9 ha, WRI, 1990) or the amount of forest (both closed and open) which has already been cleared during human settlement (0.7×10^9 ha, Matthews, 1983). From this gross view, it appears that forestry could play a significant role, even if not a full offset, and it is appropriate to look specifically at the amount of land which might be available and the kind of growth rates which might reasonably be anticipated. This is with full recognition that carbon storage and carbon uptake are separate issues: while a forest can presumably sequester carbon indefinitely, the rate of uptake will decline eventually as the forest matures. There is only a finite time interval over which high rates of net carbon uptake can be counted on. Preserving forests to store carbon is thus a long term commitment, while utilizing growing forests to offset current emissions must be viewed as a limited term measure, i.e., its effects will continue for at most several decades. Reforesting non-forest land could slow the rate of growth of atmospheric CO_2 and provide time to assess better the actual impacts of climate change, to develop alternate energy systems, and to allow human systems to accommodate to changes in climate. Preserving forests offers an immediate benefit by reducing net CO_2 emissions.

The question then is how much of an offset could be provided by new forests. We examine the case for the U.S. even though it is hardly a typical country in this respect. It is characterized by high per-capita emission rates for fossil-fuel derived CO_2, a large land resource base, high land-rental rates, high labor costs, and a temperate climate. Marland (1991) has provided some preliminary indications of land requirements for a group of tropical nations with mostly low rates of emissions and anticipated high growth rates for trees. Of course, as with CO_2 releases, for global change it really is immaterial where in the world CO_2 is sequestered. It might turn out to be most beneficial for some nations to establish reservoirs outside their own national boundaries.

Moulton and Richards (1990) have assembled a large amount of U.S. Forest Service data which suggest that the U.S. could offset fully 56% of current CO_2 emissions with an intensive tree planting and forest management program on 142 x 10^6 ha of land identified as "economically marginal and environmentally sensitive croplands and pasture lands and non-Federal forest lands." The incremental cost of this carbon collection ranged from 5.80 $/ton C for the first 123 x 10^3 tons (for active forest management on 39 thousand ha in the northern plains) to 47.75 $/ton C for the last 302 x 10^3 tons (for passive forest management on 467 thousand ha in the northeast). The total cost would run to $19.5 billion, an average of 26.70 $/ton C. The Moulton and Richards analysis used real data for rates of bolewood accumulation and expanded this to total ecosystem carbon accumulation, including below ground tree parts and soil organic matter, using site and climate specific multipliers. Less optimistic targets from Moulton and Richard's work suggest that 10% of U.S. CO_2 emissions could be offset on 28.7 x 10^6 ha at a cost of 13.25 $/ton C. Other analyses (e.g., OTA, 1991) suggest that this is more nearly the amount of land that might realistically be available for tree planting in the U.S. over the next few decades. In Figure 8.4, we show Moulton and Richards' cumulative curve, starting at the lowest cost carbon, of the amount of carbon that could be sequestered and the amount and type of land that would be involved. The marginal cost of this sequestration is shown in Figure 8.5.

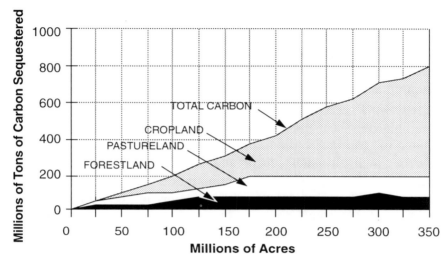

Figure 8.4   Potential annual carbon storage through reforestation in the U.S., by land type and area. (From Moulton and Richards U.S.D.A., GTRW0-58, 1990.)

Some people feel that the Moulton and Richards analysis is optimistic on both the rate of carbon sequestering (specifically the ratio between carbon in bolewood and total ecosystem carbon) and land rental rates. The U.S. forest industry certainly has extensive experience with reforesting large land areas but the scale of this enterprise is unprecedented and there is a need for more data on what realistic

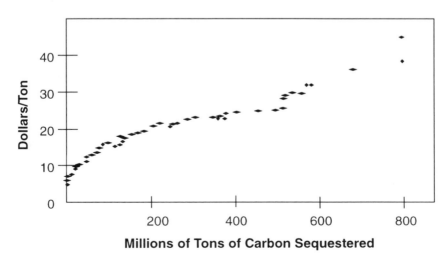

Figure 8.5   Marginal cost of carbon storage in the U.S. through reforestation as a function of total annual carbon storage. (From Moulton and Richards U.S.D.A., GTRW0-58, 1990.)

rates of net carbon uptake might be with optimal management over decadal time scales.

A logical extension of contemplating large expanses of new tree plantings is consideration of the ultimate fate of the trees. The rate of carbon uptake will decline with time as the trees mature, but the carbon will stay in storage away from the atmosphere so long as the trees are not harvested. A frequent assumption is that the trees must be left in place to maintain carbon storage, but other analysts assume that harvesting will ultimately occur and that the objective should be to ensure that the maximum possible amount of harvested material ends up in long-lived products like construction lumber. According to Harmon et al. (1990), when old growth forests in the Pacific Northwest are harvested only 42% of the carbon in bolewood ends up in products with a life span greater than 5 years, and up to 250 years can be required for the total mass of sequestered carbon to reach the pre-harvest level. This suggests: first, that on CO_2 considerations alone it does not make sense to harvest mature trees to establish fast-growing, younger stands; and, second, that carbon sequestering is seriously compromised if the trees are ultimately harvested for short duration wood or fiber products. On the other hand, sustained-yield plantations operated for timber or pulp production should theoretically operate at zero net CO_2 flux except for fossil-fuel inputs required for plantation management.

We can envision, however, a set of circumstances where wood harvest makes sense in the greenhouse gas context. If wood could be harvested and used efficiently as a replacement for fossil-fuel combustion, it would be possible to design a system in which carbon from biomass is recycled through the atmosphere but there is no net emission as would follow from fossil-fuel burning. In this case the tree crop would be harvested at regular, and fairly short, time intervals designed to maximize the mean annual carbon uptake (biomass production) and the plantation would never grow old enough to contain very much carbon in storage. Figure 8.6 provides a cartoon of net carbon fluxes as a function of time to suggest that, over the long term, the system that results in the least carbon emission is one based on a sustainable biomass energy crop. This diagram does not convey the magnitude of the land resource necessary to supply significant quantities of usable energy nor does it quite capture the idea that it depends on being able to get large quantities of biomass fuel with low internal

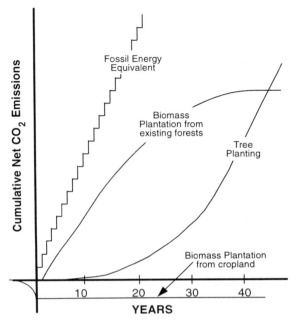

Figure 8.6 Schematic illustration of cumulative CO_2 emissions from alternative uses of forests to offset emissions from a power plant or to produce power. Path A describes reference emissions from a fossil-fuel power plant. Along path B, forests are planted sufficient, initially, to offset completely the emissions in path A. However, as the forests mature, the rate of uptake declines. Line a-b indicates the amount of carbon ultimately stored in mature trees. Along path C the power plant is fueled by wood from a mature forest which is turned into a plantation forest. Line o-c represents the difference in carbon between the original mature forest and the succeeding plantation. Finally, along path D a sustainable fuel-wood plantation is established at the time the power plant is designed, and in an area not previously forested. In this case line o-d represents the net carbon stored in the plantation. (From Marland, G. and Marland, S., *Water, Air, Soil Pollut.,* 64, 181, 1992. With permission.)

energy demands and high conversion efficiencies (Marland and Marland, 1992). Consideration of these latter two points suggests that, since carbon cycling rather than carbon storage is the prime consideration, fast growing species other than trees might be appropriate. Trees, however, have the very considerable advantage that they store large amounts of carbon over a number of years and, thus, integrate over a number of growing seasons and insulate against short term variability in growth factors like weather.

This discussion could quickly evolve into a renewable fuels discussion except for one important point. Focus on carbon flows emphasizes that, for environments with high growth rates and easy access for harvest and markets, it makes sense to contemplate planting trees (or perhaps other fast-growing crops) to recycle carbon and replace fossil fuels, whereas, for environments with slower growth rates and higher energy costs for harvest and marketing, it makes sense to contemplate planting (or protecting) trees to take up and store carbon over long periods of time.

The challenge for research is to establish quantitative bounds on which lands are appropriate for sequestering carbon and which are most appropriate for energy crops. For energy crops, technology is just evolving for efficient harvest and conversion to electricity and/or liquid fuels. For either carbon sequestering or energy crops there is much research yet to be done to learn how to optimize tree selection and management regimes to maximize benefits with respect to climate change in a manner compatible with aesthetic and other objectives. We need more data on carbon storage in soils and forest litter, on nutrient cycling, and on the chemical inputs necessary for plantation success. We need to know the implications for biodiversity, the risks and options to monocultural plantations. And, over longer time scales, we need to know the potential offered by genetic selection and engineering.

8.4.3 Organic Carbon Storage in Soils

Estimates (IPCC 1990; Tans et al., 1990; Post et al., 1991; Sarmiento and Sundquist, 1992) of the amount of CO_2 released by fossil fuel burning and deforestation exceed those for the sum of the amount appearing in oceans and the atmosphere (over the period 1850–1990) by a factor of 1.5 (with large uncertainty, perhaps ±0.4). The biosphere (trees, soils, marine humus, and marine sediments) is an obvious candidate as a reservoir for the remaining carbon storage, often referred to as the missing carbon sink. However, reservoirs and processes involving the biosphere on a global scale are exceedingly uncertain, and are usually invoked by inference, rather than by direct measurement. Prior to the last century the CO_2 content of the atmosphere remained almost constant for hundreds of years, so, in their natural state, one assumes that the various carbon reservoirs must also have remained nearly constant. If so, then any recent change in carbon storage by the biospheric reservoirs must have been induced by humans. Until the sinks for carbon are identified, their role in the future response of the system cannot be properly understood. Will they continue to remove the same amounts of CO_2 or will they gradually saturate? For geoengineering the issue is: could these reservoirs be engineered to our advantage?

Recently, Keeling et al. (1989) and Tans et al. (1990) pointed out that the interhemispheric gradient of atmospheric CO_2 is only about half the expected value. Since roughly 95% of human CO_2 sources are in the Northern Hemisphere, the atmospheric buildup should be larger in the Northern than in the Southern Hemisphere, but the magnitude of the difference depends on transport. Keeling et al., suggest that a combination of north to south transport of CO_2 through the ocean and a Northern Hemisphere biospheric sink account for the small difference. Tans et al., reject ocean transport and call entirely on uptake by the Northern Hemisphere biosphere. The magnitude of the biospheric uptake demanded by Tans et al., is roughly equivalent to the apparent imbalance in the global CO_2 budget. Putting aside the differences between the interpretations offered by these two groups, the small magnitude of the interhemispheric CO_2 gradient seems to require that there must be a major biospheric CO_2 sink currently operating on the Northern Hemisphere continents.

If the missing carbon is being stored in the biosphere above ground, the fractional increase in these reservoirs appears to be unreasonably large. Direct measurements show that the amount of CO_2 in the atmosphere has risen since 1850 from 288 to 353 ppmv (or 140 GtC). Estimates of carbon stored in trees, including changes in forest cover, suggest a decrease of about 170 GtC. Total CO_2 released by fossil-fuel burning is about 210 GtC. About 110 GtC has entered the ocean. Clearly, estimates for the observed increase of anthropogenic CO_2 in the ocean and atmosphere (250 GtC) do not match the amount released by fuel burning and deforestation (380 GtC). We refer to the difference, roughly 130 GtC, as "missing carbon." For reference we adopt the following values for carbon storage today: 640 GtC in forests (above ground), and 1600 GtC in soils, both values are quite uncertain.

The missing carbon equals ~20% of the biomass in trees in the remaining forests. Hence, if trees store the missing carbon, then today's forests have 1.2 times more biomass than preindustrial forests.

If trees are becoming more massive, then the most likely cause is fertilization by the increased concentration of atmospheric CO_2. (However, note that most of the atmospheric growth has occurred relatively recently: Keeling's Mauna Loa record shows 80 GtC growth in atmospheric CO_2 since 1957.) No biologist has proposed such a large growth enhancement from CO_2 fertilization of trees. Also, there is little direct evidence, e.g., from tree ring measurements, to support CO_2 growth enhancement in forest trees. It does not seem possible that trees by themselves are the missing sink.

By contrast, the amount of missing carbon corresponds to only about 8% of the carbon in the humus reservoir. If one assumes that the amount of humus rises with the input of new organic material to the soil, and that this input rises with growth in atmospheric CO_2, then the situation appears more promising. Since other reservoirs appear to be even less likely candidates as the site of missing carbon, a new look should be taken at the role of soil humus. Not only is it a prime candidate for the so-called missing sink, but it is also a promising candidate for engineering removal of CO_2 from the atmosphere, by changing agricultural or silvicultural practices.

8.4.4 Enhancement of Carbon Storage in Oceans

CO_2 exchange between the atmosphere and ocean is proportional to the difference in partial pressure of CO_2 in the air and the surface, mixed layer of the ocean. In turn the partial pressure of CO_2 in the mixed layer is controlled by the solubility of CO_2 (especially depending on temperature and alkalinity), by the rate of mixing of surface waters with deeper layers, and by removal of carbon by biological processes. Sarmiento and Sundquist (1992) estimate that the net oceanic uptake of CO_2 has risen to 2 $GtCyr^{-1}$, following the rising atmospheric partial pressure of CO_2. However, estimates are highly uncertain (see Tans et al., 1990, Watson et al., 1990).

Surface waters of the ocean are mixed with the deep layers primarily by deep water formation at high latitudes and by upwelling at low latitudes. While the mechanisms for deep water formation are not fully understood (Hoffert, 1991), there are concepts of how to alter deep water formation (see below). However, a doubling of the average global upwelling velocity, for example, is not expected to have an effect on atmospheric CO_2 concentrations over time intervals shorter than 100 years (Kheshgi, 1989). The solubility of CO_2 in ocean surface waters could be increased by increasing ocean alkalinity, or by biologically reducing the amount of dissolved inorganic carbon in ocean water.

As the partial pressure of CO_2 has risen from its preindustrial value, the pH and carbonate ion concentration of the surface waters have fallen. If the alkalinity of the oceans were increased, higher levels of surface water pH and carbonate ion concentration could be maintained and additional CO_2 could be dissolved in the oceans without altering the ocean biosphere. The difficulty of this approach is in enhancing the source of ocean alkalinity. While the natural feedback of increased weathering of limestone due to higher atmospheric concentration of CO_2, and possibly higher rainfall amounts, is expected to increase the runoff rate of elements responsible for oceanic alkalinity, the present riverine runoff rate would have to increase by a factor of 30 to compensate for CO_2 emissions at the current rate. The weathering rate of limestone could be accelerated artificially to some extent, and the effectiveness of methods for doing so should be considered. Alternatively, other alkaline minerals could be gathered and dissolved in seawater. However, reserves of minerals that would both affect ocean alkalinity and dissolve readily in seawater are limited. For example, worldwide recoverable reserves of Na_2CO_3 (soda ash) are estimated (Kostick, 1985) to be 43 Gt; if all the soda ash reserves were dissolved in the oceans, this would allow an equal number of moles of CO_2 to be dissolved in ocean water, only 5 GtC. It is theoretically conceivable to increase ocean alkalinity at a rate sufficient to stabilize the growth of atmospheric CO_2, while preserving the marine biosphere, but the means for doing so at a rate comparable to the rate of human CO_2 emissions has yet to be discovered.

Marine photosynthesis reduces carbon with net primary production at about 30 GtCyr^{-1} (Sundquist, 1985). This rate is about half the rate of primary production in the terrestrial biosphere. However, as in the terrestrial biosphere, nearly all of the organic carbon is soon oxidized; oxidation occurs throughout the water column and on the ocean floor. Rapid settling of organic particles and subsequent oxidation effectively pumps dissolved CO_2 from surface waters to deeper ocean layers. A small fraction of the settling particles are buried in sediments, trapping carbon in the form of organic carbon and $CaCO_3$ at rates of 0.05 and 0.2 GtCyr^{-1}, respectively (Sundquist, 1986). To have a significant geoengineering impact the marine biosphere would have to be altered enough to increase the rate of oceanic CO_2 absorption to levels comparable to fossil-fuel emissions. Changes in the marine biosphere that might have such an impact are, for example, increased fertilization, increased ratio of carbon to nutrients in organic particles (Redfield ratios), and decreased decay or more rapid settling of organic particles. Society has no experience cultivating the oceans on the grand scale that would be required. Likely such impacts could only be achieved by controlling species present in the marine biosphere and would, perhaps, entail genetic engineering.

Considerable attention has been focused on enhancing carbon storage in the deep ocean by iron fertilization of the Antarctic ocean. Recently, Martin et al. (1990) discovered that the addition of a very small amount of iron to the otherwise nutrient (nitrate and phosphate) rich Antarctic surface waters leads to increased primary productivity. Settling organic particles caused by higher productivity in the Antarctic could pump carbon from the surface waters, and hence from the atmosphere, to the deep ocean while also sinking nutrients.

Further analysis of the iron fertilization mechanism (Peng and Broecker, 1991; Joos et al., 1991) shows that if the nutrient content of Antarctic waters could be depleted, as is seen in the surface waters of the low latitude oceans, then the projected uptake of CO_2 would be only a fraction of fossil-fuel CO_2 emissions. As an example, assume that totally successful iron fertilization were carried out for a full century, and that without fertilization the atmosphere's CO_2 content would reach 600 ppmv, then, with totally successful iron fertilization, model results of Peng and Broecker (1991) show that CO_2 levels would still reach at least 540 ppmv. In their dynamic model the flow of CO_2 into the Antarctic rapidly drives the CO_2 content of its surface waters back toward the atmospheric value. However, if the growth of CO_2 could be reduced by even 60 ppmv (40 year's accumulation at today's rates) at reasonable cost, and with minimal side effects, such an approach might be a useful element of a comprehensive approach to CO_2 control.

The amount of iron needed to do the job may have been greatly underestimated. Iron is a highly reactive metal which rapidly becomes attached to particulate matter. The residence time of iron in surface waters of the Antarctic is probably no more than a few weeks. It is important to consider the competition between nonbiologic and biologic pathways for the added iron. Furthermore, release of iron from ships or tethered moorings would lead to a streaky distribution. To compensate, enough iron would have to be added to maintain the area between the streaks at an iron content high enough to accomplish the biologic goal. Finally, since this approach works by allowing the biospheric pump to operate more rapidly, the iron addition must be continued year after year. If at some point nutrient addition stopped, much of the CO_2 sequestered in Antarctic deep water would be released back to the atmosphere, largely undoing any gains that were achieved by the fertilization.

Even with the highest biological expectation (c.f. Berger, 1991), iron fertilization would not draw the nutrient content of Antarctic waters down to near zero levels even during the summer months. It would be surprising if light or some other factor did not place a limit on the degree of drawdown. These considerations suggest that the 60 ppmv reduction found by Peng and Broecker (1991), should be reduced, perhaps by a factor two.

While it is unclear what the possible ecological side effects of this process might be, there is at least one geochemical side effect that must be considered. The 1.6 m/kg of PO_4 in surface water, which is the target of iron fertilization, would, when converted to organic matter, create an oxygen demand of 280 m/kg. It is possible that parts of the thermocline would become anaerobic. As our knowledge of the depth distribution of the oxidation of organic matter produced in the Antarctic is limited, it is currently not possible to quantify this impact reliably, even if the dynamics of mixing and overturn were well constrained.

Thus, while research on iron fertilization and other approaches to stimulation of the marine biosphere should still be considered, the potential for reducing atmospheric CO_2 may be less than it first appeared. Iron fertilization, alkalinity enhancement, or other means to fix CO_2 using the marine biosphere may be elements of a comprehensive strategy to control the concentration of atmospheric CO_2.

8.5 OPTIONS FOR GEOENGINEERING THE CLIMATE SYSTEM

8.5.1 Modifying the Earth's Albedo with Tropospheric Aerosols

One possible method of climate control is to use the properties of sulfate or other aerosols and their potential influence on the Earth's radiation budget. Any aerosol-based intervention strategy should first assess the potential effects on climate in detail. To do this reliably, improved knowledge of current aerosol effects is a necessary prerequisite. As a basis for speculation on future aerosol effects (both intentional and unintentional), therefore, we first review present understanding as a basis for a crude indication of the sensitivities of the climate system to aerosols.

Tropospheric aerosol particles influence radiative processes in at least two ways. Their dominant "direct" effect is to scatter solar radiation, with backscattering acting to increase albedo. Aerosol particles may also have an "indirect" radiative effect by influencing cloud albedo through their role as cloud condensation nuclei (CCN). In both instances, increased amounts of submicrometer aerosol particles cause a decrease in the amount of solar radiation absorbed by the surface-troposphere system. There are other mechanisms by which aerosols can potentially influence climate, including absorption of shortwave solar radiation (e.g., by soot), absorption of longwave (terrestrial) radiation (e.g., by silicates), alteration of the lifetime of water vapor in the atmosphere, the extent and duration of rain, and the occurrence of ice in clouds. The absorption of both short and longwave radiation is generally thought to be small except near specific sources of soot (e.g., Kuwait) and dust (e.g., deserts), and the climatic role of aerosols remains largely an open question. Effects on precipitation and water vapor lifetime are also open questions, and ice nucleation is probably the least understood of all.

Sulfate ($SO_4^=$) aerosol particles from the oxidation of anthropogenic SO_2 are of major importance to the direct forcing, and may provide an important indirect effect. They may be important factors influencing precipitation and water vapor lifetimes. Also, they act as cloud condensation nuclei but are probably not important ice nuclei. Enough is now known to allow quantitative assessments of the direct radiative forcing by $SO_4^=$ aerosol particles, while their indirect forcing has only been demonstrated qualitatively. The direct forcing of anthropogenic $SO_4^=$ aerosol has been quantified both theoretically and experimentally using well-known data on its mass scattering coefficient as a function of relative humidity. Uncertainties do exist, leading to an overall uncertainty in the calculated forcing of a factor of about two.

The sensitivity of the direct forcing to the amount of $SO_4^=$ aerosol can be given in various ways. One common index of the amount of the effect is the aerosol optical depth, δ, which is defined by Beer's Law for the transmission of solar radiation through the atmosphere (neglecting Rayleigh scattering, O_3, and water absorption, etc.). In a highly simplified form of the equation:

$$I/I_o = e^{-\delta}$$

(Eq. 8.1)

where I_o and I are the incident and transmitted amounts of solar radiation. The optical depth, δ, is directly proportional to the mass of $SO_4^=$ aerosol in the vertical column above the Earth,

$$\delta = \alpha B_{SO_4}^=$$

(Eq. 8.2)

where $\alpha = 8.5$ m^2g^{-1} is a quantity arrived at by empirical correlation. Thus, a concentration of $SO_4^=$ of 10 μg m^{-3} (the approximate annual average in Ohio) over an altitude range of 3000 m (a typical depth of the planetary boundary layer) yields $B_{SO_4}=$ 30 mg m^{-2} and $\delta = 0.25$. This is about five times the average value for clear marine air in the Southern Hemisphere (Charlson et al., 1991).

Because about 20% of the loss of solar energy is due to backscatter to space, the additional loss of sunlight from sulfate aerosols is about 5%. If the average solar flux incident on the aerosol layer is about 200 W m^{-2}, this local forcing would be -10 W m^{-2} or about six times as large as, but opposite in sign to, the current radiative forcing by anthropogenic greenhouse gases.

Chemical models allow the extension of this calculation to the Northern Hemisphere (NH) and to the globe, where δ due to anthropogenic $SO_4^=$ is estimated to have averages of about 0.07 and 0.035 for the NH and globe, respectively, i.e., the effect is nearly entirely concentrated in the NH. When included in a more complete radiative transfer calculation that includes fractional cloud cover and global geometric considerations, the resultant spatial average radiative forcing, at current emission levels of SO_2, due only to anthropogenic $SO_4^=$ are -1.1 and -0.6 W m^{-2} for the NH and globe, respectively. Shine et al. (1990) estimate that the increase of atmospheric CO_2 between 1765 to 1990 has increased radiative forcing by 1.5 Wm^{-2}. Though far more uncertain, these estimates for sulfate forcing are of a magnitude sufficient to offset CO_2 forcing appreciably.

Penner et al. (1991, 1992) suggest that smoke particles emitted by human burning of biomass may produce aerosols in amounts with a global effect on radiative forcing of -2 W m^{-2}. (The contribution from the direct effect, -0.8 W m^{-2} is estimated far more reliably than the -1.2 W m^{-2} from indirect effects.) Again the estimated effect is comparable to the estimate of radiative forcing associated with CO_2 increases since pre-industrial times (indeed it is larger, and opposite in sign). This source of biomass aerosol is concentrated largely in the tropics. As with sulfate aerosols, aerosols from biomass may produce indirect effects, but these are much more difficult to quantify.

While this comparison of anthropogenic aerosols and CO_2 forcing demonstrates that aerosols cannot be neglected in climate models, it is potentially misleading because it might suggest that man-made aerosols directly compensate for greenhouse forcing. However, the two forcings have radically different geographical and altitudinal distributions. The meteorological consequences of geographically non-uniform forcing are quite unknown and must be left as high priority topics for research. Also, the history of forcing with time is certainly quite different for the various sources.

Sensitivity of cloud albedo to the number population of CCN is the basis of the so-called indirect effect. An increase of CCN of 30% only over the oceans is calculated to yield a global radiative forcing of ca. -1 W m^{-2}, again a quantity that would be comparable in magnitude but opposite in sign to forcing by anthropogenic CO_2. It is well known that air masses with enhanced anthropogenic $SO_4^=$ also have higher CCN populations than do unpolluted air masses. Typical remote marine CCN concentrations at 0.5% supersaturation are 50–100 cm^{-3}, while polluted air masses often have 1000 cm^{-3}. Observations of pollution sources (plumes of aerosol from ships and from inhabited areas of land) that influence marine stratiform clouds show that enhanced aerosol concentrations do indeed cause increased albedo, in qualitative agreement with theoretical calculations (Radke et al., 1990). However, at present there is no means to extend these qualitative, local observations to a quantitative, global assessment. There certainly is significant potential for substantial radiative forcing by anthropogenic CCN. Again, the non-uniform geographical distribution of anthropogenic CCN implies that the forcing and meteorological consequences must also be non-uniform. Uncertainties exist in the estimated direct forcing, but these are amenable to immediate improvements through both observation and modelling. This is due in large measure to the already well established, internally consistent body of data and theory linking chemical and physical aerosol properties and processes, and to the relative simplicity of the effect being proportional to the mass of $SO_4^=$ in the atmosphere. By comparison, the indirect effect still has major gaps of fundamental understanding. Chemists can predict with some confidence the mass concentration of anthropogenic $SO_4^=$. However, while there is little doubt of the sign of the effect on cloud albedo, there is as yet no means for reliably predicting the number concentration of CCN, and, hence, for predicting the magnitude of the albedo effect.

Besides these uncertainties, there are many other unknowns in the general area of aerosol influences on climate with regard to both natural and man-made aerosol. Tropospheric and stratospheric aerosols are known to vary naturally but the role of this variability in climate is still in need of exploration. The effects of anthropogenic aerosol other than SO_2 derived $SO_4^=$, such as soot, organic aerosol and dust, also should be investigated from both a regional and global perspective. A critical issue is to understand better the relation between sources of aerosols, and their ultimate distribution. In particular, this calls for an improved understanding of the lifetime and transport of atmospheric aerosols and their precursors.

8.5.2 Do Human Aerosols Already Contribute to Climate Modification?

The idea that sulfate aerosols could be an important agent in climatic change has recently undergone a revival. Twenty years ago, when the quantification of potential climate forcing factors was even more rudimentary than it is today, some climatologists suggested that there could be compensating effects from human releases of aerosols and greenhouse gases. The desire to explain the record of global mean temperature changes stimulated this speculation. Up to the 1940s there was a warming trend and the thought that this might be due to the increase in atmospheric CO_2 from fossil-fuel combustion was briefly popular. Updated observations showed, as early as 1946, that the warming trend had been replaced by a cooling trend, but it was not for some time that this cooling became prolonged enough to deserve an explanation. In the eyes of some, aerosols were the answer. The late J. Murray Mitchell considered the compensating aerosol and greenhouse effects more thoroughly than most (e.g., Mitchell, 1975). He suggested that, by the late 20th century, the greenhouse effect would once more dominate.

Aerosols have become important again today partly because of a better understanding of the processes involved. There are a number of separate issues: the sources of aerosol, anthropogenic versus natural, with the latter enhanced by feedbacks; the mechanisms, clear sky vs. cloud albedo effects; the implications for past and future climatic change; and detection of aerosol effects, including defining the spatial character of the signal.

The two primary sources considered in the literature are SO_2 derived from fossil-fuel combustion (see, e.g., Schwartz, 1988) and dimethyl sulfide (DMS) derived from marine organisms (Charlson et al., 1987). DMS plays an important role in the natural CCN cycle, and its concentration may be dependent on hypothesized feedback mechanisms, e.g., global warming leads to changes in ocean mixed layer temperature and mixing, leads to changing physical and chemical (i.e., nutrient) conditions, leads to changes in speciation and productivity, leads to changing DMS flux, leads to a reinforced (or diminished?) climate change. Although there is some evidence that this feedback process may operate on the time scale of ice ages, its influence in possibly modifying the enhanced greenhouse effect is highly speculative.

Sulfate aerosols have a direct effect on the radiation balance of the atmosphere under clear sky conditions (recently reassessed by Charlson et al., 1992), and may noticeably affect the albedo of marine stratiform clouds, an indirect effect (Charlson et al., 1987, 1992). The form of the relationship between sulfur source strength and the consequent radiative forcing perturbation will almost certainly differ according to the mechanism of sulfur introduction (see, e.g., Wigley, 1991, for a preliminary analysis). Charlson et al. (1990) point out that there are greater difficulties in quantifying the cloud albedo effect than the clear sky effect. The latter involves the step from total sulfur mass loading to cloud condensation nuclei number, currently quantifiable but with minimal confidence. The various uncertainties mean that the indirect effects could be anywhere from insignificant to extremely important.

It is possible that human-induced aerosol changes, related to emissions of SO_2 and to biomass burning, have influenced climate on a global scale, possibly offsetting some of the warming due to the enhanced greenhouse effect. The effects would be larger in the Northern Hemisphere (NH), since this is where most SO_2 is emitted, but interhemispheric heat and mass transport would ensure a "spill over" signal in the Southern Hemisphere (SH), at least on decadal and longer time scales. Wigley (1989) and Schlesinger et al. (1992) have shown that even a small aerosol effect would improve the agreement between models and data. The early idea that the global-scale (but largely NH) cooling between the 1940s and the 1970s was an aerosol effect is still a possibility; but there are other tenable (and equally speculative) explanations [e.g., a change in the thermohaline circulation (Watts, 1985; Wigley and Raper, 1987; Watts and Morantine, 1991)].

If biomass and SO_2-related aerosol effects are significant, then they will likely modify future greenhouse gas-related changes in climate. Strong efforts to reduce acid rain problems by cutting down SO_2 emissions or to decrease biomass burning could lead to a small warming effect. Furthermore, since a reduction in emissions of CO_2 would probably cause a parallel reduction in SO_2 emissions, and because of the widely differing response times of the CO_2 and SO_2 systems, reduced CO_2 emissions could, under extreme scenarios, lead to a counter-intuitive warming rather than cooling (Hansen and Lacis, 1990; Wigley, 1991). More probably, SO_2 effects could slow down the response of the climate system to attempts to reduce the enhanced greenhouse effect.

There is no evidence of DMS-related effects in the instrumental climate record, and future effects depend on highly speculative feedback mechanisms. Nevertheless, important effects could occur, given the large magnitude of expected changes in climate and the likelihood of major changes in ocean climate.

The dominance of NH SO_2 sources implies a greater cooling effect in the NH. Although the climate data show some evidence for such a differential, it could still be attributed to natural variability. In other words, we cannot claim to have detected any SO_2-related aerosol effect at the hemispheric scale. On the other hand, natural variability could obscure quite a large SO_2 effect. Wigley (1989) estimates that a differential radiative forcing between the hemispheres of up to 1.5 W m^{-2} over the past 100 years would be undetectable. A 1.5 W m^{-2} negative forcing would offset roughly 60% of the enhanced greenhouse forcing in the NH (Although this estimate was based on the cloud albedo effect, it applies equally to the total, clear-sky-plus-cloud, effect). Lack of detection is not surprising given that we cannot yet claim to have unequivocally detected the enhanced greenhouse effect.

Detection may be facilitated by using multivariate, or "fingerprint" techniques (Wigley and Barnett, 1990). SO_2 emissions and biomass burning are strongly regional, so they ought to produce a distinct spatial signal pattern (see, e.g., Charlson et al., 1991; Penner et al., 1991). Since multivariate methods have not yet succeeded in detecting the enhanced greenhouse effect, one might expect even greater difficulty in detecting aerosol effects. This may not be the case, however, since the greenhouse signal pattern is probably more difficult to separate from the signals of other climate change processes than the aerosol signal pattern. There is a pressing need to quantify better the spatial details of both the direct and indirect aerosol signals.

8.5.3 Engineering External Climate Forcing Factors

In evaluating the potential applicability of the options suggested below, it is helpful to recall the magnitudes of important fluxes. Radiation models estimate that doubling the atmospheric carbon dioxide concentration would increase net trapping of infrared radiation at the tropopause by about 4.4 W m^{-2}. Shine et al. (1990) estimate that increases in concentrations of greenhouse gases since 1765 have increased the trapping of IR radiation by about 2.45 W m^{-2}. For geo-engineering schemes to be effective, they must be capable of counterbalancing fluxes of comparable amounts. For reference here, schemes will be evaluated based on their ability to counterbalance 2.2 W m^{-2}, or the equivalent of half of the flux change due to a doubling of the CO_2 concentration.

The primary external climate forcing factor is solar radiation. One advantage of dealing with solar radiation is that experience with the seasonal and diurnal cycles and with volcanic eruptions provides some insight into what would need to be done and what possibilities exist.

8.5.3.1 Space

From an energetic consideration, reducing incoming sunlight in space is most attractive because the flux is greatest and, therefore, the size of the intervening mechanism can be least. However, implementing such a scheme requires lofting and sustaining the required materials in orbit. Schemes suggested include:

Orbiting Mirrors

The NAS (1992) proposed that 55,000 mirrors, each with a surface area of 100 km^2, be lofted into near Earth orbit. Assuming randomly oriented orbits with the mirrors oriented parallel to the surface, this would give total coverage of 5.5 x 10^6 km^2, which is about 1% of the Earth's surface. As a result, the mirrors would intercept 1% of the incoming solar radiation. If aligned perpendicular to the incoming radiation, or placed preferentially in low latitude orbits, about 2% of the incoming solar radiation could be intercepted. The large number of mirrors would permit incremental implementation. Such mirrors would, however, create shadows on the surface roughly equivalent to eclipses; such events would be quite frequent and probably troubling, even accounting for diffraction effects. A greater number of smaller mirrors could be used to alleviate this problem, but the effort to launch and to keep separated so many mirrors becomes larger. Removing the mirrors (or the debris resulting from colliding mirrors) from orbit would require a significant effort, making this intervention difficult to reverse.

Orbiting Layer of Absorbing or Scattering Particles

One could place in orbit a cloud of soot particles that would absorb solar radiation. Assuming a specific absorption of 10 m^2/g and requiring coverage of 5 x 10^6 km^2 in order to absorb 1% of the incoming solar radiation would require 5 x 10^8 kg to be placed in a dispersed orbit; for scattering particles, the amount would be somewhat larger. Injection could be performed incrementally as greenhouse concentrations rise. Keeping the particles in orbit might be difficult, however, due to the solar wind and other effects. In addition, such particles would significantly pollute near-Earth orbits. There would also be virtually no means for reversing the process; the particles would move on their own until being removed through contact with the atmosphere. See the discussion of various space-based solar screens by Mautner (1991). Inspired by Hoyle (1957), Kahle and Deirmendjian (1973) considered the consequence on atmospheric conditions of a relatively thick, absorbing layer that affected only solar radiation. However, soot particles, in particular, also emit IR, and this may offset some of the reduction in solar heating.

Solar Deflector at the Lagrange Point

Early (1989) proposed that a 2000 km diameter solar shield made of lunar materials be placed at the first (L1) Sun-Earth Lagrange point (1.5×10^6 km from Earth). (See also the discussion of space parasols invarious orbits by Hudson (1991).) The shield would deflect 2% of the radiation incident on the Earth (deflection instead of reflection to reduce the impact of the solar wind) and would be essentially unnoticeable from the Earth due to its relatively small apparent size at that distance. Early (1989) estimated that installation of a 10-μm thick shield weighing 10^{11}kg would cost from 1 to 10 trillion dollars. Moving the shield would be relatively inexpensive, so the phenomena is reversible; the incremental effect could be achieved by tilting of the deflector to reduce cross-sectional area. A major problem withthis proposal is the installation cost before any effect is created. Orbits at L1 are unstable, even in the restricted three body system, and the dynamics here are more complex, with an eccentric Earth-sun orbit and perturbations from Earth-moon interactions. So, maintaining alignment would require active steering.

All of these schemes would also reduce the UV flux, which would decrease stratospheric ozone. Since this would tend to cool the stratosphere, it would enhance the effectiveness of solar blocking. Other chemical effects would also require consideration.

8.5.3.2 Atmosphere

The prime difficulty with modifying the solar radiation fluxes in the atmosphere is keeping the modifying material aloft. Because large structures would require large amounts of energy to keep them aloft, resort must be made to either very small particles (~0.1 μm) that have very slow removal times or to systems based on balloons that are evacuated or filled with hydrogen, helium, or other buoyant gases.

Reflective Stratospheric Aerosols

The suggestion to inject sulfate aerosols goes back to at least SMIC (1971), Budyko (1974), and Dyson and Marland (1979). Such a human volcano could lead to significant backscattering of solar radiation. Broecker (1985) estimated the amount of sulfur that needed to be carried aloft by special aircraft, and how this sulfur could be added as a component of existing jet fuel and emitted by commercial aircraft. NAS (1992) evaluated various lofting schemes for an aerosol dust, including balloon systems and launch by artillery pieces. Alternately, sulfur could be emitted at the surface as COS and allowed to mix upward and be transformed to $SO_4^=$. For injection altitudes in the lower stratosphere, the particle lifetimes would likely be less than a year (based on lifetimes of volcanic aerosols), so the effect would be naturally reversible and could be implemented incrementally. (Fossil fuel combustion by commercial aircraft currently injects sulfur into the upper troposphere, but this would not be a preferred method for geoengineering, since sulfates are rapidly removed at this altitude.) Because much of the scattering is forward, the sky would become more white, with direct solar radiation reduced substantially even when total radiation is reduced only modestly. The large diminution of direct solar radiation would negatively affect solar energy sources relying on direct radiation and astronomers would also likely object to the increased scattering, although the public might appreciate the enhanced color of sunsets due to the scattered radiation. The amount of sulfur injected would be far less than that now causing "acid rain" and would be dispersed rather than concentrated, and thus would not create problems when rained out. Stratospheric aerosols would likely be a sink for ozone. Consequently, such aerosols would lead to additional cooling by cooling the stratosphere, but effects on surface UV would need to be considered.

Absorptive Stratospheric Aerosols

Much in the manner of "nuclear winter" aerosols (Turco et al., 1983), injection of soot into the stratosphere would lead to cooling of the lower atmosphere, although account would need to be taken of the increased downward IR flux. Again, the implementation could be incremental, but the modification of the temperature structure might lengthen removal times. Heating of the stratosphere would also perturb stratospheric chemistry in ways deserving attention. To reduce downcoming energy by 2.2 W m^{-2} would require absorption of about 2% of the solar radiation (accounting for the effect of altered stratospheric IR flux). For soot with an absorption efficiency of 10 m^{2}g^{-1}, offsetting the solar radiation would require roughly 0.5 Mt to be kept aloft. Compared with sulfate, soot would absorb solar radiation, thus creating a slight dimming of the sky, but not a whitening. It was such an effect that Sagan and Turco suggested might result from the

Kuwaiti oil fires if the smoke had been lofted into the stratosphere without scavenging. (Smoke particles and solar absorption in the troposphere would warm the climate. This effect is small because the lifetimes of smoke particles are generally very short, although fossil-fuel combustion may emit sufficient carbon particle amounts to create some effects).

High Altitude Balloons

NAS (1992) suggests stratospheric injection of small (few meter diameter), thin-skinned, helium-filled aluminum balls to reflect solar radiation. Larger balloons could be used, but might impose flight hazards. The implementation could be incremental, but the removal rate would be determined by the lifetime of the balloon (and its lifting capability) rather than by natural removal processes. The non-selective directional scattering of the balloons could be made more efficient by using corner reflectors to enhance the effect of the lower sphere of the balloon (see Canavan and Teller, 1991); in this case incoming solar radiation would be reflected back to space without scattering, which would eliminate several problems associated with aerosols and spherical balloons. Both shapes, unless specially designed, would backscatter (or absorb and reradiate) radiation outside the visible, including natural IR and radio transmissions by humans. This could be an advantage or disadvantage, depending on how the balloons were designed and other factors (corner reflectors with angles slightly different from 90° would be needed to prevent interference with transmitters on the surface; spherical shapes might be of great assistance for ham radio, etc.). The balloons could alternatively be filled with hydrogen (except that escape would lead to stratospheric water vapor) or might be evacuated, which would require construction and insertion at altitude and ultralight construction materials (e.g., aerogels) to avoid weight penalties for an evacuated chamber. A question that must be resolved is to determine the optimum altitude (or altitudes) or mechanism that would limit the advective accumulation of balloons in polar areas (at the proper level, the circulation changes as seasons evolve may ensure adequate distribution).

Reflective Tropospheric Aerosols

Preliminary calculations indicate that current emissions of SO_2, largely from fossil fuel combustion, and aerosols related to burning of biomass are exerting a cooling influence on the climate (Charlson et al., 1990, Penner et al., 1991). Enhancing such emissions would therefore appear to offer the potential for greater cooling. The injected aerosols would need to be highly efficient at scattering because contamination by even small amounts of absorbing aerosols (e.g., soot) would generally tend to change the cooling effect to a warming effect. Additional SO_2 could be injected by additional releases from existing power plants (although if regionally concentrated, "acid rain" and visibility degradation become concerns) or could be specially injected into the Southern Hemisphere troposphere so that their effects and impacts on visibility and acid deposition would occur primarily over unpopulated ocean areas. To be effective, such emissions should be lofted as much as possible; smoldering combustion to create aerosols at the surface would be less effective, as well as undesirable due to the injection of other pollutants. Because tropospheric aerosols are generally removed from the atmosphere in days to weeks, the injections would have to be continuous (although one might inject only in low latitudes and only during the summer in mid-latitudes when solar radiation is most intense) and in rather large amounts. (To achieve a Southern Hemisphere burden of 0.01 g/m^2, assuming a five-day lifetime, would require an injection rate of about 180 million tons of sulfur per year, compared with current emissions from fossil fuel combustion amounting to ~75 million tons of sulfur per year!) It is also not certain how the atmosphere would respond to such changes in forcing; cloud and humidity changes might well result.

Overall, the schemes involving injection of aerosols, while relatively straightforward and generally only amplifying inadvertent and/or natural forcings that are already occurring, have a number of difficulties. The potential effectiveness of corner-reflecting stratospheric balloons has yet to be thoroughly evaluated.

8.5.3.3 Surface
Energetically, the surface is the least efficient location to modify the global radiation balance because only about half of the solar radiation reaches the surface; the relatively low energy requirement to deploy the system would provide a counter-balancing advantage were it not for the extensive use of the surface for other purposes.

Whitening the Land Surface

The four largest land use types, in order of increasing albedo, are forests, grasslands, deserts, and snow and ice covered areas. Since the start of agriculture, human actions have been changing surface land use and its albedo. These inadvertent changes have surely affected climates, but only on a local scale are they likely significant. In considering potential advertent actions, preserving high albedo snow-covered areas is clearly desirable (but will be difficult with warming temperatures), but transforming forests or grassland to deserts would have many detrimental side effects. An alternative approach may be to increase the albedo of vegetation (via genetic engineering or substitution), an approach that may be feasible without reducing overall productivity in regions (e.g., equatorial regions) where available sunlight exceeds the amount now being used. Albedo varies from species to species and within species according to age of the leaf, turgidity, presence of waxes or other materials on surfaces. A set of experiments in Nebraska demonstrated that the albedo of soybeans could be increased by coating the crop with reflecting clays and diatomaceous earths and by breeding for leaves approximately three times as pubescent as normal. Both methods increased reflectance of short and long wave radiation, reducing energy absorption and, hence, evapotranspiration (as described in Rosenberg et al., 1983). Although modification of the global radiation balance over the land surface is problematic, modification of the local land surface to reduce urban heating (and therefore, reduce emissions associated with air conditioning demand) appears to merit further investigation.

Whitening the Ocean Surface

The ocean covers about 70% of the Earth's surface and has a low reflectivity typically ranging from 5 to 10%. Increasing the albedo of the ocean surface via films, foams (Jones, 1990), or biological organisms would have many other important effects (e.g., altering the carbon and nutrient cycles and evaporation); making sea ice in high latitudes (by taking advantage of the cooling potential of the polar winter temperatures) and then hauling the sea ice to lower latitudes may be possible for limited areas.

Covering large areas of the ocean (roughly 10% per CO_2 doubling) with highly reflective floating chips or spheres is perhaps conceptually feasible. By reducing solar warming of the ocean this approach would also reduce the rate of sea level rise. For a uniform effect,

Table 8.3   Estimated solar flux components.

Solar Flux Components		
Solar insolation	342	Watts/m^2
(top of the atmosphere)		
Solar radiation reflected to space	100	Watts/m^2
(29% albedo)		
• Clear sky fraction (50%)		
– Surface (13% albedo)	32	Watts/m^2
– Clouds	0	Watts/m^2
– Atmosphere (backscattered 7%)	24	Watts/m^2
Total	**56**	**Watts/m^2**
• Clear sky fraction (50%)		
– Surface	0	Watts/m^2
– Clouds	133	Watts/m^2
– Atmosphere (above clouds 44%)	12	Watts/m^2
Total	**145**	**Watts/m^2**
Solar radiation	78	Watts/m^2
(absorbed by atmosphere — 23%)		
Solar radiation	164	Watts/m^2
(absorbed by surface — 48%)		
Total allocation of solar insolation	**342**	**Watts/m^2**

[1] Cloudy-clear sky effects, when averaged over the planet, gives a cloud effect of 44 Watts/m2, in agreeement with EBBE results.

[2] A 1% change in the solar constant is equivalent to:
.01 x 340 x (1 – 0.29) = 2.4 Watts/m^2. This is about half a CO_2 doubling.

reflectors would need to be dispersed widely and uniformly. In contrast to other options, this approach would probably be aesthetically unacceptable since the floating materials would likely accumulate on ocean shores, rather than maintaining reasonable coverage. Another difficulty is that accumulation of biological materials on surfaces might degrade reflectivity.

An alternative, more natural, approach might involve moving of icebergs from high to low latitudes, but this would require moving roughly the entire Arctic icesheet each year to have a significant effect.

Means of reducing the absorption of solar radiation by the surface-atmosphere system have been suggested that would have their effect in space, in the atmosphere, and at the surface. Table 8.3 provides estimates of various terms in the solar radiation budget for use in estimating the magnitude of change required by geoengineering schemes.

8.5.4 Engineering Climate Feedbacks and Processes

Although the underlying cause of the greenhouse-induced climatic change is the modification of the infrared radiation balance by the increasing concentrations of greenhouse gases, it is the response of atmospheric, oceanic, cryospheric, and land surface processes and feedbacks that determines the actual changes that take place. Thus, modifying the processes that are internal to the climate system offers a potential path to controlling or influencing the greenhouse response. The most important and, perhaps, overwhelming difficulty with this approach is the very close coupling of many processes in ways that are only poorly understood, so that even if modification were possible, predicting the consequences and side effects is very difficult and uncertain. Nonetheless, there appear to be several potential modifications to which the climate system might be sensitive.

8.5.4.1 Atmospheric Processes

Two possibilities to gain control of the climate by intervening in atmospheric processes include cloud modification (see NAS, 1992) and reduction of the atmospheric water vapor burden. An increase in the sulfate aerosol burden of the atmosphere may brighten clouds, particularly marine stratus decks, by increasing the number density of cloud droplets, although this may also lengthen the lifetime of water vapor in the atmosphere and enhance the water vapor greenhouse effect. Koenig (1974) conducted a model simulation to investigate inadvertent changes on a regional scale; more recently, Charlson et al. (1991) evaluated global scale effects of current sulfur emissions.

It is often thought that increasing evaporation rates, for example in low latitude ocean or tropical land areas, may increase cloudiness (thereby increasing solar reflection), but cloud amount and location are typically controlled by relative humidity, vertical stability, and atmospheric circulation, rather than by absolute humidity. As a consequence, attempts to increase evaporation (for example by placing floating wicks in the ocean) may have little effect. It may actually be more effective to reduce the atmospheric water vapor burden and the water vapor greenhouse effect. This might be attempted by creating a monomolecular slick on the ocean to inhibit evaporation. Inhibiting evaporation would, of course, lead to reduced rainfall, which is generally viewed as detrimental, and might not reduce the concentration of atmospheric water vapor, if its lifetime increases. Such an increase in the lifetime of water vapor might occur unless the atmospheric cloud condensation nuclei burden is also reduced. In addition, any reduction in water vapor and/or cloudiness would allow more solar radiation to reach the surface, where it has the largest impact on surface heating.

8.5.4.2 Oceanic Processes

The ocean transports heat from low to high latitudes and buffers climatic change by taking up heat. A large global ocean conveyor belt (Broecker, 1991) carries heat and chemical constituents from the upper to deep oceans and back again. Heat transport and release in the North Atlantic, for example, helps warm Europe in the winter while cooling lower latitudes. Enhancing the strength of the conveyor belt would thus warm high northern latitudes during the cool seasons while moderating any warming in lower latitudes. Because the conveyor belt is apparently driven by a salt excess (or fresh water deficit) in the North Atlantic basin, this situation could be enhanced by either damming the Bering Strait (where relatively fresh water enters) or by diverting rivers (e.g., in Siberia; see Lamb, 1971) which carry fresh water into the Atlantic so that their moisture is carried to the Pacific (e.g., by evaporating the water as a side effect of its use for irrigation). Reducing glacial calving and drainage from Greenland (e.g., via preservative coating) would have a similar effect as well as moderating sea level rise. Enhancing the conveyor belt circulation might also help moderate sea level rise by reducing downward heat transport in the ocean gyres.

8.5.4.3 Cryospheric Processes

Sea ice, snow cover, glaciers, and polar ice sheets play several important roles. These include reflection of solar radiation, limits on cooling of the polar ocean by the insulating effect of sea ice, protection of land from freezing as deeply when under snow cover, inhibition of evaporation of moisture (which limits cloud formation in polar regions), and storage of water above sea level. Although somewhat surprising, the present seasonal cycle of sea ice seems to exert a warming effect on the climate system by insulating the polar ocean in the winter (thereby retaining its heat) and by a lowered albedo in the summer (thereby allowing increased solar absorption). If this cycle could be reversed, this warming effect of sea ice could be reversed.

One way to do this would be to use pumps to carry water from below the sea ice and to spray it such that it comes out as snow or ice particles on the top of the sea ice. This process would bypass the insulating effect in the winter and provide thicker sea ice with its higher albedo to reflect summer radiation. Although such pumping would require energy, it may be possible to use ice-water temperature gradients to derive the energy. One potential question to be assessed is what would happen to the salt; the expectation is that it would melt through the ice, which might affect sea ice structure.

8.5.4.4 Land Surface Processes

Although the land surface exerts important effects on the atmosphere, altering existing interactions to affect global climate would require significant changes over very large areas. Aside from potential albedo changes, which were discussed earlier, changing the surface moisture balance might be considered (and has been carried out since the start of irrigation). Increased evaporation (e.g., via irrigation) is known to reduce local surface temperature.

Overall, modifying climate system processes and feedbacks involves a complex set of interactions that are very difficult to evaluate and assess.

8.5.5 Engineering Climate System Response

An alternative to preventing climatic changes is to geoengineer the responses to the change. The responses can generally be classified as temperature, precipitation and storms, and sea level.

8.5.5.1 Temperature

Global temperature modification by means other than those already mentioned does not seem feasible. However, local actions to adjust the temperature via albedo or moisture changes can be effective. Alternatively, air conditioning has become the primary way to protect members of society from high temperatures, though the energy required contributes to CO_2 emissions when electricity is produced from fossil fuels.

8.5.5.2 Precipitation and Storms

Although model projections of a warmer Earth are uncertain, it appears that the intensity and tracks of storms may change. In addition, it has been suggested that severe storms such as hurricanes may increase in frequency and/or intensity (Emanuel, 1987). Development of storm modification and rainfall enhancement activities, which are not now possible, may be able to provide the capability for amelioration of severe situations.

8.5.5.3 Sea Level

Projections suggest a potential sea level rise of up to about 1 m by 2100. Areas below sea level that could be flooded with seawater are too limited to displace more than a very small amount of the projected rise. In fact, reductions in groundwater (due to pumping) and reservoir volumes (due to siltation) probably play a small role in increasing sea level. Melting of mountain glaciers could contribute a significant fraction to the projected sea level rise, so that preserving (or enhancing) such mountain glaciers would be helpful. Similarly, protecting the Greenland and Antarctic icecaps would be helpful.

An active approach to counteracting sea level rise from greenhouse gases could be undertaken by pumping ocean water and spraying it as snow onto the East Antarctic icecap. This would require pumping of up to about 3 trillion tons of ocean water per year up several kilometers and onto the icecap, which would add about 0.3 m depth (as water) per year to the East Antarctic icecap. Such active movement of water would require substantial amounts of energy (possibly costing more than the damages from the rising sea level). However, it may be possible to enhance natural transport of ocean water up onto East Antarctica by adjusting (increasing, decreasing, or changing the shape) sea ice extent around the continent, which might more favorably reposition the storm tracks for snow buildup.

Overall, there is some potential for moving actively to moderate the impacts of the climate system, but these appear more feasible on local rather than global scales because energy requirements for intervening are large and side effects may be important.

8.6 FUTURE RESEARCH ACTIONS

Preliminary inquiry suggests that geoengineering options might be technically able to counteract inadvertent human impacts on the Earth's climate system. It is also clear that our understanding of the

Earth system is insufficient to predict the range of consequences either from an enhanced greenhouse effect or from geoengineering. Nonetheless, the potential negative impacts of climate change and the possibilities for effective, low cost geoengineering solutions indicate that further research seems appropriate.

Research to investigate geoengineering responses to climate change should be considered as long term, strategic insurance for society. An essential basis for pursuit of geoengineering options must be an improved understanding and predictive capability for climate change itself. Essential to that goal, and also to better design of response options, must be understanding of the fundamental processes that influence climate, including the behavior of clouds, aerosols, ocean dynamics, and the operation of numerous physical and, especially, biospheric interactions that influence the carbon cycle. Of course, this research is essential for society to understand and respond to concerns about global change in any event.

Already at this stage it would be useful to acquire information and inventories of the distribution of likely sites, and the potential of various countries and regions to effect programs in forestry, and the availability of geological reservoirs as sites for storage of greenhouse gases. Such studies would help to define the potential and costs of those options.

Geoengineering projects would require integrated, comprehensive, systems analysis to assess options for design, costs, and evaluation of associated impacts. Also, research and planning would require consideration of implementation strategies broadly, including authorization and response to ethical issues that will surely be raised around any approach. Clearly there are distinctions between inadvertent and advertent climate change. These issues should be addressed by groups involving expertise outside the normal engineering and scientific disciplines.

Finally, given the long term strategic nature of the issues, research should address and support options that might draw from related disciplines, such as materials science, robotics, self-organizing systems, and the potential for genetic engineering and biosystems studies to contribute. Ultimate implementation of strategies, such as loading the atmosphere with albedo enhancing materials, likely may require systems that operate automatically and rely on natural sources for materials and energy.

REFERENCES

Albanese, A. S., M. Steinberg, Environmental Control Technology for Atmospheric Carbon Dioxide, Final Report, DOE/EV-0079, U.S. Department of Energy, Washington, D.C., 1980.

Baes, C. F. Jr., S. E. Beall, D W. Lee, G. Marland, The collection, disposal, and storage of carbon dioxide, in *Interactions of Energy and Climate*, W. Bach, J. Pankrath, J. Williams,(Eds.), pp. 495–519, D. Reidel Publishing Co., 1980.

Berger, W. H., Climate change: no change down under, *Nature*, 351, 186–187, 1991.

Breuer, G., Can forest policy contribute to solving the CO_2 problem?, *Environ. Int.*, 2, 449–451, 1979.

Broecker, W.S., T.-H. Peng, *Tracers in the Sea*, Eldigio Press, Lamont-Doherty Geological Observatory, Palisades, NY, 1982.

Broecker, W. S., *How to Build a Habitable Planet*, Eldigio Press, Lamont-Doherty Geological Observatory, Palisades, New York, 1985.

Broecker, W. S., The great ocean conveyor, *Oceanography*, 4, 79–89, 1991.

Budyko, M. I., The method of climate modification, *Meteorol. Hydrol.*, 2, 91–97 (in Russian), 1974.

Canavan, G., E. Teller, Distributed remote sensing for defense and the environment, Los Alamos National Laboratory Report LA-UR-91-1169 (Rev.), Los Alamos, NM., 1991.

Charlson, R. J., J. Langner, H. Rodhe, Sulphate aerosol and climate, *Nature*, 348, 22, 1990.

Charlson, R. J., J. Langner, H. Rodhe, C. B. Levoy, S. G. Warren, Perturbation of the northern hemisphere radiative balance by backscattering of anthropogenic sulfate aerosols, *Tellus AB*, 43, 152–163, 1991.

Charlson, R. J., J. E. Lovelock, M. O. Andrea, S. G. Warren, Oceanic phytoplankton, atmospheric sulphur, cloud albedo, and climate, *Nature*, 326, 655–661, 1987.

Charlson, R. J., S. E. Schwartz, J. M. Hales, R. D. Cess, J. A. Coakley, Jr., J. E. Hansen, D. J. Hofmann, Climate Forcing by Anthropogenic Aerosols, *Science*, 255, 423–430, 1992.

Dale, V. H., R. A. Houghton, C. A. S. Hall, Estimating the effects of land-use change on global atmospheric CO_2 concentrations, *Can. J. Forest Res.*, 21, 87–90, 1991.

Decker, L. C., Sheep Mountain CO_2 line fills out West Texas EOR grid, *Oil Gas J.*, p. 57, May 19, 1986.

Dyson, F.J., Can we control the carbon dioxide in the atmosphere?, *Energy*, 2, 287–291, 1977.

Dyson, F. J., G. Marland, Technical fixes for the climatic effects of CO_2, in *Workshop on the Global Effects of Carbon Dioxide from Fossil Fuels*, Miami Beach, FL, March 7–11, 1977, pp. 111–118, W. P. Elliott, L. Machta, (Eds.), U.S. Department of Energy CONF-770385, UC-11, May, 1979.

Early, J. T., Space-based solar screen to offset the greenhouse effect, *J. Brit. Interplanetary Soc.*, 42, 567–569, 1989.

Emanuel, K. A., The dependence of hurricane intensity on climate, *Nature*, 326, 483–485. 1987.

Golomb, D., H. Herzog, J. Tester, D. White, S. Zemba, Feasibility, Modeling, and Economics of Sequestering Power-Plant CO_2 Emissions in the Deep Ocean, MIT-EL 89-003, Massachusetts Institute of Technology, Energy Laboratory, Cambridge, MA., 1989.

Hall, D. O., H. E. Mynick, R. H. Williams, Carbon Sequestration vs. Fossil Fuel Substitution: Alternative Roles for Biomass in Coping with Greenhouse Warming, PU/CEES Report No. 255, Center for Energy and Environmental Studies, Princeton University, NJ, 1990.

Hansen, J. E., A. A. Lacis, Sun and dust versus greenhouse gases: an assessment of their relative roles in global climate change, *Nature*, 346, 713–719, 1990.

Harmon, M. E., W. K. Ferrell, J. F. Franklin, Effects on carbon storage of conversion of old-growth forests to young forests, *Science*, 247, 699–701, 1990.

Hendriks, C. A., K. Blok, W. C. Turkenberg, Technology and Cost of Recovering and Storing Carbon Dioxide from an Integrated Gasifier, Combined Cycle Plant, Lucht 92, Ministry of Housing, Physical Planning, and Environment and Energy Study Centre, Leidschendam, The Netherlands, 1990.

Herzog, H., D. Golomb, S. Zemba, Feasibility, modeling and economics of sequestering power plant CO_2 emissions in the deep ocean, *Environ. Prog.*, 10, No. 1, 64–74, 1991

Hoffert, M. I., Climatic change and ocean bottom water formation: are we missing something?, in *Climate-Ocean Interaction*, M. E. Schlesinger, (Ed.) Kluwer, Dortrecht, The Netherlands, 1991.

Horn, F. L., M. Steinberg, Possible Storage Sites for Disposal and Environmental Control of Atmospheric Carbon Dioxide, BNL-51597, Brookhaven National Laboratory, Upton, NY, 1982.

Houghton, J. T., G. J. Jenkins, J. J. Ephraums (Eds.), *Climate Change, The IPCC Scientific Assessment*, Cambridge University Press, Cambridge, U.K. (IPCC), 1990.

Hoyle, F., *The Black Cloud*, Harper and Brothers, New York, 1957.

Hudson, H.S., A space parasol as a countermeasure against the greenhouse effect, *J. Brit. Interplanetary Soc.*, 44, 139–141, 1991.

Jones, D., Firefighting foam, *Nature*, 348, 396, 1990.

Joos, F., Sarmiento, J. L., U. Siegenthaler, Estimates of the effect of southern ocean iron fertilization on atmospheric CO_2 concentrations, *Nature*, 349, 772–775, 1991.

Kahle, A. B., D. Deirmendjian, The black cloud experiment, Rand Corporation Report R-1263-ARPA, Santa Monica, CA, 1973.

Kheshgi, H. S., The sensitivity of CO2 projections to ocean processes, in Third International Conference on Analysis and Evaluation of Atmospheric CO_2 Data, Extended Abstracts Report E.P.M.R.D. No 59, 124–128, World Meteorol. Organ., Geneva, 1989.

Koenig, L. R., A numerical experiment on the effects of regional atmospheric pollution on global climate, Rand Corporation Report R-1429-ARPA, Santa Monica, 1974.

Kostick, D. S., Soda ash and sodium sulfate, in Mineral Facts and Problems, U. S. Bureau of Mines, pp. 741–755, 1985.

Lamb, H. H., Climate-engineering schemes to meet a climatic emergency, *Earth Sc. Rev.*, 7, 87–95, 1971.

Manabe, S., R. T. Wetherald, On the distribution of climate change resulting from an increase in CO_2 content of the atmosphere, *J. Atmos. Sci.*, 37, 99–118, 1980.

Marchetti, L., On geoengineering the CO_2 problem, *Climatic Change*, 1, 59–68, 1977.

Marchetti, C., On geoengineering and the CO_2 problem, International Institute for Applied Systems Analysis, Laxenburg, Austria, 1975.

Marland, G., The prospect of solving the CO_2 problem through global reforestation, DOE/NBB-0082, U.S. Department of Energy, Washington, DC, 1988.

Marland, G., Why should developing tropical countries plant trees? (A look to biomass fuels to approach zero net CO_2 emissions), *Climatic Change*, 19, 227–232, 1992.

Marland, G., Global climate change: some implications, opportunities, and challenges for U.S. forestry, pp. 61–16, *Proceedings of the 21st Southern Forest Tree Improvement Conference*, Knoxville, TN, June 17–20, 1991.

Marland, G, S. Marland, Should we store carbon in trees?, *Water, Air, and Soil Pollution*, 64, 181–195, 1992.

Martin, J. H., Gordon, R. M., S. E. Fitzwater, Iron in Antarctic waters, *Nature*, 345, 156–158, 1990.

Mautner, M., A space-based solar screen against climatic warming, *J. Brit. Interplanetary Soc.*, 44, 135–138, 1991.

Matthews, E., Global vegetation and land-use: new high-resolution data bases for climate studies, *J. Climate Appl. Meteorol.*, 22, 474–487. 1983.

Mitchell, J.M., Jr., in *The Changing Global Environment*, S.F. Singer, (Ed.), pp. 149–173, Reidel, Dordrecht, The Netherlands, 1975.

Moulton, R.J., K.R. Richards, Costs of Sequestering Carbon Through Tree Planting and Forest Management in the United States, U.S. Department of Agriculture, Washington, DC, GTR WO-58, 1990.

Moulton, R.J., K. Andrasko, Policy options: reforestation, *EPA J.*, March/ April, 14–16, 1990.

National Academy of Sciences, National Academy of Engineering, and Institute of Medicine, *Policy Implications of Greenhouse Warming: Mitigation, Adaptation, and the Science Base*, National Academy Press, Washington, D.C., 1992.

Office of Technology Assessment, U.S. Congress, Changing by Degrees: Steps to Reduce Greenhouse Gases, OTA-0-482, U.S. Government Printing Office, Washington, D.C. 1991.

Peng, T-H., W. S. Broecker, Dynamical limitations on the Antarctic iron fertilization strategy, *Nature*, 349, 227–229, 1991.

Penner, J. E., R. E. Dickinson, C. A. O'Neill, Effects of aerosol from biomass burning on the global radiation budget, *Science*, 256, 1432–1434, 1992.

Penner, J. E., S. J. Ghan, J. J. Walton, The role of biomass burning in the budget and cycle of carbonaceous soot aerosols and their climate impact, in *Global Biomass Burning*, pp. 387–393, J. S. Levine, (Ed.), MIT Press, Cambridge, MA, 1991.

Post, W. M., T-H. Peng, W. R. Emanuel, A. W. King, V. H. Dale, D. L. DeAngelis, The global carbon cycle, *Am. Sci.*, 78, 310–326, 1991.

Radke, L. F., J. A. Coakley, Jr., M. D. King, Direct and remote sensing observations of the effect of ships on clouds, *Science*, 246, 1146—1149, 1990.

Rosenberg, N. J., B. L. Blad, S. B. Verma, in *Microclimate: the Biological Environment*, pp. 404–406, J. Wiley & Sons, NY, 1983.

Sarmiento, J. L., E. T. Sundquist, Revised Budget of the Oceanic Uptake of Anthropogenic carbon dioxide, *Nature*, 356, 589–593, 1992.

Schlesinger, M. E., X. Jiang, R. J. Charlson, in *Climate Change and Energy Policy: Proceedings of the International Conference on Global Climate Change: Its Mitigation through improved Production and Use of Energy*, edited by L. Rosen, R. Glasser, American Institute of Physics, New York, pp. 75–108, 1992.

Schwartz, S. E., Are global cloud albedo and climate controlled by marine phyto-plankton?, *Nature*, 336, 441–445, 1988.

Shine, K. P., R. G. Derwent, D. J. Wuebbles, J-J. Morcrette, Radiative forcing of climate, in *Climate Change, The IPCC Scientific Assessment*, J.T. Houghton, G. J. Jenkins, J. J. Ephraums, (Eds.), Cambridge University Press, Cambridge, U.K., pp. 41–68, 1990.

Solar Energy Research Institute, The Potential of Renewable Energy: An Interlaboratory White Paper, SERI/TP-260-3674, Golden, Colorado, U.K., 1990.

Steinberg, M., H.C. Chen, F. Horn, A Systems Study for the Removal, Recovery, and Disposal of Carbon Dioxide from Fossil Fuel Power Plants in the U.S., DOE/CH/00016-2, U.S. Department of Energy, 1984.

Steinberg, M., An Option for the Coal Industry in Utilizing Fossil Fuel Resources with Reduced CO_2 Emissions, BNL 42228 (Rev. 5/89), Brookhaven National Laboratory; Upton, NY, 1989.

Steinberg, M., Biomass and Hydrocarb Technology for Removal of Atmospheric CO_2, BNL 44410 (Rev. 2/91), Brookhaven National Laboratory; Upton, NY, 1991.

Study of Man's Impact on Climate (SMIC), Man's Impact on Climate, W. H. Matthews, W. W. Kellogg, G. D. Robinson (Eds.), MIT Press, Cambridge, MA, 1971.

Sundquist, E. T., Geological perspectives on carbon dioxide and the carbon cycle, in The Carbon Cycle and Atmospheric CO_2: Natural Variations Archean to Present, *Geophys. Monogr.* Ser. 32, E. T. Sundquist, W. S. Broecker, (Eds.), pp. 5–59, AGU, Washington D. C., 1985.

Sundquist, E. T., Geologic analogs: their value and limitations in carbon dioxide research, in *The Changing Carbon Cycle*, J. R. Trabalka, D. E. Reichle, (Eds.) Springer-Verlag, New York, pp. 371–402, 1986.

Tans, P. P., Fung, I. Y., T. Takahashi, Observational constraints on the global atmospheric CO_2 budget, Science, 247, 1431–1438, 1990.

Turco, R. P., O. B. Toon, T. P. Ackerman, J. B. Pollack, C. Sagan, Nuclear winter: global consequences of multiple nuclear explosions, *Science*, 222, 1283–1293, 1983.

Watson, R. T., H. Rodhe, H. Oeschger, Siegenthaler, Greenhouse gases and aerosols, in *Climate Change, The IPCC Scientific Assessment*, edited by J.T. Houghton, G. J. Jenkins, J. J. Ephraums, Cambridge University Press, Cambridge, U.K., pp. 1–40, 1990.

Watts, R. G., M. C. Morantine, (1991). Is the Greenhouse Gas-Climate Signal Hiding in the Deep Ocean?, *Climate Change*, U.K., pp. iii–vi.

Watts, R. G., (1985). Global Climate Variation due to Fluctuations in the Rate of Deep Water Formation, *J. Geophys. Res.*, pp. 8067–70, Vol. 90.

Wigley, T. M. L., Possible climate change due to SO_2-derived cloud condensation nuclei, *Nature*, 339, 365–367, 1989.

Wigley, T. M. L., Could reducing fossil-fuel emissions cause global warming?, *Nature*, 349, 503–506, 1991.

Wigley, T. M. L., T. P. Barnett, Detection of the greenhouse effect in the observations, in *The IPCC Scientific Assessment*, J. T. Houghton, G. J. Jenkins, J. J. Ephraums, (Eds.), pp. 239–255, Cambridge University Press, Cambridge, U.K., 1990.

Wigley, T. M. L., S. C. B. Raper, Thermal expansion of sea water associated with global warming, *Nature*, 330, 127–131, 1987.

Wolsky, A. M., E. J. Daniels, B. J. Jody, Recovering CO_2 from large and medium-sized stationary combustors, *J. Air Waste Manage. Assoc.*, 41,449–454, 1991.

World Resources Institute, World Resources: 1990–1991, World Resources Institute, Washington, D.C., 1990.

Wright, L. L., A. R. Ehrenshaft, Short Rotation Woody Crops Program: Annual Progress Report for 1989, ORNL-6625, Oak Ridge National Laboratory, Oak Ridge, TN, 1990.

Chapter 9

ENGINEERING RESPONSE TO GLOBAL CLIMATE CHANGE

Author: Robert G. Watts

9.1 INTRODUCTION

The scientific community is in near unanimous agreement that we will be faced in the next century with climate change due to increased greenhouse gas forcing. The magnitude of the climatic change, its distribution, the rate at which it will occur, and the impacts it will bring about are not known with certainty. How shall we respond?

Mark Twain once said that learning how to play two small pairs in the game of poker is as important as a college education, and costs about as much. Roger Price (1970) elaborated: "Poker, I've always thought, should be taught in colleges because it is an accurate analog to the larger realities of life. It teaches pragmatism and educates the player to the inevitable cause and effect. When you have two small pair and a large bet has been made, you cannot say, 'I'll get back to you later,' and you cannot take your money back. You must make a decision, and, of course, if you want to win, you must play. You must draw cards, an act which implies the possibility of losing. The (ordinary person) is a non-player. His fear of losing has destroyed his desire to win. He is the eternal kibitzer."

The business of the engineer is to design in the face of uncertainty. We do not, in fact, know what is in the cards for the future. We do, however, know a good deal about the possibilities and the probabilities of greenhouse induced climatic change, and we know much more about the current inefficiencies of energy use and the need for some kind of non-fossil fuel energy source for the long-term future. There is enough information available for the engineering community to begin to move forward toward helping to create a sustainable world. It is a daunting task. With the emergence of the possibility of the greenhouse effect, a certain sense of urgency has now entered the picture. It may be 10 to 20 years before an unequivocal signal indicating that climatic change is occurring emerges from the global data sets. It will in all likelihood be at least that long before climate models are able to predict future climate change with assurance, especially on regional scales. Yet if we continue to operate as we have in the past, we are essentially betting the livelihoods of future generations either:

1. that greenhouse warming will not be substantial, or
2. that it will not be so rapid or so extreme that we cannot adapt to it, or
3. that the change will be either beneficial or benign.

The consequences of losing such a bet are considerable. We must not be eternal kibitzers, waiting until climatologists can tell us precisely what the cards hold. If we decide to allow the climate to change as it will and to adapt to these changes as they materialize, this should be an acknowledged and informed choice, not the result of procrastination caused by our inability to make a decision. We need to begin now to plan in a responsible way for an uncertain future. Prudence requires that these plans account for the possibility of global warming, yet this should not and need not be the only consideration. The technological infrastructure is designed today subject to human needs and to certain constraints that are determined by economics, safety, and convenience, along with the laws of physics. In the future, environmental concerns, including potential global climate change, must become an additional constraint.

The possibility of greenhouse induced global warming brings with it a certain urgency for dealing with problems that we should be dealing with for other reasons. For example, decreasing the rate at which we use fossil fuels delays the arrival of the time when we will inevitably run out of these important resources. In many cases, efforts to improve the efficiency with which we use energy can actually improve the economy and security of the nation and the world. The development of renewable and long-term energy sources such as solar energy and fusion will eventually be necessary as fossil fuel resources are depleted.

It is also clear that the Earth's population cannot be allowed to increase without bounds. We should strive toward levels of human population that the Earth can support with a reasonably comfortable lifestyle. Research aimed at understanding the Earth as a system is important in its own right. In short, research and development inspired by the prospect of greenhouse gas induced climatic change will benefit humanity regardless of whether climate change actually occurs.

The prospect of climate change prompts us to recommend that certain actions be undertaken sooner rather than later. The development of renewable energy sources such as solar and geothermal energy and long-term options such as fusion must be accelerated in order to be able to replace depleted fossil fuel resources in due time.

In Chapter 1, we noted that Darwin felt that what he termed the Fifth Revolution would be brought on by the exhaustion of the Earth's reserves of fossil fuels, and would lead inevitably to the inability of the Earth's resources to sustain its human population. The view expressed here is a more optimistic one. If we plan for the future in an intelligent way, the Fifth Revolution can become a transition from the use of fossil fuels to the use of renewable energy sources. Indeed, the current forms taken on as a result of each of Darwin's first three revolutions can be altered through application of the Fourth Revolution: the Scientific Revolution. As discussed in Chapter 4, the concept of the First Revolution must be broadened to accept the fact that there are other sources of energy besides fossil fuels. The practice of modern agriculture allows much more food to be produced per hectare than even a few years ago. More changes are suggested in Chapter 7. The current design of urban and suburban areas is such that transportation by private automobile is encouraged. These areas are not designed well to accommodate energy efficient transportation. Each of the first four revolutions outlined by Darwin has served humanity well. These have largely taken the forms of social and technological advances. Technology has only recently been identified as a destructive force, and this is largely because of the reluctance of the society as a whole (not just industry and government) to adopt environmental concerns as a design constraint. Darwin stated that the nature of the Scientific Revolution is the "discovery that it is possible consciously to make discoveries about the fundamental nature of the world, so that by their means man can intentionally and deliberately alter his way of life." The application of appropriate current technology and the invention of new technologies is surely the hope of the future. It is the only way we have of promising an adequate lifestyle in the future. But how can this be accomplished?

9.1.1 The "Cascade of Uncertainty": Shall We Begin?

Much has been said about the uncertainty in our ability to predict and describe the details of climate change with climate models, as well as the uncertainty in our ability to state with assurance that climate change due to the greenhouse effect has already occurred. Even if we could predict the future climate with certainty, we are not yet able to understand the impacts that climate change would have on the ecosystem. Furthermore, we cannot predict the effects that these impacts might have on the economic and political systems of the world. This unfortunate set of circumstances has been described by Bill Clark, who calls it "the cascade of uncertainty." Outcomes predicted by a series of uncertain models, each with ever more uncertain inputs become increasingly less certain until making decisions based upon the final predictions appears hopeless. The truth of this proposition is undeniable, of course. The implication that it renders decision making impossible is, however, dangerous. Some policy makers may be tempted to conclude that we must proceed with "business as usual" until some definitive future scenario can be predicted, or at least until a large part of the uncertainty is erased. Uncertainty is, of course, as integral a part of engineering and economic planning as it is of climate modeling. Nearly every decision is made within the context of its own cascade of uncertainty. The timeliness of making the decision must always be weighed against the consequences of being wrong and the consequences of waiting, which is itself a decision.

Let us examine the issue of responding to the greenhouse problem within the context of the cascade of uncertainty. The community of climatologists is in substantial agreement that if we continue to emit greenhouse gases into the atmosphere, climate change will occur and the impacts will very likely be serious. The rate of change will quite possibly be so rapid that many ecosystems will be severely stressed. The great uncertainty begins when we ask to know the details of the distribution of the future climate and its ecological and socioeconomic impacts. This is where the problem lies. It will in all probability be at least a decade, and possibly many decades, before this uncertainty begins to be cleared up. Can we wait that long before beginning to act?

We should emphasize that responses need not immediately take the form of widespread and expensive infrastructure changes or the adoption of energy sources that are not yet economically competitive.

Some responses need to take the form of intensified research both in the study of climate change and impacts and in alternative energy sources, as well as education, policy options, and economics.

9.1.2 No Regrets?

A number of options for reducing energy use are available now either at little or no economic cost or even with economic benefits. These should be adopted now even if we ignore the possibility of greenhouse induced climate change.

Such strategies are referred to as "no regrets" strategies. No one can argue convincingly against the proposition that we should be doing things that avert the possibility of global warming that we should be doing for other reasons anyway. It is a truism. (Certainly we should be doing things that we should be doing: increasing the efficiency of energy use when it has a short payback, for example.) Several examples are offered in Chapter 3. What is becoming increasingly clear is that this may not be enough. While increasing the efficiency of energy production and end use is desirable in itself, we need also to be looking to the more distant future. The "no regrets" strategy is a creative solution to the problems of potential greenhouse warming and national security for the immediate future, but it would be short sighted to presume that it alone will deal effectively with longer term problems in either national security or environmental concerns for the U.S. or for other countries. If the greenhouse problem turns out to be as serious as many climatologists believe it will, then we shall regret not having done more. Planning for the more distant future is an urgent problem with which we must deal as a world society. "No regrets" policies must be viewed in this perspective. Increased energy efficiency will contribute substantially to reducing the energy needed to sustain world requirements, but, as pointed out in Chapter 4, as the world energy requirements increase to meet the demands of developing countries we shall have to develop new sources of energy. Ultimately, we shall have to deal with a world economy that relies on renewable energy sources.

The infrastructure associated with energy production cannot be changed immediately. The more rapidly we change it, the more severe the resulting economic dislocation will be. The sooner we begin, the more deliberately and carefully we can proceed. We should also point out that the energy infrastructure in many developing countries is now being rapidly expanded, and it relies increasingly on fossil fuel. Once it is in place, it will be much more difficult to affect change.

One might compare the present situation with that of an army in the battlefield. The commander is uncertain whether the enemy will attack from the front or from the right or left flank, on the ground or from the air. Yet he does not wait until the attack begins. Instead, he makes his plans. How do we proceed?

9.1.3 The Appropriate Responses

Our strategy and tactics must be developed in the context of the current international system. Dominant political, economic, and social institutions have developed in response to and in spite of competing claims from rival actors about how authority, resources and justice should be allocated. A great deal of psychic and social energy, not to mention more material kinds of investment, has been invested in the construction of this societal infrastructure. We might usefully divide social institutions into two types: status quo and revisionist.

Status quo institutions are the currently dominant ones. Lacking cogent information about who will be affected how by global climate change, they have little incentive to make any changes that might threaten their dominance. When levels of uncertainty are high they will predictably take a conservative stance. Revisionist institutions are critics and/or competitors of the dominant ones. Sometimes they have not reached the level of institutionalization, but are looser, less coherent entities we call social or political movements. The radical environmental and human rights movements are examples. With zeal, they seize on evidence of any wrongdoing, lack of vision, or injustice on the part of status quo institutions (e.g., government agencies, largest banks) to support their demands for reform, or change. Naturally enough, they tend to see predictions about climate change in terms of how it might affect their goals vis-a-vis the dominant institutions.

At the international level, the dominant institution is the sovereign nation-state. Sovereignty is the possession of ultimate authority over a given population in a given territory. Although often honored in the breach, state sovereignty remains the fundamental principle of the international system. There is no overarching, authoritative world (or even regional) government to establish order and justice. Each state must look to itself in the final analysis. Given this institutional structure, achieving cooperation among sovereign states in a frequently hostile world is a major problem for those who would approach issues from a perspective that relies on transnational cooperation.

Because the effects of climate change will probably be diverse, impacting different regions in different ways, and because of the high levels of uncertainty in our understanding of these effects and their consequences, not all institutions are on the same page at the same time. Any prescription for change involves some redistribution of wealth or of power. States and other dominant institutions, obeying the systemic imperative to look out for their own interests (whether social, economic, or political) first, are inclined to cooperate only when it is in their perceived self-interest. Thorough-going altruism is a rather scarce commodity in the world. Given the relatively short time spans of politics, altruism is usually seen by political actors as a rhetorical device. Acting altruistically in the real political world is typically seen as dangerous.

Such systems will not change overnight. How much time it will take for cooperation to become the rule rather than the exception depends on the level of redistribtion demanded and the cogency of the information that science can provide about global climate change. For science, the problem is that, normally, considerable time is needed to obtain information that will be consensually accepted, and thus serve as a rationale for policymaking that requires redistribution of wealth and power. Good science proceeds carefully and cumulatively to test novel insights. Unless urgency is patent, significant shifts of time, effort, and material resources are not to be expected.

If such shifts are driven by political forces, the types of information that will be acceptable to decision makers will be significantly shaped by political factors. Politics, one old definition says, is about who gets what, where, and how. Arguments over the distribution of valued goods are likely to dominate the political arena, so that other considerations are lost or depreciated. Political urgency is not the best foundation for good science.

With all this in mind, let us now turn to some specific ideas about what might realistically be done. The factors with which we deal are energy, time, and information.

9.2 ENERGY, TIME, AND INFORMATION

It has been pointed out by Spreng (1978) that time and information can be traded for energy in a number of ways. The energy requirement for performing a task often depends strongly on how fast the task is performed. The time required to perform various household tasks, for example, is decreased when washing/drying machines, dishwashers, and vacuum cleaners are used. Similarly, the time required to commute is often decreased as the maximum speed and acceleration of the chosen mode of transportation increase. Time is also an important factor in the design and operation of energy systems through considerations of the longevity of the required infrastructure and the resulting amortization costs. If the infrastructure can be designed so that it lasts a very long time, the amortization costs can be very low, resulting in less expensive energy. However, putting in place an essentially immortal infrastructure must be contemplated with care because of our inability to predict future needs and constraints with certainty (as pointed out by Marland and Weinberg (1988)). This points to the important role of information.

Information enters the scene in at least three forms: technical knowledge, uncertainty about the future, and the use of electronic devices as a substitute for energy or as a means for improving energy efficiency. For example, information exchange through telecommunications can allow us to avoid work related travel altogether, at least theoretically. Technical know-how is directly related to energy efficiency in that it allows designers and builders to use the best available

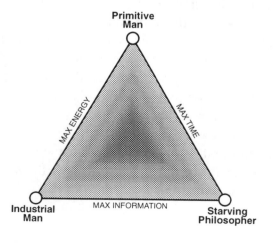

Figure 9.1   Spreng's idea relating time, energy, and information.

technology. On the other hand, our limited ability to predict the future leads to over-design and redundancy in energy systems design. Spreng's ideas are illustrated in Figure 9.1.

The corners of the triangle represent the extreme cases of the starving philosopher (who has much information and time but uses little energy), primitive man (who has little information and must use much time and energy), and industrial man (who accomplishes things quickly by using much information and energy).

By making different choices one can move a state point around inside the triangle. For example, by making use of the best available technology, one can save energy by moving closer to the maximum available information line, but frequently with a sacrifice of time. Time is required to obtain information (e.g., by reading labels and researching which alternatives actually result in saving energy). Similarly, a given task, for example commuting, might take longer if smaller, slower cars with low acceleration are used. If carpools are used, one might spend more time waiting (either for a late ride or for a late rider), or it may take longer to stop to pick up passengers than to go directly to a final destination.

Acquiring more knowledge through research can result in efficiency improvement, but time is lost in waiting. Reducing uncertainties about the future allows for proper discounting and for eventually building a more permanent infrastructure that will probably be less expensive when amortization costs are considered. Uncertainty about the future necessitates flexibility in design strategies and the design of a flexible infrastructure that can be modified or added to without undue difficulty. These ideas are discussed further in the following subsections.

9.2.1 Time and Energy

Time can be substituted for energy in a number of obvious and perhaps not so obvious ways. We have already alluded to some of these. We all recognize that time as well as energy is an important resource. The time that we have at our disposal in our lifetimes is finite, and we constantly seek ways to conserve it. In modern societies the things that we call labor saving devices are often really time saving devices. We wash our dishes in automatic washers and we use clothes dryers instead of taking the time to hang clothes on clotheslines to dry in the Sun. We drive personal automobiles when we could use public transportation (with savings of energy) not only because it is convenient, but because it saves time. Commuting time to our work places is often minimized by the use of personal automobiles by avoiding waiting time for public transportation, or for waiting for a late ride or a late rider when carpooling is used. We use automobiles with high acceleration, and we drive at speeds higher than the speed for which maximum efficiency is attained in order to minimize the time required to reach our destinations. Of all of our resources, many would rank time as one of our most precious commodities. In some sense impatience is an enormous waster of energy.

Time also enters into our considerations of energy policy through the concept of a time preference function. The usefulness of any capital stock such as energy depends not only on its present yield, but also on future prospects. The overall value of a given energy policy would then depend not only upon the current costs and availability of certain energy sources but also on future costs and availabilities and on how we choose to discount the future. We might choose to ignore future considerations almost entirely, in which case we would use a very high discount rate. On the other hand, if we take the viewpoint of an extreme preservationist, we would choose a very low discount rate. It is important to understand that discount rates are not functions of economics alone. In the real world, all markets are influenced by economics, politics, and engineering within a social setting. Morals, ethics and personal preferences enter the equation along with scientific and political possibilities and the perception of the future. If building a future with little environmental insult is of primary concern, discount rates will tend to be low, while the perception of an insecure future generally leads to high discount rates. On the other hand, developing countries often discount possible future environmental problems in favor of near-term economic growth. Obvious examples are the destruction of forests and the increasing interest in soft coal as an energy source, especially in China, where it is an abundant resource.

Time can often be a consideration in personal choice of the best and most efficient energy technology because many consumers of energy do not consider the cost of energy over a product's lifetime.

Several examples of how energy can be saved by using technologies that are already available are presented in Chapter 3. Major advances have been made recently in efficient lighting technologies, window systems, appliances, and building control systems. Higher efficiency lighting and low emissivity windows can result in large energy savings, often at negative cost, but their market penetration is presently low. Major barriers to market penetration of these new technologies are lack of information and lack of incentives to overcome higher first costs. Builders often are either unaware of the best available technology or lack the incentive to employ it (because they do not pay for future energy bills), or both. Likewise, consumers are often unaware of efficiency differences between products or are driven by capital costs instead of total (life cycle) costs.

Finally, time must enter our considerations through the idea of investment and amortization costs. Investment in a particular infrastructure is a powerful incentive not to change. Energy producers are unlikely to look kindly upon the idea of abandoning a coal-fired power plant before it has reaches retirement age. Careful planning is necessary before building an energy infrastructure with a long expected lifetime.

9.2.2 Information and Energy: Technical Knowledge

Education at all levels is a key element in promoting the use of currently available energy saving technologies. Students, practicing engineers and related professionals, and government planners, should be educated on the cost effectiveness of technologies and encouraged to include environmental impacts as a design constraint, using the estimated costs of externalities. While this may be accomplished relatively easily in engineering schools, it will be considerably more difficult to retrain practicing engineers and professionals.

The penetration of some technologies into the market may require incentives in the form of government standards. The U.S. federal government has mandated energy-efficiency standards for some residential appliances and for fluorescent ballasts. The recent Energy Policy Act of 1992 (EPAct) sets standards on additional products such as lamps, motors, commercial heating, cooling, and water heating equipment, plumbing products, and distribution transformers. While initial savings will be modest from the first generation of standards, several of these standards will receive periodic updates.

In order to capture more potential energy savings beyond standards, it is necessary to overcome consumer emphasis on minimizing first cost by educating consumers on life-cycle costs (considering energy costs over a product's lifetime). Labeling and information programs for windows, lamps, and luminaries are included in EPAct. Residential energy efficiency ratings and energy efficient mortgages are part of the legislation. Funding for energy efficient lighting and building energy centers, as well as federal energy efficiency information programs, are included.

Other provisions of EPAct relating to energy efficiency include state adoption of commercial and residential building energy efficiency standards, federal building energy efficiency programs, district heating and cooling studies, grants to state regulatory agencies, assistance to state energy conservation and low-income programs, encouragement of efficiency investment by electric and gas utilities, advanced appliance development, studies of early retirement of inefficient appliances, and industrial energy efficiency programs. These are good first steps toward comprehensive federal incentives toward U. S. energy efficiency in all sectors of the economy.

Many items for research and development leading to efficiency improvement are identified in Chapter 3. Opportunities exist in the manufacturing sector through a combination of options including enhanced end-use efficiency, fuel switching, co-generation of heat and power, and process conversion and redesign. Some of these involve implementation of known and proven technologies. Fuel switching opportunities in the manufacturing sector involve replacing coal-fired boilers with those that use oil and gas. Care must be taken, however, to evaluate total system performance in light of greenhouse warming potentials or some more precise criterion. If natural gas leakage is relatively high the advantage in terms of greenhouse warming may be lost. It appears, however, that these losses can be made relatively small with good engineering design. Nevertheless, appropriate precautions are essential. Co-generation can displace fairly large amounts of CO_2 emissions by using waste heat. Several studies have indicated that the potential for displacing significant quantities of fossil fuel use in some industries with electricity through process-wide modifications is large.

Gas turbines have traditionally been used only for peaking power in electric generating facilities, mostly due to their low efficiencies. Following intensive research in aircraft turbines by the military, gas turbines have recently become both highly efficient and low in cost. When these turbines are used in gas-steam combined cycles, efficiencies as high as 50% can be attained, compared to the average efficiency of 35% for modern coal fired power plants.

Other opportunities exist in the manufacturing sector for both waste minimization through process optimization, and waste utilization through recycling, especially in the production of paper, glass, organic chemicals, and petroleum refining.

Significant opportunities should emerge over the next decade in these areas to adapt process design to accept larger volumes of recycled materials as input feed stocks. Likewise, pervasive process redesign across industries can result in important inroads in waste minimization and avoided CO_2 emissions resulting from the lost energy in those wastes. In some instances, additional energy can be saved in disposal and treatment of those wastes.

In the transportation sector, fuel use by personal automobiles represents a large portion of world energy use. The technological structure for increasing personal automobile efficiency requires both creative research and policy actions. For example, consider the use of express lanes for high occupancy vehicles. This is a case where a policy action results in decreased energy use and time saved. High efficiency automobiles are already available, but often require the use of expensive, state of the art materials. Consumer choice exerts an important influence on the average passenger miles per gallon of personal automobiles as well as the choice between single passenger use, car pooling, and the use of mass transit. There is disagreement on the relative impact of fuel prices and regulations on the development of fuel efficient cars.

The U.S. Government's corporate average fuel efficiency (CAFE) regulations have resulted in increased fuel efficiency of new cars in this country. High gasoline prices in Europe are accompanied by the use of smaller, more efficient cars. However, while gasoline costs more than twice as much in Europe as in the U.S., fuel efficiency is only about 20% greater, indicating that the demand for automobile fuel may be highly inelastic.

Fuel switching to compressed natural gas for automobiles is technologically feasible. Public acceptance of alternately fueled vehicles awaits the availability of fueling stations, among other things. A potentially more serious problem with cars fueled by natural gas is that relatively small losses may cause a more serious greenhouse problem than that resulting from CO_2 production. Losses in well designed pipelines can probably be held to a fraction of 1%, but preventing losses during refueling and from the automobile itself may present a more formidable engineering design problem. Moreover, estimating how much loss is permissible will require a more realistic measure of the relative importance of methane than the currently used greenhouse warming potential.

The use of public mass transit systems could result in decreased fuel use in the transportation sector, but public acceptance and use is a problem. The costs of private car use in the industrialized countries would probably have to be raised to a high level given the current infrastructure and the lack of convenience of public transportation in most urban and suburban areas.

9.2.3 Information and Energy: Developing Countries

Since global warming is an international concern, the U.S. (emitting 25% of the world's CO_2) and the other OECD countries (emitting another 25% of the world's CO_2) cannot adequately respond to the challenge alone. However, the policies of the industrialized countries play the largest role in current greenhouse gas emissions and can exercise significant influence on reducing future emissions. Planned international negotiations involve many complex issues, including important ethical and equity questions. Developing countries will be most likely to participate in efforts to combat global warming if the industrialized countries take the first steps, and if international agreements provide developing countries with the necessary financial and technical resources to support their own programs.

Industrialized countries have been responsible for the majority of greenhouse gas emissions from energy consumption in the past. In 1985, about 48% of total carbon from CO_2 was emitted by the US/OCED, 36% by the USSR, Eastern Europe and China, and 16% by the developing countries (IPCC, 1990).

Much attention has been focused recently on developing countries because of the impending growth in their emissions. While energy consumption is still 5 to 100 times lower per capita than in OECD countries, the rate of growth has been 7 times higher over the past two decades (Levine et al., 1991). If standards of living as well as population grow as projected, developing country greenhouse gas emissions could grow three- to fourfold by 2025, and their contribution to global warming would then rise to 30% of the total, with the US/OECD at

34%, and the USSR, Eastern Europe and China at 37% (IPCC, 1990).

Developing countries are not necessarily bound to follow the development path that the industrialized countries have pursued over the past 50 years. As shown in Figure 3.2 in Chapter 3, a great discrepancy exists in the historical relationship between GDP and commercial energy consumption between the U.S. and Japan, for example.

Developed countries can contribute toward helping developing countries "leapfrog" the early inefficient stages in several ways. The first is by "good example." Developing countries cannot be expected to curtail greenhouse gas emissions while developed countries continue to consume energy at substantially higher rates while enjoying a much higher standard of living. New efficient technologies and renewable energy sources will be much more widely used in the developing world if they have already been commercialized in the US/OECD, reducing prices and increasing their appeal as "state-of-the-art."

Energy efficiency can be especially important for developing countries to reduce costs and environmental damage and to increase energy security. Significant energy savings, on the order of 20–25%, are available with present technologies at a payback period of two years or less (Levine et al., 1991). Substantial energy efficiency savings opportunities exist in OECD countries similar to those described for the U.S. in Chapter 3.

Solar energy and hydropower may already be cost effective or nearly so in many countries, particularly developing countries, as discussed in Chapter 4. Arid and other unproductive land that is of low value for other purposes could be exploited for solar energy. If this could be accomplished in countries like Mexico and India, for example, where the daily solar flux at the Earth's surface in at least some regions is large throughout most of year, solar energy could conceivably become a major export industry, helping the economies of these countries.

Since solar energy is only available when the sun is shining, a solar energy economy requires some sort of energy storage. A variety of enabling energy technologies are discussed in Chapter 4. Research in these technologies needs to proceed at a sufficiently rapid pace that they become economically attractive before developing countries become committed to a large fossil fuel infrastructure.

The portion of atmospheric CO_2 increase that can be attributed to deforestation comes mostly from developing countries, mainly from tropical forests. Deforestation takes place for a variety of reasons: fuel wood use, clearing for agricultural use and pasture, and extraction of timber resources, for example. From the point of view of developing countries such practices are perfectly appropriate to promote development, at least in the short term. If these practices are to decrease, and if reforestation is to increase, it must be demonstrated that it is in the best interests of the particular countries in question.

Important barriers exist to realization of savings described above for developing countries. As elaborated in Chapter 3, world capital constraints will make financing of projected supply difficult. International lending policies, which have favored large capital-intensive energy projects, must shift toward more cost-effective investments in energy efficiency. Lack of trained manpower, adequate delivery infrastructure, and access to efficient technologies must be overcome by technology transfer and assistance from the developed countries.

The structure of international trade also makes regulation of greenhouse gas emissions complicated. Treaties and trade practices should be devised that prevent emissions and foster environmental protection, rather than weakening existing regulations. Emissions trading is another policy that may be promising if logistics can be worked out and enforcement provisions established.

Another way that the developed world can aid the developing world is through technology transfer. If long-term success is to be achieved, efforts must be aimed at producing talent and appropriate infrastructure in these countries, and be disseminated from within as far as possible. Professional and technical training programs and access to technology and programs are excellent investments in reducing global warming as well as developing international human capability.

9.2.4 Information and Energy: Uncertainty About The Future

The need to reduce uncertainties about future climate and its impacts is necessary for planning for the long-term future. It seems certain that neither the general public nor policy makers will become convinced that greenhouse induced climatic change is a real problem until climatic change and its impacts can be predicted with more confidence, or unambiguously detected. Moreover, responses, especially mitigative responses, depend strongly upon our ability to predict regional changes. Therefore, research in climate change and the resulting impacts should certainly go forward, probably with increased urgency. This will require continued efforts in data collection and monitoring as well. The same is true of agricultural and biolgical systems (Chapter 7).

An essential component of global change research lies in understanding the global biogeochemical cycles. The role of trace gases in the chemistry of the atmosphere, as well as the role of the biosphere in producing and consuming trace gases, is currently poorly understood. The hydrologic cycle is currently parameterized rather crudely in general circulation models. The interaction of vegetation with the hydrological cycle is also poorly understood. Ocean biogeochemical processes undoubtedly both respond to and affect climatic change. These are areas of active research that need to be encouraged.

Research in some renewable energy technologies that can quite possibly have a large impact on the energy sources of the future needs to be funded in order to reduce uncertainties concerning their future potential. Among those with uncertain futures are fusion, fission with breeders, solar space power, and storage and transmission technologies. Funding for renewable technologies and the related enabling technologies such as fuel cells, hydrogen, high temperature superconductors for storage and transmission, and nontoxic batteries, has decreased in recent years. This trend needs to be reversed if the U.S. is to become competitive in these technologies in the future.

If climatic sensitivity is close to that deemed most likely by the IPCC report, no reasonable mitigative response will prevent some climatic change. It therefore seems likely that some adaptation will ultimately be necessary.

Impact assessment, which has been largely neglected in climate change research, is an essential component necessary to prepare for appropriate and effective adaptation. Since there are many uncertainties in the quantitative assessments of sea level rise, for example, emphasis should be given to robust engineering solutions. Engineering research could be conducted in beach nourishment or structures that can be effective for a range of changing conditions.

Design under uncertain constraints has been the business of the engineer for many decades or centuries. However, engineers need to be constantly apprised of the changing needs of society and made keenly aware of the environmental constraints associated with climatic change. The assumption of a stationary climate, or at least the existence of climate with predictable and limited variability based on historical experience, is ingrained in engineering training, design, and operation. Professionals must be provided with the analytical tools to consider climatic shifts outside of traditional expectations.

The possibility of climatic change is not the only uncertainty that must be dealt with, of course. The future population and its distribution, the political state of the world, the future needs and desires of the various population groups, as well as the availability of the various energy sources are also uncertain.

The maintenance of stable shorelines at their present locations in environmentally sensitive areas may only be possible by engineering means including beach nourishment, accretion of fine material by active protection, armoring the forelands in front of sea dikes and strengthening them by revetments. These may affect or be viewed as affecting the environment in a negative manner through modification of the natural development of wetlands and marshes, loss of wildlife habitats by erosion, and development towards more shoreline armoring. For such cases, engineering solutions should be sought that meet the protection goals as well as the environmental quality desired. Acknowledging that this approach will not be possible in every case, definitive and well-monitored experiments must be carried out to address environmental concerns. They must lead to prognoses with respect to the development of the total affected system. Public education on coastal protection as well as environmental aspects are necessary to develop an adequate answer from the affected society whether to retreat or to maintain the existing protection system (Chapter 6).

Water resource planning is currently based on historical records of precipitation, stream flow, and temperature along with future demands projected from past trends. A key assumption is that climate is constant and thus, variability and extremes are somewhat predictable and manageable. A changing climate violates this underlying assumption. Current planning and design practices may not be adequate to deal with a changing climate. Moreover, existing water resource systems may not perform as planned. It is necessary to develop tools to analyze both new and existing water resource projects for risk/reliability relationships and uncertainty (Chapter 5).

Flexibility of design implies that designers of energy systems must keep in mind that as information about energy efficiency and other constraints are identified design constraints must change accordingly. Designed flexibility, on the other hand, means that the energy infrastructure must be designed in such a way that it can be easily adapted to new environmental conditions. For example, the transportation infrastructure should be designed in such a way that it is easily adaptable to new modes of transportation or new social structures as they arise. On the other hand, engineers must constantly keep abreast of current new energy efficient technologies as they arise. This places an enormous burden on practicing engineers and technologists as well as educators to see that information transfer is efficient and effective. Once again, this points to the fact that the efficient transfer of information is a cornerstone of effective energy policy.

9.2.5 Planning for the Long Term: Future World

What are our options? The nature of the future is generally the result of either good or poor planning in the present. According to Gregory (1931) it is said that "in China a road is good for seven years and then bad for 4000." Judging from the technological progressions in the past 100 years, it will be very difficult to guess what the future holds. For example, the nature of private and mass transport will likely be very different 100 years from now, perhaps in ways that are difficult for us to imagine. The best and most creative minds must explore what the future may be like and what we would like it to be like. In a world that is becoming stressed by population increase it will not be adequate to wait until events unfold and react to them. We need to understand that, equipped with the Fourth Revolution, we can, with creativity and determination, decide upon our future.

The various estimates of human population increase indicate that it will reach 10 to 20 billions by the end of the 21st century, or two to four times the current population. Since the carbon dioxide production from fossil fuel burning is the production per capita times the population, the increase in population would have immense implications for resource consumption and CO_2 emissions. Furthermore, since the largest increases in population by far will likely be concentrated in the developing countries, it will be important to pay close attention to how these countries develop, and to aid them in developing along environmentally conscious paths. An example will perhaps illustrate this point.

Many of the conventional scenarios of energy growth rely to some extent upon extrapolations of trends in industrialized countries. As we have pointed out in Section 9.2.3, as well as in Chapter 3, this need not be the case. The most rapidly growing sector of energy use in both developed and developing countries is transportation, particularly the personal automobile. The need for the personal automobile in urban areas of the U.S. and other industrialized countries is intimately tied to the way in which urban areas are designed. Just as the advent of the personal automobile made possible easy commuting to suburbs, the current pattern of a central urban area surrounded by suburban sprawl makes it difficult to design effective and inexpensive mass transport. This pattern has not yet developed in many developing countries.

Although improvements in fuel efficiency and the like are likely to make future automobiles less environmentally destructive, the industry must be viewed as a relatively mature one. It seems likely that in the rather distant future, that is, in 50–100 years, the nature of ground transportation will be quite different from what it is today. The advent of the personal automobile using fossil fuel as we have known it may have passed its apex. The question that we must ask is "What technology is likely to replace it?" It may be replaced by electrically powered vehicles or by hydrogen fueled personal automobiles. If new urban areas are sensibly designed, however, mass transport might greatly reduce inner city use of personal automobiles. Similarly, the development of high speed magnetic levitation trains could largely replace short to medium distance airplane trips.

The implementation of new technologies requires putting in place infrastructures that are frequently expensive. Hence, a great deal of careful thought and planning must precede putting in place an infrastructure that is potentially long lived. However, new technologies will undoubtedly emerge in the future that essentially replace those of today. It is essential that we maintain enough flexibility in both our minds and in our infrastructure to take advantage of opportunities for a transition to new technologies as they present themselves.

9.3 THE SOCIAL CONTEXT OF ENGINEERING

It is possible to think of technical fixes to the greenhouse problem if we ignore social, political, and economic constraints. "All" one has to do is build more efficient cars, improve building insulation, etc.

Fossil fuel use could be eliminated by producing electricity from nuclear or solar sources. Implementation of these purely technical advances is the hard part. In the book *A Field of Dreams* the main character hears a voice that tells him that "If you build it, they will come." The very practical question that engineers and the industries that employ them must ask is, "If we build it, will they buy it at a price that will produce a profit?" This is a very real constraint on whether a particular conceived technological fix can be implemented in practice. The process of implementing a particular technology involves the consideration of many variables within the triad that consists of industry (the producers of goods and services), society (the consumers of goods and services), and government, which acts, among other things, as a regulator. Industry normally reacts to profit. If a given industry does not make a profit, at least in the long run, it cannot survive. Individual members of society generally react to perceived self interest. Governments (democratic governments, at least) react both to the will of the people and to the pressures of various interest groups. Each is a stockholder in the sense that it gains or loses depending upon the outcome of decisions and the events that follow.

9.3.1 Values, Economics, and Decisions

Value is the set of constructs upon which a given society or a member of that society makes decisions. Economists naturally like to think of values in purely analytical terms because this makes them, at least theoretically, quantifiable. Accordingly, the value of a given item or service is precisely the monetary value for which one could purchase or sell it. Value, then, is rooted in utility. In his book, *Theory of Political Economy*, William Stanley Jevons wrote that "Repeated reflection and inquiry have led me to the somewhat novel opinion, that value depends entirely upon utility." Jevons was quite clear about what he meant by utility. This is evident in his statement that "persons are thus led to speak of such a nonentity as intrinsic value." Utility is that "abstract quality whereby an object serves our purposes, and becomes entitled to rank as a commodity." A little reflection on recent events, however, reveals that there are circumstances in which intrinsic values become a very important part of decision making, at least on a personal level. Utilitarian philosophy would discount the Spotted Owl absolutely in favor of jobs for lumbermen. In the same sense, the replacement of virgin forests by forest plantations has no utilitarian cost unless the destruction of biodiversity can be accounted for quantitatively.

Utilitarian economic theory is a highly simplistic notion in this sense, but it may be all that we have to work with within a theoretical framework. It is important, however, if we want our plans to work, always to keep in mind that intrinsic value is real in the minds of people, and that the intangibles associated with it must be considered. Intangibles are, by definition, difficult or impossible to quantify, but they must be considered in any intelligent decision making strategy.

Even if the set of values that are important to consider in making a given decision could be accurately listed, quantification would be a very difficult problem. Fischoff (1991) has discussed the general problem of value elicitation. He discusses the spectrum of value elicitation within the continuum spanning from what he calls the philosophy of articulated values to the philosophy of basic values. The philosophy of articulated values holds that people will pursue their own best interests, and that they will make value judgments based solely on the question at hand. On the other hand, the philosophy of basic values holds that people lack well-differentiated values on specific questions unless they are very well acquainted with the associated problems. They tend to derive valuations from a set of basic values through some inferential process. Fischoff gives many examples indicating the difficulty of assessing human values. In the real world people probably make value judgments on the basis of a philosophy somewhat between the two extremes.

Generally speaking, economists expect people to pursue their own best interests, making choices that reflect their values through their decisions. The consumer is assumed to evaluate his or her basic concerns, which are called attributes. An attribute represents a quality that the consumer considers in making a decision. The relative importance that the consumer places on all of the attributes can then be captured in a multi-attribute utility function. The consumer makes a decision based on examining this utility function in the light of market availability. As an example, consider a number of attributes in the decision to purchase an automobile: price, comfort, safety, gas mileage, style, etc. To simplify the discussion, and to facilitate making certain points later, we limit the utility function to two dimensional parametric space and consider three attributes: cost, gas mileage, and safety. The trade-offs among these attributes are shown in Figure 9.2. At a certain price the buyer is willing to trade safety for gas mileage. At a higher price he expects a safer automobile or better gas mileage or both. These curves are called indifference curves because the buyer is indifferent to the combination of safety and gas mileage at a given cost. Under certain reasonable assumptions the curves can be shown to be monotonic decreasing and concave as shown in the figure. The slope of a given curve will depend upon the relative importance a particular consumer places on the attributes.

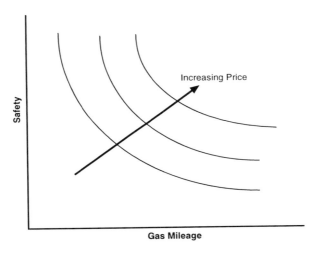

Figure 9.2 Indifference curves.

All technology has to be affordable to the consumer. This alone is not enough, however. The consumer must value the new technology enough to purchase it. Finally, the new technology must be available. Put simply, absent collusion by external forces (such as government regulations) producers must be able to deliver the new technology at a profit and consumers must be willing to purchase and use it.

9.3.2 The Environmental Constraint

The introduction of environmental concerns as a design constraint will often (although not always) increase the cost of an item. Purchasers must therefore place a value on the perceived environmental costs that is large enough to offset the added costs. In some cases, as in "no regrets" strategies, it is a matter of educating the public to total life cycle costing. Purchasing efficient lights entails a larger first cost, but in the long term results in net savings. This has been discussed in Chapter 3 and in the present chapter.

In order for the public to responsibly place a value on environmental concerns, it must be educated about the prospects of environmental degradation due to energy production, including possible climatic change. There is, however, a danger here that must be recognized at the outset. Education is not the same as indoctrination. In his book *Extraordinary Popular Delusions and the Madness of Crowds*, Charles MacKay recalls Schiller's dictum: "Anyone taken as an individual is tolerably sensible and reasonable — as a member of a crowd, he at once becomes a blockhead." We need to avoid "crowd thinking" when we seek solutions to problems like global warming. There are few guidelines as to how to do this. Climate change is a long-term problem, so the public must be constantly aware of its possible dangers. Faced with the more immediate and visible problems of unemployment, poverty, famine, and war, the public tends to quickly tire of hearing about the longer term prospects of less certain, and certainly less visible problems like global warming. On the other hand, confronting problems such as the prospect of global warming, like those associated with the problem of reducing the national debt in the U.S., can only be effectively dealt with in a democratic society if the constituency (the public) is willing to confront the problem and endure the personal sacrifices that are necessary. The best choice is probably to present what we know in appropriate classes and popular articles and books, clearly separating what is known from what is suspected, in a non-apocalyptic manner. "Crowd thinking" tends to have a short lifetime.

But if matters are clearly and impassionately presented to the public, we must be prepared to accept the will of the people. This presents the scientific community with an enormous responsibility, perhaps unlike any that we have had in the past. This is particularly true of the engineering community, which has, in the past, used only design constraints that insure such attributes as economy of sales and the immediate safety of the consumer, for example, to prevent lawsuits. The environmental constraint must include global concerns that are to some extent intangible, and the public must accept these as real concerns. Engineers must be willing and able to explain to the public why these concerns are important to them and how they personally might be affected by their decisions to purchase items that are environmentally destructive.

In the next section we discuss a theoretical construct for approaching this problem scientifically. There are certain opportunities that make sense for immediate implementation. Others present challenges that might be possible in the near future, but require research before they can be implemented economically. Finally, the long-term potential solutions to the problem of attaining a sustainable energy future whose futures are somewhat uncertain are termed possibilities. These are the sources that will require extensive research to see whether they are possible and practical.

9.4 OPPORTUNITIES, CHALLENGES, AND POSSIBILITIES

Federal support of research and development in the U.S. in the post-World War II era has officially been centered around the proposition advanced by the Vannevar Bush report (NSF, 1945) urging that basic science should be viewed as a "public good," and therefore be supported by public funds. Nevertheless, a large part of research and development funding has gone to the support of technological development through "mission-oriented" research in the areas of space and defense as well as medicine and energy. More recently, significant attention has been given to precommerical technological development that is essential to the long-term health of the nation's economy and to national security.

Heaton et al. (1992) have defined environmentally critical technologies as "those that can reduce environmental risk substantially through significant technical advance." They further state that "technological developments can be considered environmentally critical if:

- their use brings about large, cost-effective reduction in environmental risk;
- they embody a significant technical advance;
- their adoption involves a favorable ratio of social to private returns."

The general idea of the Heaton report is to identify appropriate criteria and to list the environmentally critical technologies that are deserving of private and public support. In the private sector, many companies are beginning to recognize the creation of environmentally friendly technologies as a business opportunity. Nevertheless, although environmental research and development funding by private companies has increased somewhat, it is still not sufficient to meet the challenges of the future. There are a number of legitimate reasons for this. It is often difficult for private companies to recoup their investments in research and development in new technologies. Owing to the competitive nature of private industry, the results of research in new technologies are often kept as proprietary information. This results in duplication of effort, delays in publication of conclusions, and delays in the use of the new technology by competing industries. Also, private research and development tends to emphasize technologies that can potentially yield a near-term profit, rather than solving longer term problems. Emphasis is most often placed on marketing new applications of products already developed or nearly developed. Even so, market penetration of improved technologies such as efficient lighting and appliances has been slower than would be desirable. Research and development concerning longer term environmental problems has been inadequate.

Private investment in environmentally friendly agricultural practices has been underemphasized for a number of practical reasons. For example, the use of modern information and monitoring systems can lead to optimal use of water, fertilizers and pesticides, but the benefits are mostly environmental, and may not substantially increase profits by individual farmers.

Hydrology and coastal management have generally been viewed as being in the public domain, and research and development in these areas has been typically funded by federal and state governments. However, changes in climate and its variability have not normally been considered as design parameters in the past.

Public support of research is warranted whenever the research cannot be expected to be supported by the private, for profit, sector and when the research may be deemed "in the public good." In the next subsection we attempt to delineate some rather subjective measures for selecting what research associated with climatic change and the engineering response falls into this category.

9.4.1 The Valuation Criterion Space

The methodology for choosing which research deserves public support is necessarily subjective. The members of our workshop spent considerable time discussing methods of evaluating different research agendas. Generally, we look for some sort of balance between "costs" and "benefits." Unfortunately, valuation of both costs and benefits is somewhat elusive. Both costs and benefits involve, to some degree, intangibles. For example, both costs and benefits may be valued differently by different social groups, even within a given country. This point is particularly acute among the lesser developed countries, where rapid economic development is often viewed as much more important than environmental concerns. We offer here a "valuation criterion space" in which we recognize both cost and benefit as vectors involving multiple components, and for which selection of appropriate research and development recommendations is somewhat subjective. Components of the chosen vectors are shown in Table 9.1.

Table 9.1   Valuation criteria vectors.

Cost/Difficulty	Benefits
• Cost of materials and labor	• Reduced environmental risks
• Technical difficulty	• Significant technical advance
• Public acceptance resource	• Potential to meet long-term demand
• Time to implement	• Potential for global implementation
• Intangibles	• Intangibles

Cost and difficulty are lumped together, and include real labor and material costs as well as intangibles, technical difficulty, public acceptance, and time required for implementation. The benefits vector includes components somewhat similar to, but not identical to those listed by Heaton et al. (1992). These are reduced environmental risks (we are mainly concerned with the risks of climatic change), significant technical advances, the potential to meet long-term resource demand (energy, food and fiber, water-resources, etc.), and the potential for global implementation.

The valuation criterion space is illustrated qualitatively in Figure 9.3, and outlined in Table 9.2. The *No Regrets* space contains those research, development, and deployment options that produce benefits with a net economic gain (negative costs). An example is energy efficient lighting, which is cheaper when total life cycle costing is considered, is already technically developed, and which can be deployed immediately, although public acceptance may be a problem. Options that can be implemented on a relatively short time frame but at some cost occupy the *Opportunities* space. An example is wind power, which is currently competitive with other energy sources on a cost per kilowatt basis, but only in areas where the wind speed is consistently reasonably high, and with some time required to establish the needed infrastructure. The potential to meet a large fraction of long-term global energy demand is low.

If research is likely to lead to cost effectiveness on a short to moderate time frame we place an option in the *Challenges* space. An example is solar photo-voltaic energy production. While the cost of this energy source is still not competitive, it has been decreasing fairly rapidly. The potential for meeting long-term global energy demand once the technology is mature is high provided that enabling technologies involving distribution and storage are also developed.

In the *Possibilities* space we place those options that offer potentially very large benefits, but whose future is uncertain due either to high cost or technological difficulties. Obvious examples are fusion technologies and solar power satellites. Both of these options offer practically unlimited sources of energy. The future success of both technologies is, however, uncertain.

We have emphasized throughout this chapter that choosing whether a particular research, development and deployment option falls within the *Opportunities, Challenges,* or *Possibilities* sectors of the valuation space, or falls within the *No Regrets* sector, is somewhat subjective. We offer what we interpret as the consensus view of our workshop. Other groups would no doubt come to a different consensus. The strength of our group lay in its diversity and in the fact that

Table 9.2   The valuation criteria space.

No Regrets Space
Deployment on a short time frame with net economic gain.

Opportunities Space
Options that can be implemented in a relatively short time but at some cost and with some technological difficulty. Moderate potential to meet long-term resource needs.

Challenges Space
Research likely to lead to movement into the Opportunities Space on a short to moderate timeframe.

Possibilities Space
Uncertain future due to high cost or difficulty, but with potentially great benefits.

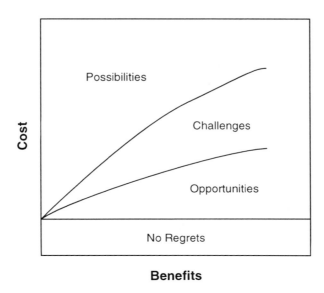

Figure 9.3   The valuation criteria space.

engineers and physical scientists who are actively involved both in climate research and in impacts and associated engineering activities were participants. In this spirit we offer the concluding quantifications.

The list of items that we have presented speaks volumes about the idea of creating national energy plans for one, four, or even eight years or longer. The problem of supplying energy to the people and nations of the world is a long-term problem and one that crosses continents and national boundaries. What is needed is a global energy plan that spans decades or centuries. We do want to emphasize the notions presented in this chapter concerning both design flexibility and flexibility of design. While a long-term plan for energy policy seems necessary, as research leads to insights about the future of the human interaction with the health of the planet, and as values and technology change, as they will, we need to be prepared to alter our goals for a long-term plan.

9.5 A RESEARCH AND DEVELOPMENT AGENDA

NO REGRETS

Delivering Efficiency
- Expand the mandate for utilities to act as energy services. Profits to utilities from investments in conservation [Chapter 3].
- Establish test criteria for standards and labels for all major energy using end uses [Chapter 3].
- Improve building energy efficiency by developing and using energy efficient technologies: new building design, efficient lighting, smart windows, day lighting, water heating, HVAC, modification of existing buildings [Chapter 3].

Institutional Water Policy
- Develop improved tools for integrated systems analysis, including indices for economic value and evaluation of tradeoffs in multiple use systems [Chapter 5].
- Modify institutional constraints to increase water resource projects' operational flexibility [Chapter 5].
- Implement institutional changes making water a commodity responding to market forces while providing for basic needs [Chapter 5].
- Develop water conservation and efficient water use practices [Chapter 5].

Education
- Train building operators in efficient buildings operation and maintenance practice [Chapter 3].
- Educate engineers regarding efficient lighting design [Chapter 3].
- Improve the perception of the potential role of engineering in responding to sea level rise through education [Chapter 3].

Material R & D
- Develop and test foam blowing agents and refrigerants that are not greenhouse gases [Chapter 2].

OPPORTUNITIES

Design Under Uncertainty
- Emphasize robust engineering solutions for possible sea level rise due to global climate change [Chapter 6].
- Incorporate risk/reliability and uncertainty analyses in planning and design procedures for water resources [Chapter 5].
- Establish criteria for evaluating, identifying, and setting priorities for vulnerability and resiliency of existing water resources systems and infrastructure [Chapter 5].
- Develop flexible designs that provide acceptable performance over the full range of probable water resource conditions and specified extreme events [Chapter 5].

Water Use
- Develop and employ sensors and systems for more efficient agricultural water use [Chapter 7].
- Develop new construction materials and methods for rehabilitation/retrofit of existing water resources infrastructure [Chapter 5].
- Apply petroleum industry extraction techniques to extend ground water harvesting [Chapter 5].

Methane Management
- Close leakages of natural gas [Chapter 2, Chapter 8].

Transportation
- Continue development of new materials for transportation applications that are stronger and more durable and that generate less greenhouse gas emissions [Chapter 3].
- Continue development of advanced propulsion systems for transportation vehicles including adiabatic diesel engines, ceramic gas turbine engines, advanced electronics, and sensors to maximize vehicle efficiency and performance [Chapter 3].

Economic and Social Valuation
- Facilitate improved valuation of shoreline properties and estimation of response costs [Chapter 6].

- Promote equity considerations in assessing responsibilities for mitigating effects of sea level rise [Chapter 6].
- Promote research in culture theory as a means of including intangibles in decision making [Chapter 9].

Modeling and Systems Analysis
- Develop improved predictive capability in modeling sea level change [Chapter 6].
- Develop refined set of global warming potential (GWPs [Chapter 2].
- Evaluate possible changes in episodic sea level forcings [Chapter 6].
- Refine coupled climate models (ground water, surface water, climatic, land use, natural resources, economic, and environmental) [Chapter 1, Chapter 5, Chapter 7].
- Develop improved understanding/description of physical, chemical, and biological processes and their interactions [Chapter 5, Chapter 7].
- Develop models that couple to direct environmental measurement [Chapter 2].
- Develop more refined mesoscale models to produce useful hydrologic/climatic data at 250 km^2 spatial scale and daily temporal scale [Chapter 5].

Structural Response to Sea Level Rise
- Improve the design of dikes and evaluate effects of sea level rise and extreme events in dikes [Chapter 6].
- Conduct world-wide study of areas of relative sea level rise and resulting system responses and engineering approaches [Chapter 6].
- Improve coastal structures and construction methodology (including port facilities) [Chapter 6].

Hydro-Based Power Generation
- Increase hydropower generator and turbine efficiencies [Chapter 5].

Delivering Efficiency
- Develop solar house, zero energy, passive solar and other technologies [Chapter 3].

Monitoring Technologies

- Improve global climate and biological systems monitoring network [Chapter 2, Chapter 7].
- Improve deep ice core drilling technology for CO_2, N_2O, CH_4, O_2 [Chapter 2].
- Global monitoring of soil temperature and moisture [Chapter 2].
- Integrate and make existing databases available [Chapter 2].
- Develop and implement technology for monitoring remotely and periodically the world's shorelines [Chapter 6].
- Update on a periodic basis (5–10 years) objective evaluation of reliable water level data [Chapter 6].
- Develop measurement technology and make measurements for world-wide flux and reservoirs (inventories) of RIAC (gases, terrestrial carbon, marine fluxes) [Chapter 2].
- Improve instrumentation to measure and monitor trace gas fluxes [Chapter 2].

Fossil Fuel Efficiency

- Improve fossil fuel power plant efficiency (coal, oil, natural gas), roughly 50% near term [Chapter 3].

Resource and Ecosystem Management

- Co-location of industry for efficient use of resources — turn waste streams into product streams [Chapter 3].
- Redesign of manufacturing processes to accept increased volumes of recycled input feed stocks in paper, glass, chemicals, etc. [Chapter 3].
- Accelerated RD & D of electroprocess technologies that yield high processed fossil displacement rates, foster waster minimization, improve on site treatment of stack gas effluents and other residuals [Chapter 3].

International Energy Technology Transfer

- International technology transfer of efficient building design and operation and industrial technology [Chapter 3].
- International lending policies favoring investments in energy efficient technologies [Chapter 3].

Technologies for Reducing GHG Emission
- Eliminate N_2O from catalytic converters and industrial processes [Chapter 2].
- Develop and implement technology to reduce net emissions of CO_2 and SO_2 from cement manufacture [Chapter 2].

Decision Making Methodology
- Develop knowledge-based tools for flexible operations, based on real-time data, of water resources projects [Chapter 5].
- Develop rational decision making methodologies in environmentally sensitive coastal areas [Chapter 6].
- Improve the design of lowland drainage [Chapter 6].

Sediment Management
- Develop and apply procedures for evaluating excess/deficit conditions of dynamic beach profiles [Chapter 6].
- Establish effects of sea level rise on inlet response [Chapter 6].
- Improve the design and maintenance of navigational channels [Chapter 6].

CHALLENGES

Water Use
- Develop methods for identifying new water supplies by approaches such as weather modification, desalination, improved conservation, storm and waste water reuse, etc. [Chapter 5].

Carbon Sequestration in the Biosphere
- Protect and increase carbon storage in the terrestrial biosphere [Chapter 2, Chapter 8].
- Develop management techniques to increase carbon in soils [Chapter 8].

Carbon Separation
- Coal gasification (remove, recover carbon, fuel cells) [Chapter 3].

CO_2 Management

- Optimize industry capability to recapture and recycle materials and greenhouse gases (recycle carbons stocks, CH_4 from coal mines, natural gas systems, and landfills; use animal waste and crop residues for energy) [Chapter 3, Chapter 7].

Methane Management

- Reduce agricultural methane emissions [Chapter 7].
- Develop genetic engineering to decrease CH_4 production from rice, etc. [Chapter 2, Chapter 7].
- Develop waste water treatment that minimizes methane and carbon dioxide production [Chapter 5].

Biotron

- Design and build a biotron laboratory [Chapter 7].
- Crop adaptation avoidance to/of environmental stresses [Chapter 7].
- Rhizosphere enhancement [Chapter 7].

Transportation

- Accelerate rate of telecommunications/computer innovations to mitigate future growth in travel and to replace many current uses for paper throughout the economy [Chapter 3].
- Develop attractive mass transportation options including new light rail and high speed rail systems [Chapter 3].
- Develop transportation vehicles that can utilize alternative fuels including electricity and biomass derived fuels as well as compressed natural gas and, in the long term, hydrogen [Chapter 3, Chapter 4].
- Ensure that safety issues related to vehicles, alternative fuels, and battery systems are carefully addressed [Chapter 2, Chapter 3].

Renewable Technologies

- Demonstration plants for solar thermal energy conversion [Chapter 4].
- Photovoltaics (manufacturing technology, regenerate and recycle, new-multi band gap) [Chapter 4].

Biomass Energy Production and Use
- Recycle carbon through energy plantations [Chapter 7, Chapter 8].
- Reduce or stop large-scale biomass burning [Chapter 2].
- Energy production from crops and waste [Chapter 7].

Delivering Efficiency
- On-farm storage, processing, packaging [Chapter 7].

Monitoring Technologies
- Developing space-based instrumentation for measurements of new constituents (OH, H_2SO4, O_3, etc.) [Chapter 2].

Wetland Management Technologies
- Develop technology for enhancement and management of coastal wetlands [Chapter 6].

Resource (Ecosystem) Management
- GBISC "Globally Biological Integrated Sustainable Communities" [Chapter 7].
- Integrated farming systems (recycling nutrients, water, and carbon — new and mixed species production systems [Chapter 7].

Material R & D
- Develop high temperature materials for adiabatic engines [Chapter 4].
- Improved turbines for oil/natural gas higher temperature at turbine inlet [Chapter 3, Chapter 4].
- Develop new materials requiring less energy and GHG emissions [Chapter 2].

Sediment Management
- Develop and implement procedures for effective sediment management and stabilization/accretion (including reservoirs and river geomorphology) [Chapter 6].
- Improve cost effective sediment transfer and delivery system [Chapter 6].

Enabling Energy Technologies
- Fuel cells (low, medium, high temperature) [Chapter 4].
- Better nontoxic batteries [Chapter 4].
- Compressed natural gas/liquefied natural gas [Chapter 3, Chapter 4].
- Hydrogen Technologies [Chapter 4].
- Storage (heat, electric-SMES, chemical, hydrogen, flywheels — spinning reserve) [Chapter 4].

Fission Reactor Technology
- Develop nuclear technologies for long term-breeders [Chapter 4].
- Decommissioning of nuclear facilities [Chapter 4].
- Improve existing LWR technology. [Chapter 4].
- Improved uranium isotope separation processes [Chapter 4].
- Demonstrate licensable radioactive waste management techniques [Chapter 4].
- Improved institutional arrangements for reducing the likelihood of nuclear proliferation [Chapter 4].
- Improved efficiencies of gas-cooled reactors [Chapter 4].
- Passively safe and proliferation resistant liquid metal reactor [Chapter 4].
- Heavy water reactors [Chapter 4].
- Modular high temperature gas-cooled reactors [Chapter 4].

POSSIBILITIES

Carbon Sequestration in the Biosphere
- Sequester carbon by increasing ocean alkalinity [Chapter 8].
- Micro-biological system, soil environment, waste recycling to increase net carbon in soils [Chapter 7, Chapter 8].
- Forest and halophyte management systems to sequester CO_2 [Chapter 7].
- Sequester carbon by intentional fertilization of the marine biosphere [Chapter 8].

Carbon Separation
- Covert fossil fuel to coal and hydrogen and burn only the

hydrogen (Co-process fossil fuel and biomass to sequester carbon and produce methanol) [Chapter 8].
- Coal/biomass/oil/gas (store carbon, mine mouth), (C, methanol, H_2 and CH_4) [Chapter 8].

Active Removal from the Atmosphere
- Develop laser technology to destroy greenhouse gases in the atmosphere [Chapter 8].

CO_2 Management
- Remove, recover, and dispose of CO_2 from flue gases at large combustion sources [Chapter 8].
- Manage CO_2 used for (1) enhanced oil recovery, (2) depleted oil and gas wells, (3) excavated salt domes, (4) geological voids [Chapter 2].

Methane Management
- Develop biotechnology to consume greenhouse gases [Chapter 8].

Future Energy Options, Long Term
- Solar power satellites (lunar vs. SPS [geo] comparative study, demonstration, technology development transmission, power conditioning re: control, reception, structure); modeling, re: simulation studies [Chapter 4].
- Fusion — develop economic, environmentally sound, very safe fusion reactors [Chapter 4].

Hydrologic Cycle Modification
- Alter the "ocean conveyor belt" by changing Atlantic Ocean salinity (Russian river diversion, dam Bering Strait, melt Greenland) [Chapter 8].
- Pump ocean heat to winter polar atmosphere by by-passing sea ice insulation (preserves sea ice for summer or vertical mixing of stratified Arctic-delays, re: ice formation) [Chapter 8].
- Increase evapotranspiration and cloud cover in the tropics [Chapter 8].

Hydro-Based Power Generation
- Ocean energy (wave-energy density, OTEC-turbine design, cost/nutrients, fouling, hydrology, tidal-siting) [Chapter 4].
- Develop and use innovative hydropower sources such as low head, tidal, pumped storage, etc. [Chapter 4, Chapter 5].

Atmospheric Radiation Technology
- Reduce incident solar radiation with mirrors in orbit [Chapter 8].
- Put radiation scattering or absorbing particles into the stratosphere [Chapter 8].
- Put corner reflecting balloons into the upper stratosphere [Chapter 8].
- Modify local climates by changing urban area albedo [Chapter 8].
- Increase plant albedo in low latitude forested areas. [Chapter 8]
- Enhance albedo with sulfate aerosol in the marine tropical troposphere [Chapter 8].

REFERENCES

Fischoff, B., Value Elicitation, Is There Anything There?, *Am. Psychol.*, August, 1991.

Gregory, J. W., *The Story of the Road*, Alexander Maclehose Co., London, 1931.

Heaton, G. R., R. Repetto, R. Sobin, *Backs to the Future*, World Resources Institute, Washington, DC, June, 1992.

Intergovernmental Panel on Climate Change, Formulation of Response Strategies. Report prepared for IPCC by Working Group III, June 1990.

Jevons, W. S., *Theory of Political Economy*, 2nd. ed., MacMillan and Co., London, 1879.

Levine et al., Energy Efficiency, Developing Nations, and Eastern Europe, A Report to the U. S. Working Group on Global Energy Efficiency, Washington, DC, April 1991.

Levine, J. Sathaye, A. Ketoff, CO_2 Emissions from Major Developing Countries, *Energy J.,* Vol. 12 No. 1, 1991.

MacKay, Charles, *Extraordinary Popular Delusions and the Madness of Crowds*, Farrar, Strauss, and Giroux, New York, 1932.

Marland, G., A. M. Weinberg, Longevity of Infrastructure, in *Cities and Their Vital Systems. Infrastructure Past, Present, and Future,* Jesse H. Ausubel and Robert Herman (eds.) National Academy Press, Washington, D.C. 1988, pp. 312–332.

Price, Roger, *The Great Roob Revolution,* Random House, New York, 1970.

Spreng, D. T., On Time, Information, and Energy Conservation, ORAU/IEA-78-22 (R), Institute for Energy Analysis, Oak Ridge, TN., December, 1978.

AUTHOR AND CONTRIBUTOR LIST

Chapter 1
THE FIFTH REVOLUTION

Author:
Robert G. Watts
Director
NIGEC SouthCentral Region
Tulane University
School of Engineering
605 Lindy Boggs Building
New Orleans, LA 70118

Chapter 2
EMISSIONS AND BUDGETS OF RADIATIVELY IMPORTANT ATMOSPHERIC CONSTITUENTS

Author:
Don Wuebbles
University of Illinois - Urbana
Department of Atmospheric Sciences
105 S. Gregory Avenue
Urbana, IL 61801

Co-Authors:
Jae Edmonds
Battelle Pacific Northwest Laboratory
370 L'Enfant Promenade
901 D Street SW, Suite 900
Washington, D.C. 20024-2115

Jane Dignon
Lawrence Livermore National Laboratory
Atmospheric and Geophysical Sciences Division, L-524
7000 East Avenue
P.O. Box 808
Livermore, CA 94557-9900

William Emanuel
University of Virginia
Department of Environmental Science
Clark Hall
Charlottesville, VA 22903

Donald Fisher
E. I. Du Pont de Nemours, Inc.
D320/240
Du Pont Experimental Station
P.O. Box 80320
Wilmington, DE 19880-0320

Richard Gammon
University of Washington
School of'Oceanography
WB10
Seattle, WA 98195

Robert Hangebrauck
(retired)

Robert Harriss
National Aeronautics and Space Administration
Code YS, Headquarters
Washington, D.C. 20546-0001

M. A. K. Khalil
Portland State University
Department of Physics
P.O. Box 751
Portland, OR 97207-0751

John Spence
Environmental Protection Agency
Mail Stop 84
Alexander Drive and Route 54
Research Triangle Park, NC 27711

Thayne M. Thompson
National Oceanic and Atmospheric Administration
R/E/CGI
Climate Monitoring and Diagnostics Laboratory
325 Broadway
Boulder, CO 80303-3328

Chapter 3
DEMAND REDUCTION

Authors:

Arthur Rosenfeld
Senior Advisor, Energy Efficiency, US DoE
EE-1, United States Department of Energy
Washington, D.C. 20585

Barbara Atkinson
Lawrence Berkeley Laboratory
38-40 Mailstop 90-4000
Berkeley, CA 94720

Lynn Price
Lawrence Berkely Laboratory
1 Cyclotron Road, MS 90-4000
Berkeley, CA 94720

Bob Ciliano
RCGL Hagler, Bailly, Inc.
1530 Wilson Blvd., Suite 900
Arlington, VA 22209

J.I. Mills
EG & G Idaho, Inc.
P.O. Box 1625
Idaho, ID 83415

Kenneth Friedman
International Energy Agency – OECD
2, Rue Andre-Pascal, 75775 Paris
CEDEX 16

Co-Authors:
Ed Flynn
E.I.A. Department of Energy
1000 Independence Ave.
Washington, D.C. 20250

Kenneth Friedman
International Energy Agency-OECD
2, Rue Andre-Pascal, 75775 Paris
CEDEX 16

Mark Hopkins
Alliance to Save Energy
1725 K Street NW, Suite 509
Washington, D.C. 20006

Chapter 4
ENERGY SUPPLY

Authors:
Martin Hoffert
Department of Applied Science
New York University
26-36 Stuyvesant Street
New York, NY 10003

Seth D. Potter
Department of Physics
New York University
Barney Building
26-36 Stuyvesant Street
New York, NY 10003

Contributors:
Peter E. Glaser
Vice-President
Arthur D. Little, Inc.
Acom Park
Cambridge, MA 02140-2390

Murali Kadiramangalam
Fluent Inc.
Centerra Resource Park
10 Cavendish Court
Lebanon, NH 03766-1442

Alfred Perry
8809 Cove Point Lane
Knoxville, TN 37922

Harold M. Hubbard
3245 Newland Street
Wheat Ridge, CO 80033

Carl-Jochen Winter
Deutsche Forschungsanstalk fur
Luft-und Raumfahrt e. V. (DIR)
Pfaffenwaldring 38-40
Germany

Jerry Delene
19 Windhaven Lane
Oak Ridge, TN 37803

Michael Golay
Department of Nuclear Engineering
MIT
77 Massachusetts Avenue
Cambridge, MA 02139

Myer Steinberg
Process Sciences
Brookhaven National Laboratory
Department of Applied Science
Upton L.I., NY 11973

Chapter 5
WATER RESOURCES

Author:
William H. McAnally
Chief, Waterways and Estuaries Division
USAE Waterways Experiment Station
3909 Halls Ferry Rd
Vicksburg, MS 39180

Co-Authors:
Phillip H. Burgi
Manager, Water Resources Research Laboratory
US Bureau of Reclamation
P.O. Box 25007
Denver, CO 80225

Darryl Calkins
Chief, Geological Sciences Branch
USAE Cold Regions Research and Engineering Laboratory
Hanover, NH 03755-1290

Richard H. French
Research Professor
Water Resources Center, Desert Research Institute
University and Community Colleges System of Nevada
2505 Chandler Ave., Ste. 1
Las Vegas, NV 89120

Jeffrey P. Holland
Special Assistant to the Director
Hydraulics Laboratory
USAE Waterways Experiment Station
3909 Halls Ferry Rd
Vicksburg, MS 39180

Bernard Hsieh
Research Hydraulic Engineer
Hydroscience Division
USAE Waterways Experiment Station
3909 Halls Ferry Rd
Vicksburg, MS 39180

Barbara Miller
President, Rankin International, Inc.
Drawer E
Knoxville, TN 37828

Jim Thomas
Manager, Flood Hydrology Group
US Bureau of Reclamation
P.O. Box 25007
Denver, CO 80225

Contributors:
    William D. Martin
    Acting Chief, Hydrosciences Division
    USAE Waterways Experiment Station
    3909 Halls Ferry Rd
    Vicksburg, MS 39180

    James R. Tuttle
    Chief, Water Control Branch (retired)
    Lower Mississippi River Valley Division
    U.S. Army Corps of Engineers
    3909 Halls Ferry Rd
    Vicksburg, MS 39180

Chapter 6
GLOBAL WARMING AND COASTAL HAZARDS

Author:
    Ashish Mehta
    Coastal and Oceanographic Engineering Department
    University of Florida
    Gainesville, FL 32611

Contributors:

Robert Dean
Coastal and Oceanographic Engineering Department
University of Florida
Gainesville, FL 32611

Hans Kunz
Niedersachsisches Landesamt fur Wasserwirtschaft
Forschungsstelle Kuste
An Der Muhle 5
D 2982 Norderney, Germany

Victor Law
Chemical Engineering Department
Tulane University
New Orleans, LA 70118

Say-Chong Lee
National Hydraulic Research Institute of Malaysia
KM 7, Jalan Ampang, 68000 Ampang
Kuala Lumpur, Malaysia

Zal Tarapore
Central Water & Power Research Station
Khadakwasla, Pune 411024, India

Chapter 7
AGRICULTURE AND BIOLOGICAL SYSTEMS

Author:

Norman R. Scott
Vice-President for Research and Advanced Studies
Cornell University
314 Day Hall
Ithaca, NY 14853-2801

John N. Walker
Associate Dean, College of Engineering
University of Kentucky
177 Anderson Hall
Lexington, KY 49085-9659

Gerald F. Arkin
Associate Director, Georgia Agricultural Experimental Station
University of Georgia
Griffen, GA 30223-1797

James A. DeShazer
Professor and Head, Agricultural Engineering Department
University of Idaho
Moscow, ID 83843-2040

Gary R. Evans
Special Assistant for Global Change Issues
U.S. Department of Agriculture
301 12th Street, Room 2M08
Washington, D.C. 20250

Glenn J. Hoffman
Professor and Head, Biological Systems Engineering Department
University of Nebraska
223 L. W. Chase Hall
Lincoln, NE 68583-0726

James W. Jones
Professor, Agricultural Engineering Department
University of Florida
Gainesville, FL 32611

Chapter 8
GEOENGINEERING

Authors:
Brian P. Flannery
Exxon Research and Engineering Company
Route 22 East
Annandale, NJ 08801

Haroon Kheshgi
Exxon Research and Engineering Company
1545 Route 22 East
Annandale, NJ 08801

Gregg Marland
Oak Ridge National Laboratory
Environmental Sciences Division
Oak Ridge, TN 37831

Michael C. MacCracken
Office of the US Global Change Research Program
300 D Street SW, Suite 840
Washington, D.C. 20024

Contributors:
Hiroshi Komiyama
University of Tokyo
Tokyo, Japan 113

Wallace Broecker
Lamont Doherty Earth Observatory–Columbia University
Department of Geological Sciences
Palisades, NY 10964

Hisashi Ishatani
University of Tokyo
Kaya Laboratory, Department of Electrical Engineering
7-3-1 Hongo
Bundy o-ku, Tokyo, Japan 113

Norman Resenberg
Battelle Pacific Northwest Laboratory
370 L'Enfant Promenade
901 D Street SW, Suite 900
Washington, D.C. 20034-2115

Meyer Steinberg
Brookhaven National Laboratory
Process Sciences Group
Upton, Long Island, NY 11973

Tom Wigley
University Corporation for Atmospheric Research (UCAR)
P.O. Box 3000
Boulder, CO 80307-3000

Michael Morantine
Tulane University
School of Engineering
Mechanical Engineering Department
New Orleans, LA 70118

Chapter 9
THE ENGINEERING RESPONSE TO GLOBAL ENVIRONMENTAL CHANGE

Author:
Robert G. Watts
Director
NIGEC SouthCentral Region
Tulane University
School of Engineering
605 Lindy Boggs Building
New Orleans, LA 70118

INDEX